Geometric Dimensioning and Tolerancing

basic fundamentals

by

David A. Madsen
Department Chairperson
Drafting, Manufacturing, and Building Trades Technology
Authorized AUTOCAD™ Training Center
Clackamas Community College
Oregon City, Oregon

South Holland, Illinois
THE GOODHEART-WILLCOX COMPANY, INC.
Publishers

Library of Congress Catalog Card Number 87-30967
International Standard Book Number 0-87006-673-0

4567890-88-654321

Library of Congress Cataloging in Publication Data

Madsen, David A.
Geometric dimensioning and tolerancing: basic fundamentals/by David A. Madsen.

Includes index.
1. Engineering drawings—Dimensioning.
2. Tolerance (Engineering)
I. Title.
T357.M22 1988 604.2'43--dc19 87-30967
ISBN 0-87006-673-0

NOTE

Concepts used in this book are in accordance with the national standard for Dimensioning and Tolerancing, ANSI Y14.5M-1982 published by the American National Standards Institute, 1430 Broadway, New York, NY 10018.

INTRODUCTION

GEOMETRIC DIMENSIONING AND TOLERANCING will provide you with the complete fundamentals of Geometric Dimensioning and Tolerancing (GD&T) concepts as interpreted in the American National Standards Institute document *ANSI Y14.5M-1982, Dimensioning and Tolerancing.*

A DIMENSION placed on a drawing is a numerical value used to describe size, shape, location, or related manufacturing processes. A TOLERANCE is the total amount by which a specified dimension is permitted to vary. GEOMETRIC TOLERANCING (GT) is dimensioning associated with the tolerancing of individual characteristics of a part where permissible variations relate to form, profile, radial relationship to an axis, orientation of one feature to another, and location of features.

The goal of this book is to guide you through a logical sequence of learning activities. You will move from conventional dimensioning and tolerancing into geometric tolerancing. Using clear examples, you will build a strong understanding of each concept before the introduction of new material. The concepts are covered in an easy-to-learn sequence. Learning about GD&T progresses in a format that will allow you to become comfortable with the concepts as your knowledge builds from one chapter to the next.

Flexibility is the key word when using GEOMETRIC DIMENSIONING AND TOLERANCING. This book is very successful in teaching how to read and interpret prints with GD&T symbols. It is also an excellent training tool where the emphasis is learning to interpret and draw GD&T.

- Each chapter has a test that will allow you to reinforce prior learning.
- When you have completed all of the chapter tests, a final exam is available at the end of the book that will provide a comprehensive review of the material learned.

- Drafting Problems are designed for you and are presented as real world engineering sketches. Problems may be completed after individual chapters or after you have studied the entire book. The problems range in complexity so that you may develop skills as slowly or rapidly as individual course objectives require. Drafting problems may be omitted where the course objectives are to learn GD&T print reading.
- This book is useful in Manufacturing Technology programs for machining, welding, tool and die, and other curriculums where the emphasis is GD&T print reading and not drafting. Actual prints from industry are provided along with related exercises to help you develop basic technical skills.
- Geometric Dimensioning and Tolerancing symbology is a natural for CADD applications. This book discusses how GD&T relates to use with a CADD system. GD&T/CADD templates may be designed for individual needs or purchased from software vendors ready for use.

This text has been written to conform to ANSI standards. While using the metric system of measurement, the examples have been kept as close as possible to whole units. Being in whole units, the theory, principles, and symbols can be related or applied to the decimal inch based system of measurement.

I would like to give special thanks to Robert F. Francoise, Chairman of the American National Standards Committee Y 14, and Mack Courtier, of Boeing Aerospace for their comprehensive reviews of this text; and to Cynthia B. Clark, Production Editor Technical Publishing for The American Society of Mechanical Engineers, for her permission to use specific items from ANSI Y14.5M-1982. I would also like to thank the following professionals for their technical comments: John K. Aragon, Alliance Carolina Tool and Mold; Daniel D. Eichenauer, Northwest Technical College; and Paul A. Migliaccio, Mohawk Valley Community College.

David A. Madsen

CONTENTS

Examples of automotive components manufactured from plans using geometric dimensioning and tolerancing.

Chapter 1

GENERAL TOLERANCING

This chapter will cover general tolerancing as applied to conventional dimensioning practices. Use of the term CONVENTIONAL TOLERANCING refers to tolerances related to conventional dimensioning practices without regard to geometric tolerancing practices.

Conventional dimensioning methods will provide the necessary basic background to begin a study of geometric tolerancing. This text will introduce you in step-by-step fashion to easy-to-understand concepts as you learn geometric tolerancing. In future chapters you will see conventional dimensioning and tolerancing used together with geometric tolerancing.

It is important that you completely understand general tolerancing before you begin the study of geometric tolerancing.

When mass production methods began, interchangeability of parts was important. However, many times parts had to be "hand selected for fitting." Today, industry has faced the reality that in a technological environment, there is no time to do unnecessary individual fitting of parts. Geometric tolerancing will help insure interchangeability of parts. The function and relationship of a feature of a part will dictate the use of geometric tolerancing.

Geometric tolerancing does not take the place of conventional tolerancing. However, geometric tolerancing specifies requirements more precisely than conventional tolerancing, leaving no doubts as to the intended definition. This precision may not be the case when conventional tolerancing is used, because notes on the drawing may become ambiguous.

A drafter or any person dealing with technology needs to know how to properly represent and interpret conventional dimensioning and geometric tolerancing. Generally, the drafter will convert engineering sketches or instructions into formal drawings using proper standards and techniques. After acquiring adequate experience, a design drafter, designer, or engineer will begin implementing geometric tolerancing and dimensioning on the research and development of new products or the revision of existing products.

Most dimensions in this text are in metric; therefore, a 0 preceeds decimal dimensions as in 0.25 when inch dimensions are used. The 0 need not preceed the decimal dimension.

DEFINITIONS RELATED TO GENERAL TOLERANCING

Learn the following definitions . . . you will apply these terms many times in your studies.

Dimension—A numerical value, expressed in appropriate units of measure, indicated on a drawing and in documents to define the size and/or geometric characteristics and/or locations of features of a part.

Reference dimension—A dimension, without tolerance, used for information purposes only. Shown on a drawing with parenthesis, for example: (60).

Feature—The general term applied to a physical portion of a part or object, for example: a surface, slot, tab, keyseat, or hole.

Feature of size—One cylindrical or spherical surface, or a set of two plane parallel surfaces, each of which is associated with a size dimension.

Actual size—The measured size of a feature or part after manufacturing.

Stock size—A commercial or premanufactured size, such as square, round, or hex steel bar.

GENERAL TOLERANCING CONSIDERATIONS

A TOLERANCE is the total amount by which a specific dimension is permitted to vary. The tolerance is the difference between the maximum and minimum limits. A tolerance is not given to values that are identified as reference, maximum, minimum, or stock sizes. The tolerance may be applied directly to the dimension as shown in Example 1-1, indicated by a general note, or identified in the drawing title block.

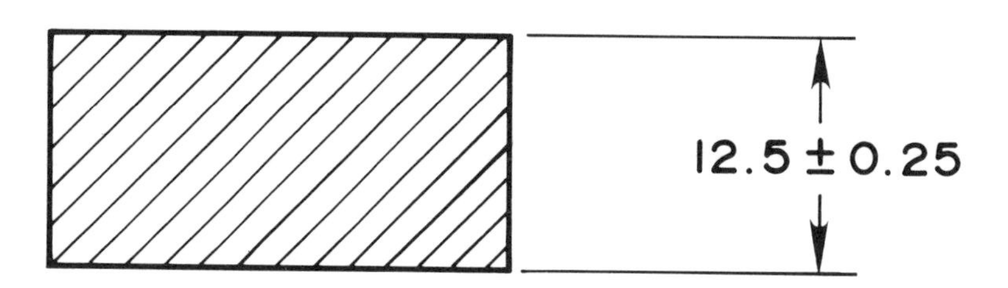

Example 1-1. Tolerance.

In example 1-1 the dimension is stated as 12.5 ± 0.25. The tolerance of this dimension is the difference between the maximum and minimum limits.

The LIMITS of a dimension are the largest and smallest numerical value that the feature can be. From Example 1-1, the upper limits are 12.5 + 0.25 = 12.75 and the lower limits are 12.5 − 0.25 = 12.25. So, if you take the upper limit and subtract the lower limit you have the tolerance: 12.75 − 12.25 = 0.5.

The SPECIFIED DIMENSION is the part of the dimension from which the limits are calculated. The specified dimension of the feature shown in Example 1-1 is 12.5 A dimension on a drawing may be displayed as in Example 1-1, or the limits may be calculated and shown as in Example 1-2. Many companies prefer this second method because the limits are shown and calculations are not required.

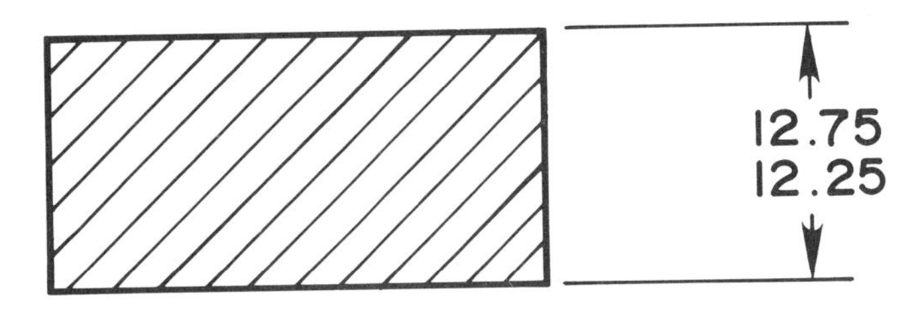

Example 1-2. Limit dimensioning.

A BILATERAL TOLERANCE is permitted to vary in both + and − directions from the specified dimension. The dimension 12.5 ± 0.25 is an equal bilateral tolerance where the variation from the specified dimension, 12.5, is the same in both directions. An example of an unequal bilateral tolerance is where the variation from the specified dimension is not the same in both directions, for example $12.5\,^{+0.3}_{-0.1}$

A UNILATERAL TOLERANCE is permitted to increase or decrease in only one direction from the specified dimension, for example

$12.5\,^{+0.1}_{\;0}$ or $12.5\,^{\;0}_{-0.1}$.

MAXIMUM MATERIAL CONDITION, MMC

MAXIMUM MATERIAL CONDITION is the condition in which a feature of size contains the maximum amount of material within the stated limits. The key words are "maximum amount of material."

An external feature is at Maximum Material Condition at its largest limit, or maximum amount of material, as shown in Example 1-3.

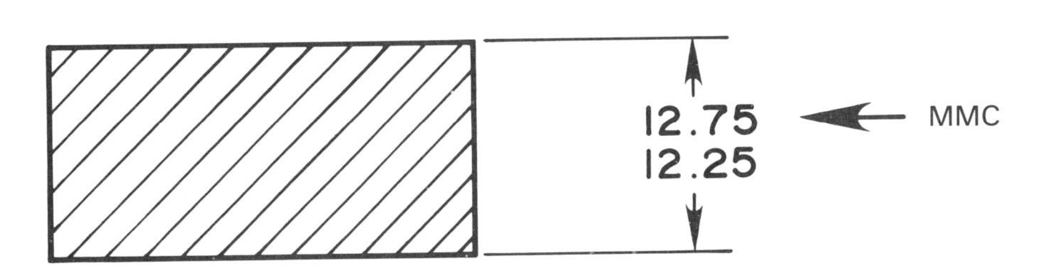

Example 1-3. MMC—External feature.

An internal feature is at Maximum Material Condition at its smallest limit, or maximum amount of material, as shown in Example 1-4.

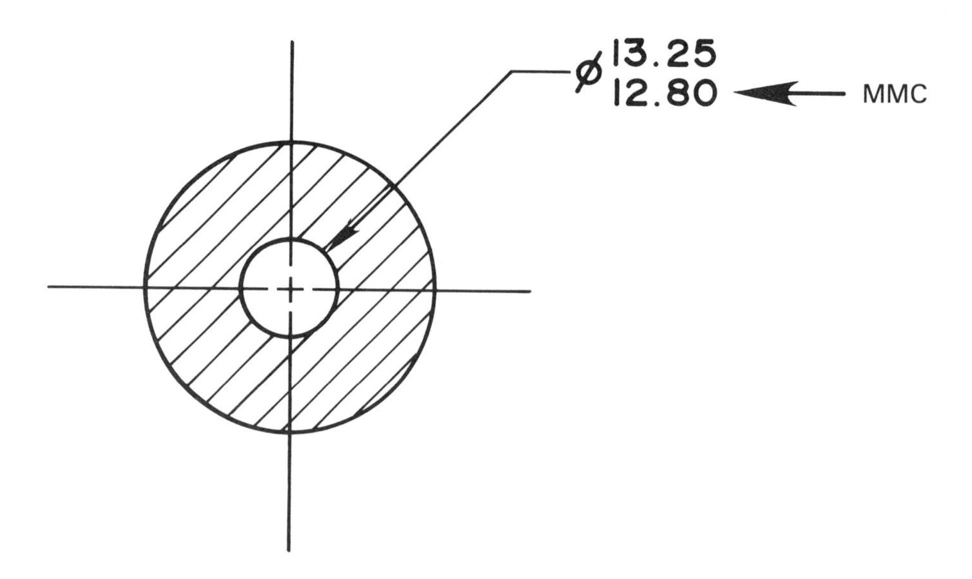

Example 1-4. MMC—Internal feature.

LEAST MATERIAL CONDITION, LMC

LEAST MATERIAL CONDITION is the condition in which a feature of size contains the least amount of material within the stated limits. The key words are ''least material.''

An external feature is at Least Material Condition at its smallest limit, least material, as illustrated in Example 1-5.

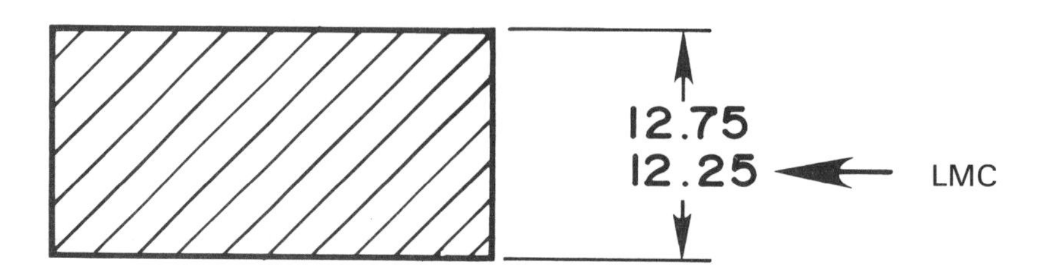

Example 1-5. LMC—External feature.

An internal feature is at Least Material Condition at its largest limit, least material, as shown in Example 1-6.

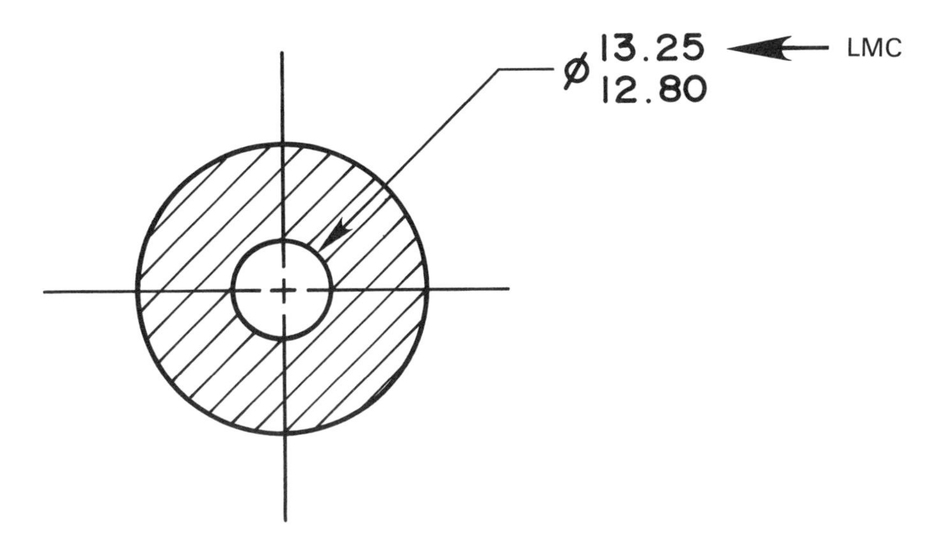

Example 1-6. LMC—Internal feature.

BASIC FITS OF MATING PARTS

Three general groups of fits between mating parts are: Running and sliding fits; Force or interference fits, and Locational fits.

Running and sliding fits (RC) are intended to provide a running performance with suitable lubrication allowance. Running fits range from RCI which are close fits, to RC 9 which are loose fits.

Force fits (FN) or shrink fits constitute a special type of interference fit characterized by maintenance of constant pressure. Force fits range from FN 1 light drive to FN 5 which are force fits required in high stress applications.

Locational fits are intended to determine only the location of the mating parts.

For more information on fits refer to the *Machinery's Handbook* under the classification of ALLOWANCES AND TOLERANCES FOR FITS, ANSI STANDARD FITS, GRAPHICAL REPRESENTATION OF LIMITS AND FITS, and STANDARD FIT TABLES.

CLEARANCE FIT

A clearance fit is shown in Example 1-7. Part 1 will fit into Part 2 with a clearance between the two parts no matter what the actual size of each part is when produced within the given tolerances.

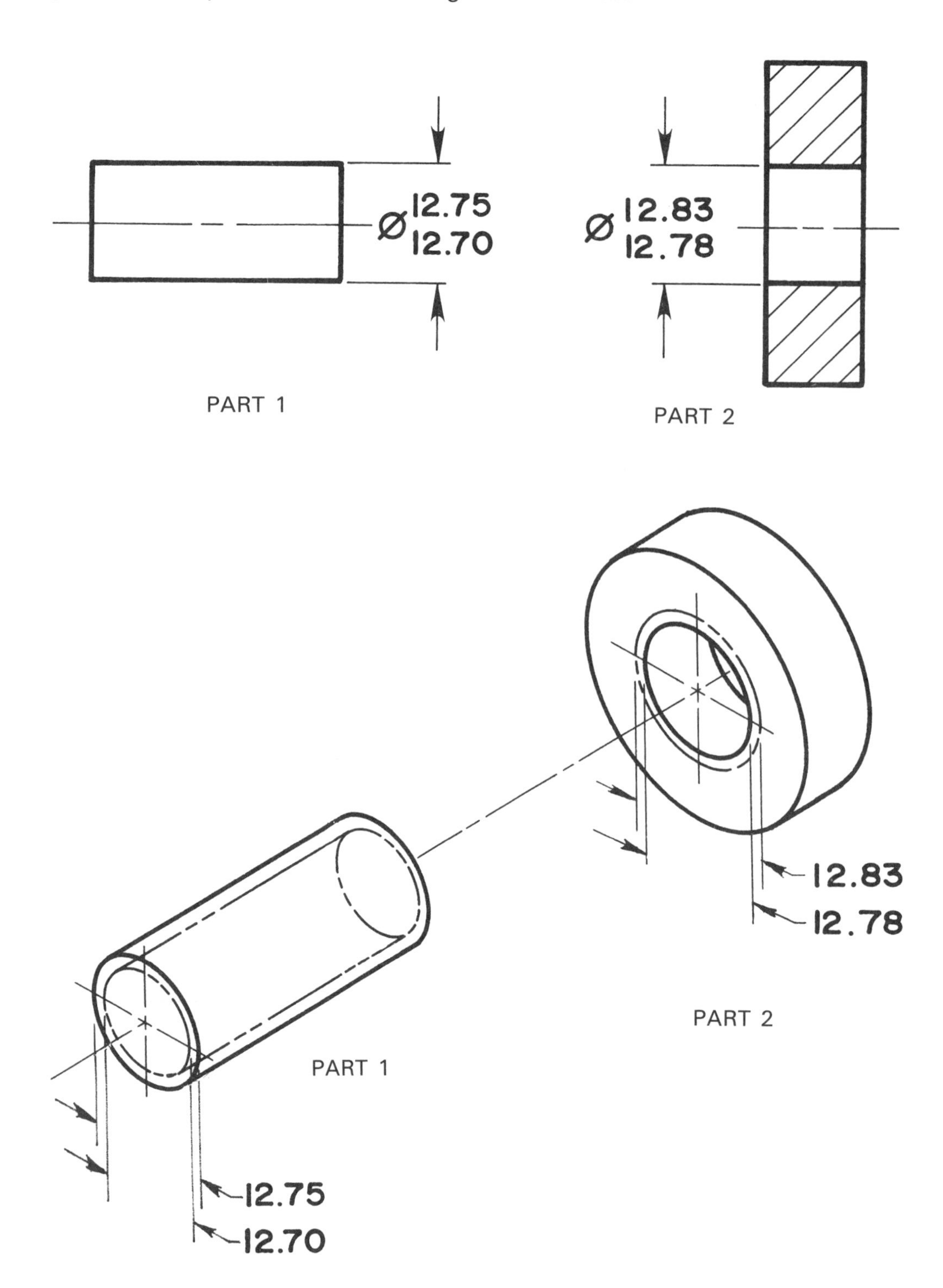

Example 1-7. Clearance fit between two parts.

ALLOWANCE

The ALLOWANCE is defined as an intentional difference between the maximum material limits of mating parts. Allowance is the minimum clearance (positive allowance), or maximum interference (negative allowance) between mating parts. Allowance can be considered to be the tightest possible fit between parts. Allowance may be calculated using the formula:

MMC HOLE
– MMC SHAFT
ALLOWANCE

Now refer to Example 1-7 as you make these calculations:

	MMC HOLE (Part 2)	=	12.78
–	MMC SHAFT (Part 1)	=	12.75
	ALLOWANCE	=	0.03

CLEARANCE

The loosest fit or maximum intended difference between mating parts is called the CLEARANCE. The clearance is calculated with this formula:

LMC HOLE
– LMC SHAFT
CLEARANCE

Refer again to Example 1-7 as you determine the clearance:

	LMC HOLE (Part 2)	=	12.83
–	LMC SHAFT (Part 1)	=	12.70
	CLEARANCE	=	0.13

FORCE FIT

A FORCE FIT is also referred to as an INTERFERENCE FIT or SHRINK FIT. This is where two mating parts must be pressed or forced together. Due to the tolerance on each part, the shaft is larger than the hole, as shown in Example 1-8.

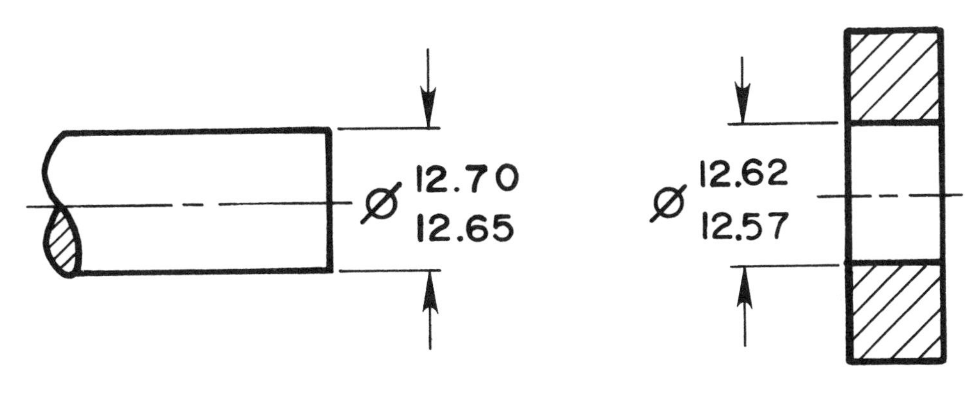

Example 1-8. A force fit between two parts.

At any produced size within the stated tolerance, the shaft will be larger than the hole.

The smallest amount of interference is:

LMC SHAFT	=	12.65
− LMC HOLE	=	12.62
MIN INTERFERENCE	=	0.03

The greatest amount of interference is:

MMC SHAFT	=	12.70
− MMC HOLE	=	12.57
MAX INTERFERENCE	=	0.13

DIMENSIONING RULES

Fundamental rules for dimensioning and tolerancing are necessary to establish the intent of dimensioning on engineering drawings and related documents. The following general rules have been taken in part from Dimensioning and Tolerancing, ANSI Y 14.5M-1982.

- Each dimension shall have a tolerance, except for dimensions specifically identified as reference, maximum, minimum, or stock. The tolerance may be applied directly to the dimension or indicated by a general note or in the drawing title block.
- Dimensions for size, form, and location of features shall be complete to the extent that there is full understanding of the characteristics of each feature. Neither scaling nor assumption of a dimension is permitted.
- Every necessary dimension shall be shown and no more dimensions than those necessary for complete definition shall be given. Reference dimensions should be kept to a minimum.
- Dimensions shall be selected and arranged to suit the function and mating relationship of a part and shall not be subject to more than one interpretation.
- The drawing should describe a part without specifying manufacturing methods, unless specifically required by the design engineer for control purposes. Specifications should be specified on the engineering documents in cases where manufacturing processes, quality assurance, or environmental information is essential to the definition of engineering requirements.
- Final dimensions shall be given on the drawing, while nonmandatory processing dimensions that provide for finish allowance, shrinkage allowance, and other requirements may be identified. Nonmandatory processing dimensions shall be identified by an appropriate note, such as NONMANDATORY, or MFG DATA.
- Dimensions should be arranged for best readability in true profile views and refer to visible outlines. Do not dimension to hidden features.
- Stock sizes manufactured to gage or code numbers shall be specified by dimensions indicating the diameter and length. Gage or code numbers may be shown in parentheses following the dimension.

- A 90° angle is implied where centerlines and lines depicting features are shown on a drawing at right angles and no angle is specified. A 90° basic angle applies where centerlines of features in a pattern or surfaces shown at right angles on the drawing are located or defined by basic dimensions and no angle is specified.
- Unless otherwise specified all dimensions are applicable at 20°C (68°F). Compensation may be made for dimensions made at other temperatures.

DIMENSIONING UNITS

Dimensions in ANSI Y14.5M-1982, Dimensioning and Tolerancing, are given in SI (International System of Units) units. The commonly used SI linear units used on engineering drawings and related documents is the millimeter. The commonly used US linear units used on engineering drawings is the decimal inch. The unit of measurement selected should be in accordance with the policy of the user. When all dimensions on a drawing are either in millimeters or in inches the general note: UNLESS OTHERWISE SPECIFIED, ALL DIMENSIONS ARE IN MILLIMETERS or (INCHES). The abbreviation ''IN.'' shall follow all inch dimensions on a millimeter dimensioned drawing, and the abbreviation ''mm'' shall follow all millimeter dimensions on an inch dimensioned drawing. Refer to ANSI Y14.5M-1982 or major drafting texts that acknowledge reference to this standard for further dimensioning practices.

Some companies, though very limited in number, prefer to use a method of dual dimensioning where both inch and millimeter dimensions are shown at the same time. One common technique is to show inch dimensions followed by millimeter equivalents in brackets, or millimeters followed by inch equivalents in brackets. When this is done the general note: DIMENSIONS IN [ ] ARE MILLIMETERS (INCHES). Another method of dual dimensioning is with inches followed by millimeters or millimeters followed by inches. When this is done the general note: MILLIMETER/INCH or INCH/MILLIMETER shall be placed on the drawing. Another practice is to show a tabulation of equivalent dimensions on the drawing. This practice eliminates the need for ''dualizing'' and cleans up the drawing. Some examples of dual dimensioning are shown in Example 1-9.

	DUAL DIMENSIONING		
POSITION METHOD	25 .984	OR	25/.984
	1.000 25.4	OR	1.000/25.4
BRACKET METHOD	25 [.984]	OR	25 [.984]
	1.000 [25.4]	OR	1.000 [25.4]

Example 1-9. Dual dimensioning methods.

CONVERTING DIMENSIONS

To convert inches to millimeters, use the formula:
25.4 × INCH = MILLIMETER.

To convert millimeters to inches, use the formula:
MILLIMETER ÷ 25.4 = INCH.

To convert millimeters to decimal inch values, use the formula:
.03937 INCH = 1 MILLIMETER.

To achieve the same degree of accuracy when converting from inches to millimeters, provide one digit to the right less for millimeter dimensions than for inch dimensions. For example:

INCH	MILLIMETER
.1	2.5
.01	0.3
.001	0.03
.0001	0.003

GEOMETRIC DIMENSIONING AND TOLERANCING FOR CADD/CAM

The implementation of geometric tolerancing and dimensioning into a mechanical drafting CADD program is practical. The geometric tolerancing (GT) symbology makes this application a bonus to the mechanical drafting system. GT/CADD symbol libraries will be introduced in Chapter 2.

Some dimensioning and tolerancing guidelines for use in conjunction to CADD/CAM are outlined in part from ANSI Y14.5M-1982 as follows:

1. Geometric tolerancing is necessary to control specific geometric form and location.
2. Major features of the part should be used to establish the basic coordinate system, but are not necessarily defined as datums.
3. Subcoordinated systems that are related to the major coordinates are used to locate and orient features on a part.
4. Define part features in relation to three mutually perpendicular reference planes, and along features that are parallel to the motion of CAM equipment.

5. Establish datums related to the function of the part, and relate datum features in order of precedence as a basis for CAM usage.
6. Completely and accurately dimension geometric shapes. Regular geometric shapes may be defined by mathematical formulas. A profile feature that is defined with mathematical formulas should not have coordinate dimensions unless required for inspection or reference.
7. Coordinate or tabular dimensions should be used to identify approximate dimensions on an arbitrary profile.
8. Use the same type of coordinate dimensioning system on the entire drawing.
9. Continuity of profile is necessary for CADD. Clearly define contour changes at the change or point of tangency. Define at least four points along an irregular profile.
10. Circular hole patterns may be defined with polar coordinate dimensioning.
11. When possible dimension angles in degrees and decimal parts of degrees; for example, 45°30′ = 45.5°.
12. Base dimensions at the mean of a tolerance because the computer numerical control (CNC) part programmer will normally split a tolerance and work to the mean. While this is theoretically desirable, one cannot predict where the part will be made. Dimensions should always be based on design requirements. If it is known that a part will be produced always by CNC methods, then establish dimensions without limits that conform to the CNC machine capabilities. Bilateral profile tolerances are also recommended for the same reason.

CHAIN VS DATUM DIMENSIONING

The difference between chain and datum dimensioning is shown in Example 1-10. In chain dimensioning each dimension is dependent on the previous dimension, while in datum dimensioning each dimension stands alone. Caution should be used when chain dimensioning because the tolerance of each dimension builds on the next. This is referred to as tolerance buildup or stacking. An example of tolerance buildup is when three chain dimensions have individual tolerances of ± 0.2 and each feature is manufactured at or near the + 0.2 limit; the potential tolerance buildup is 3 × 0.2 for a total of 0.6. To accommodate this buildup the overall dimension must have a tolerance of + 0.6. This problem does not occur with datum dimensioning.

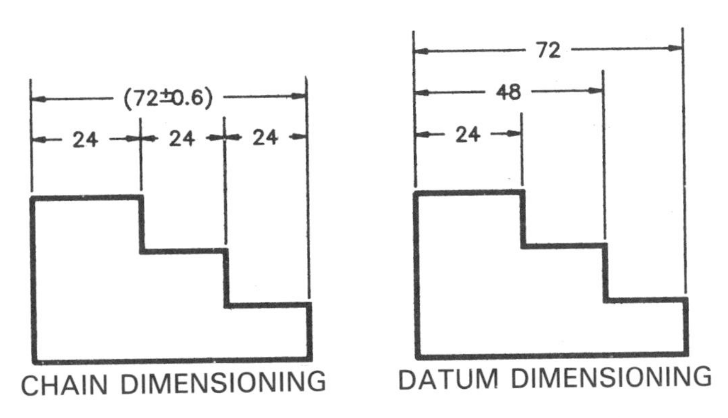

Example 1-10. Chain vs datum dimensioning.

TEST 1, GENERAL TOLERANCING **Name:**______________________

1. A ______________ is a numerical value, expressed in appropriate units of measure, indicated on a drawing and in documents to define the size and/or geometric characteristics and/or locations of features of a part.
2. ______________ is a general term applied to a physical portion of a part.
3. Define tolerance:

   ____

   ____

   ____

4. All dimensions shall have a tolerance except for dimensions that are identified as:
   a. reference.
   b. maximum.
   c. minimum.
   d. stock sizes.
   e. all of the above.
5. What are the limits of the dimension: 25 ± 0.4? ______________
6. What is the tolerance of the dimension in question 5? ______________
7. What is the specified dimension of the dimension shown in question 5?

   ____

8. Give an example of an equal bilateral tolerance. ______________
9. Give an example of an unequal bilateral tolerance. ______________
10. Give an example of a unilateral tolerance. ______________
11. Define Maximum Material Condition (MMC).

    ____

    ____

    ____

12. What is the MMC of the feature shown below? ______________

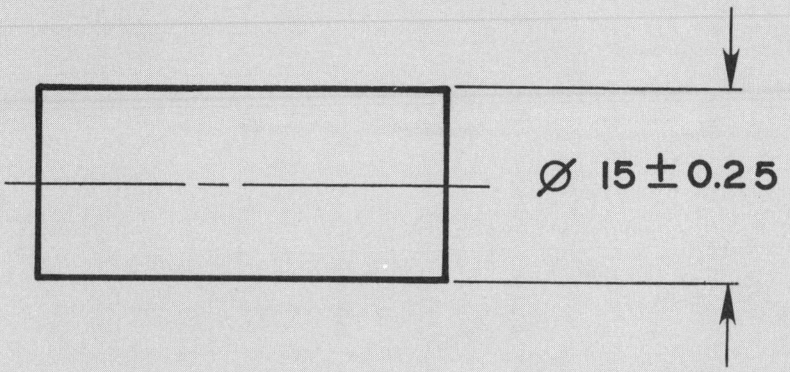

Ø 15 ± 0.25

13. What is the MMC of the feature shown below? ______________

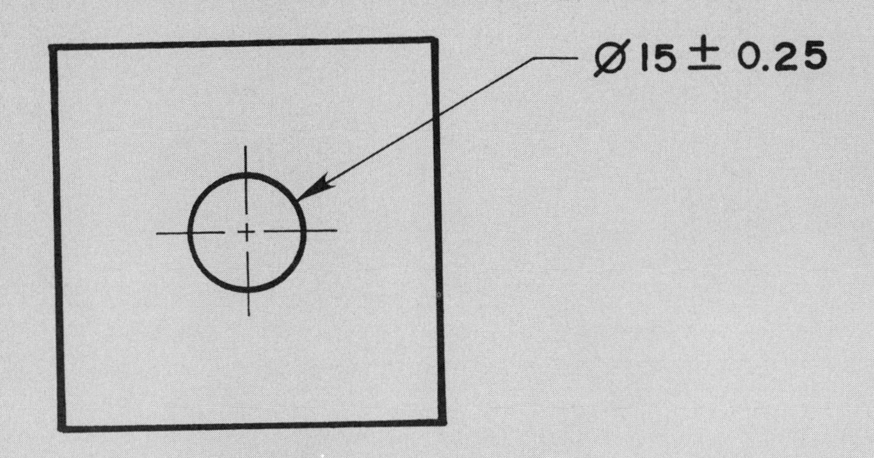

14. Define Least Material Condition (LMC).

____

____

____

15. What is the LMC of the feature shown in question 12? __________

16. What is the LMC of the feature shown in question 13? __________

17. List the three general groups related to the basic fits between mating parts.

   1. ____

   2. ____

   3. ____

18. Is the fit between the two parts shown below a clearance or a force fit?

____

Ø 19.43 / 19.18

Ø 19.76 / 19.50

19. What is the allowance between the two parts shown in question 18? Show your calculations and label each numeral.

____

____

20. What is the clearance between the two parts shown in question 18? Show your calculations and label each numeral.

____

____

21. A force fit is also referred to as
a ________________ or __________
fit.

22. Given the following information regarding the dimensions of a shaft and a collar (hole), determine the limits of the dimensions for each part. Show your calculations and label each numeral. Suggestion: review allowance and tolerance before you begin.
a. The dimension of the shaft is ϕ 14 ± 0.4.
b. A clearance fit exists between the two parts.
c. Provide an allowance 0.2.
d. The tolerance to be applied to the collar hole dimension is 0.8.

SHAFT UPPER LIMIT = ______________________________

SHAFT LOWER LIMIT = ______________________________

SHAFT LIMITS = ______________________________

MMC HOLE (UNKNOWN)
− MMC SHAFT __________
= ALLOWANCE __________

MMC SHAFT __________
+ ALLOWANCE __________
= MMC HOLE = __________

MMC HOLE __________
+ HOLE TOLERANCE __________
= LMC HOLE __________

HOLE LIMITS = __________

23. Identify the ANSI standard that is titled *Dimensioning and Tolerancing.*

____

24. What does the abbreviation SI denote?

____

25. What are the commonly used SI units on an engineering drawing?

____

26. What are the commonly used US units used on engineering drawings?

____

27. What general note should accompany a drawing to describe the predominant units used? ____________________

____

28. Define dual dimensioning. ____________________

____

29. Show an example of the bracket method of dual dimensioning.

____

30. What is the formula that may be used when mathematically converting from inches to millimeters? ____________________

PRINT READING EXERCISES FOR CHAPTER 1

Name:______________________

The following print reading exercise is designed for use in programs for machining, welding, tool and die, dimensional inspection, and other manufacturing curriculums where the objective is the reading and interpretation of prints rather than the development of drafting skills. An actual industrial print is used with related questions that require you to read and interpret specific dimensioning and geometric tolerancing representations. The answers and interpretations should be based on the previously learned content of this book. The prints used are based on ANSI standards; however, company standards may differ slightly. When reading these prints or any other industrial prints, a degree of flexibility may be required to determine how individual applications correlate with the ANSI standard.

PRINT READING EXERCISE

Refer to the Hyster Company print of PEDAL-ACCELERATOR found on page 216.

1. Are the dimensions given in inches or millimeters? ________________
2. Refer to the 0.76 x 45° BOTH ENDS dimension: What is the tolerance on the 0.76? ____________________ What is the tolerance on the 45°? ____________________
3. Refer to the ϕ4.834-4.763 dimension:
   a. What is the MMC? __________
   b. What is the LMC? __________
   c. What is the tolerance? __________

Refer to the Hyster Company print of the CASE-DIFF found on page 217.

4. Refer to the ϕ157.2-156.7 dimension:
   a. What is the tolerance? __________
   b. What is the MMC? __________
   c. What is the LMC? __________
5. Refer to the ϕ56.05-55.95 dimension:
   a. What is the tolerance? __________
   b. What is the MMC? __________
   c. What is the LMC? __________
6. Refer to the ϕ57.239-57.201 dimension:
   a. What is the tolerance? __________
   b. What is the MMC? __________
   c. What is the LMC? __________

7. Refer to the 65.1 dimension:
   a. How is the tolerance for this dimension specified? (hint: see title block)
   ____
   ____
   ____
   b. What is the tolerance? __________
   c. What is the upper limit? __________
   d. What is the lower limit? __________
   e. What is the MMC? __________
   f. What is the LMC? __________

Refer to Hyster Company print of the FLYWHEEL-DSL found on page 218.

8. Refer to the ϕ51.99-51.97 dimension:
   a. What is the tolerance? __________
   b. What is the MMC? __________
   c. What is the LMC? __________
9. Refer to the ϕ403.35 dimension in SECTION A-A:
   a. What is the tolerance? __________
   b. What is the upper limit? __________
   c. What is the lower limit? __________
   d. What is the MMC? __________
   e. What is the LMC? __________
   f. Is this a UNILATERAL or BILATERAL tolerance? __________
10. Refer to the 33.27 dimension in SECTION A-A:
    a. What is the tolerance? __________
    b. What is the upper limit? __________
    c. What is the lower limit? __________
    d. What is the MMC? __________
    e. What is the LMC? __________

Refer to the Curtis Associates print of the BRG. RETAINER found on page 219.

11. Is this drawing in METRIC or INCHES? __________
12. Refer to the ϕ2.1250-2.1245 dimension:
    a. What is the MMC? __________
    b. What is the LMC? __________
    c. What is the tolerance? __________

13. Refer to the 4 x ϕ.218 dimension:
    a. What does the 4 x denote? ______________________________
    b. Where is the tolerance specified? ______________________________
    c. What is the tolerance? ________________________
14. Interpret the 2 x R.010 MAX dimension in the GROOVE DETAIL A.

    ____

    ____

    ____

Refer to the Curtis Associates print of the MOUNTING BRACKET (WORM GEAR DRIVE) found on page 220.

15. Describe the feature associated with the ϕ2.878-2.876 dimension.

    ____

16. Refer to the ϕ1.2593-1.2587 dimension:
    a. What is the location dimension for these holes? __________________
    b. What is the MMC? ____________
    c. What is the LMC? ____________
    d. What is the tolerance? ______________

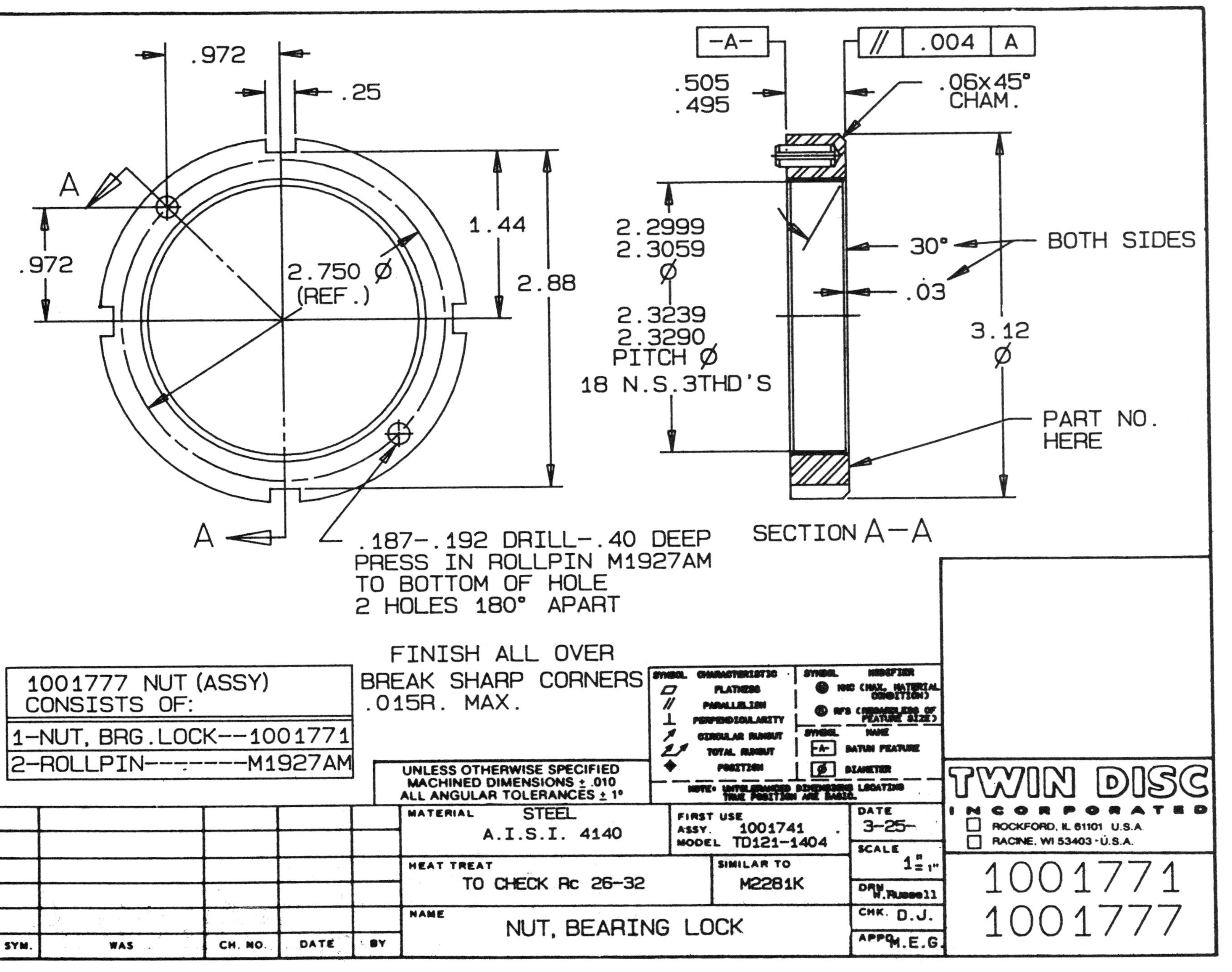

Example of typical industrial CADD drawing incorporating GT symbols.

SPECIAL SYMBOLS

All symbols are drawn recommended size based on a .125 inch lettering height.

Ø Diameter

(75) Reference Dimension

X Target Point

Dimension Origin

Conical Taper

Slope

Counterbore/Spotface

Countersink

Depth/Deep

Square (Shape)

25 Dimension Not To Scale

4 X Number of Times/Places

86 Arc Length

R Radius

SR Spherical Radius

SØ Spherical Diameter

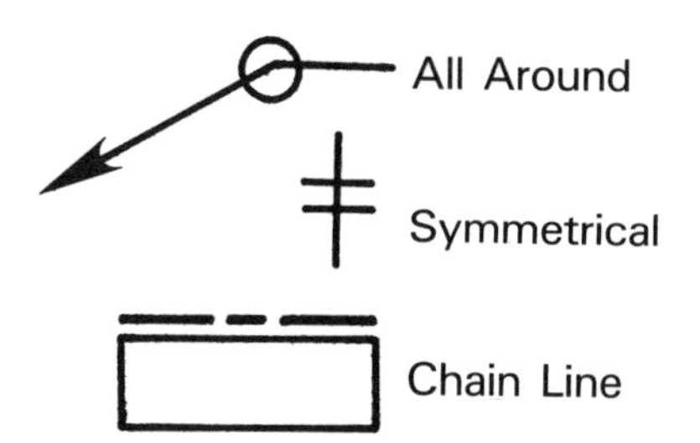

All Around

Symmetrical

Chain Line

SOME COMMON SYMBOL USES

1. 8X Ø 8.4
   Ø 12.7
   5.2

2. 6 X Ø 12 $^{+0.27}_{0}$

3. Ø 6.2 THRU
   Ø 12 X 82°

4. 45

Chapter 2

SYMBOLS AND TERMS

This chapter will aid you in the identification of symbols and terms. Your main objective is to recognize the various types of symbols with identification of name, shape, and size. Only a few terms will be defined at this time. Other terms will be clearly defined in later chapters as you learn about geometric tolerancing in an easy-to-understand format. Symbol sizes are based on drawing lettering height. Where symbols are detailed you will see the note: h = lettering height. This means that h equals the predominant lettering height on your drawing. For example, the lettering height on most engineering drawings is .125 or .156 inch depending on company standards.

Geometric tolerancing and dimensioning symbols are divided into five basic types:

1. Geometric characteristic symbols.
2. Feature control frame.
3. Material condition symbols.
4. Datum feature and datum target symbols.
5. Supplementary symbols.

When you draw symbols on test answers or in problem solutions, use clear, accurate representations. It is also recommended that an appropriate geometric tolerancing template be used for manual drafting or a symbol library for Computer Aided Design and Drafting (CADD). Geometric tolerancing symbols are drawn using thin lines which are the same thickness as extension and dimension lines.

DIMENSIONING AND TOLERANCING TEMPLATES

Dimensioning and tolerancing templates are available to help you save time when doing drawings that contain ANSI Y14.5M-1982 symbols. It is recommended that you use one of these templates when doing the exercises or tests in this book. The symbols on drawing assignments should be done using a template or a CADD system. A template or CADD presentation will help insure that all symbols are properly drawn and always uniform in size and appearance. Standard lettering height on engineering mechanical drawings is 1/8 (.125) or 5/32 (.156) inch depending on company standards and the symbols are sized based on the predominant lettering height on the drawing. Verify the lettering height specified when purchasing a template. The template shown in Example 2-1 has both geometric tolerancing and standard dimensioning symbols.

Example 2-1. Geometric Dimensioning and Tolerancing Template.

GEOMETRIC CHARACTERISTIC SYMBOLS

GEOMETRIC CHARACTERISTIC SYMBOLS are separated into five types: form, profile, orientation, location, and runout as shown in Example 2-2. The open style of arrowhead is generally preferred for CADD produced drawings because they require less machine time to complete.

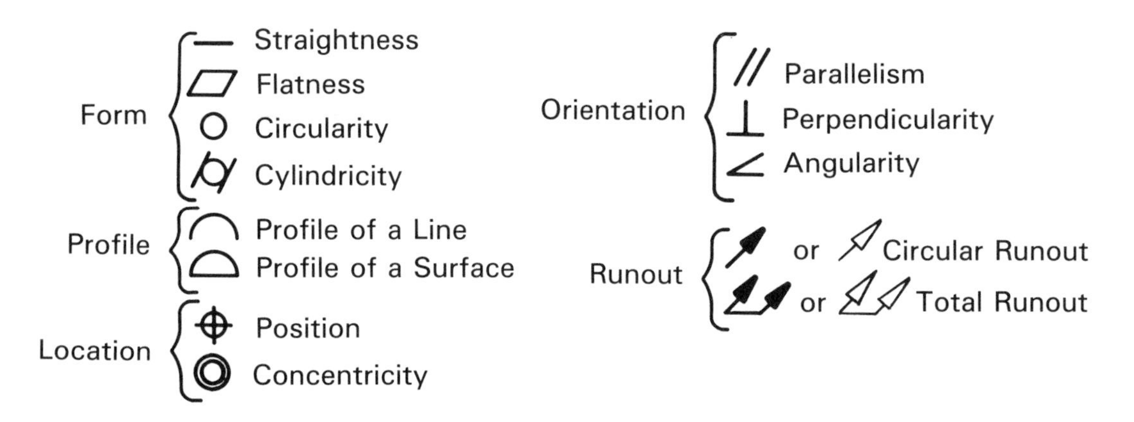

Example 2-2, Geometric characteristic symbols.

All of the symbols in Example 2-2 are drawn to the actual size and shape recommended by American National Standards Institute (ANSI) based on .125 inch high lettering on your drawing.

MATERIAL CONDITION SYMBOLS

MATERIAL CONDITION SYMBOLS are referred to as modifying symbols. These symbols are only used in geometric dimensioning applications. The symbols used in the feature control frame to indicate at maximum material condition, regardless of feature size, or least material condition are shown in Example 2-3.

Ⓜ MMC, Maximum Material Condition

Ⓢ RFS, Regardless of Feature Size

Ⓛ LMC, Least Material Condition

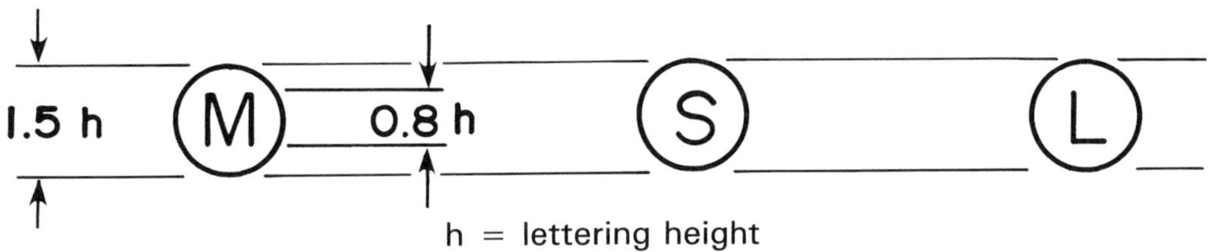

Example 2-3. Material condition symbols.

DATUM FEATURE SYMBOL

The DATUM FEATURE SYMBOL consists of a block with a datum identifying letter placed inside and a dash on each side of the letter. Any letter of the alphabet may be used to identify a datum except for I, O, or Q as these letters may be confused with numbers. Each datum feature requiring identification must have its own identification letter. On drawings where the number of datums exceed the letters in the alphabet then double letters are used starting with AA through AZ, and then BA through BZ. Datum feature symbols may be repeated only as necessary for clarity. Datum identification letters A, B, and C may be used for convenience; however, other letters are commonly used in industry. Example 2-4 shows a datum feature symbol.

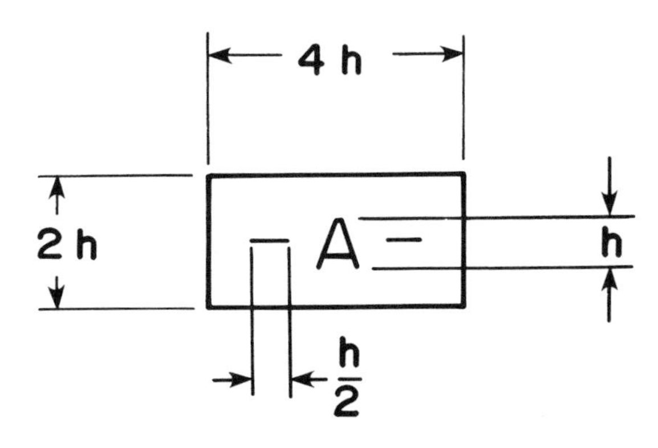

Example 2-4. Datum feature symbol.

DATUM TARGET SYMBOL

The datum target symbol is drawn as a circle with a horizontal line through the center. The top half of the circle is left blank unless the datum target symbol refers to a datum target area in which case the size of the target area is specified as shown in Example 2-5. The lower half of the circle is used to identify the related datum with the datum reference letter and datum target number assigned sequentially starting with 1 for each datum. See Example 2-5.

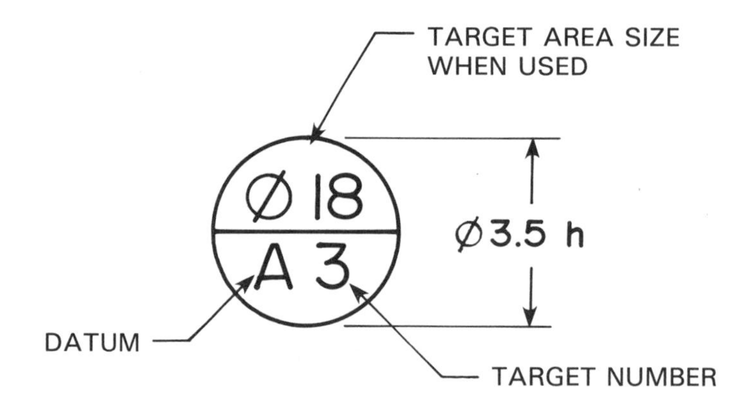

Example 2-5. Datum Target Symbol.

A radial line is used to connect the datum target symbol to the datum target point, target line, or target area shown in Example 2-6. These three examples of datum target symbols will be applied in Chapter 3 on Datums.

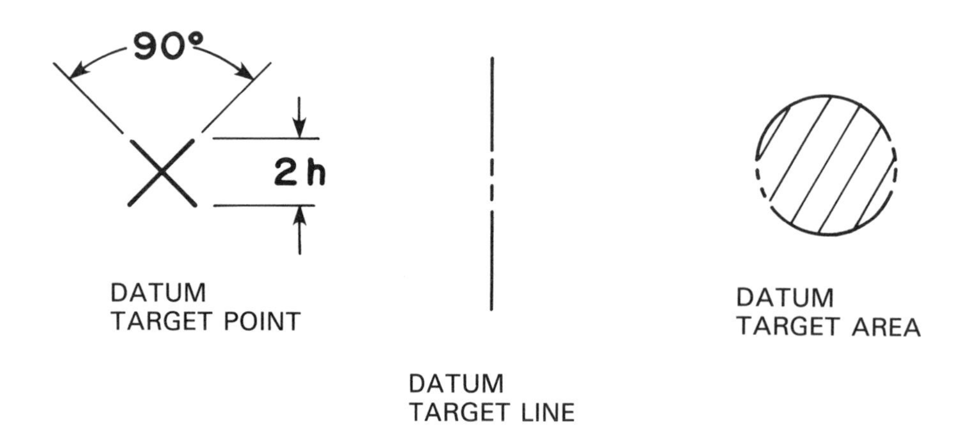

Example 2-6. Datum Target Point, Datum Target Line, and Datum Target Area.

FEATURE CONTROL FRAME

A geometric tolerance for an individual feature is specified by means of a feature control frame. The feature control frame is divided into compartments containing the geometric characteristic symbol in the first compartment followed by the geometric tolerance. Where applicable, the geometric tolerance is preceded by the diameter symbol which describes the shape of the tolerance zone, and followed by a material condition symbol. See Example 2-7.

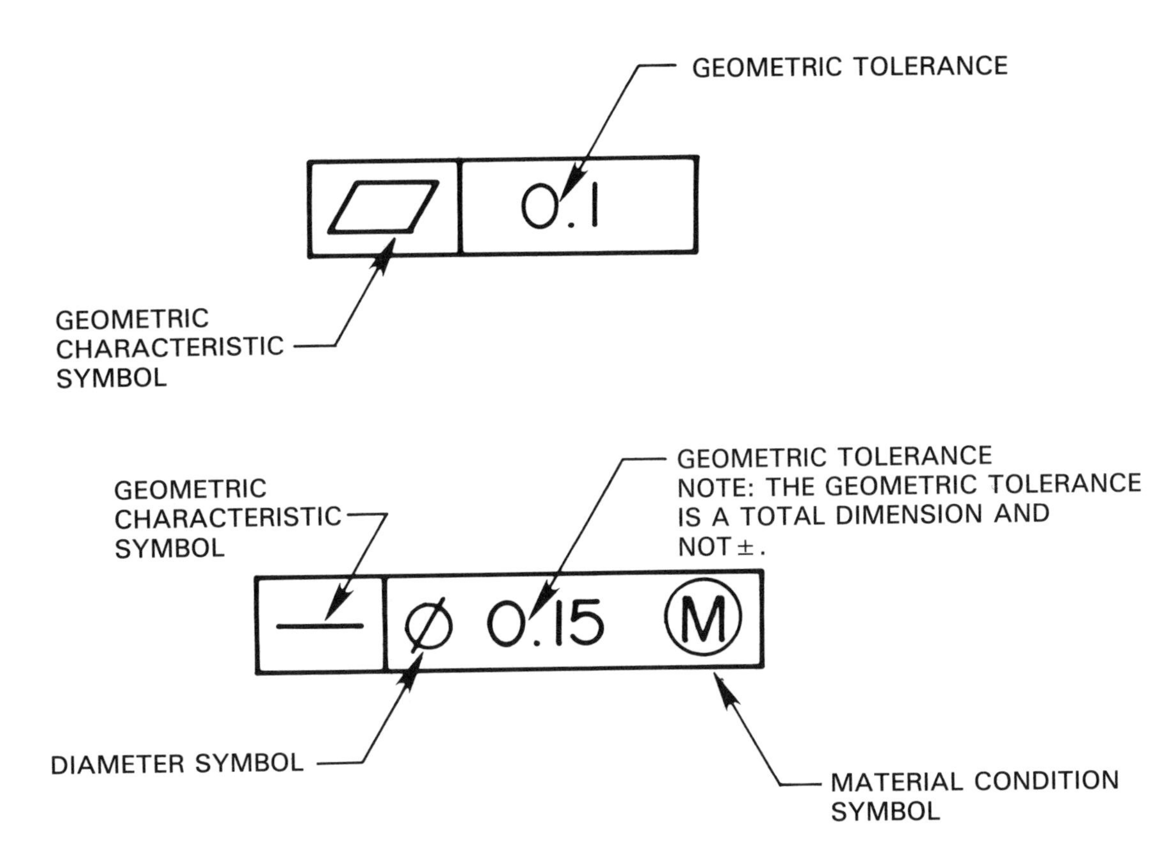

Example 2-7. Feature control frame with geometric characteristic and geometric tolerance.

Where a geometric tolerance is related to one or more datums, the datum reference letters are placed in compartments following the geometric tolerance. Where a datum reference is multiple (that is, established by two datum features—such as an axis established by two datum diameters) both datum reference letters, separated by a dash, are placed in a single compartment after the geometric tolerance as shown in Example 2-8.

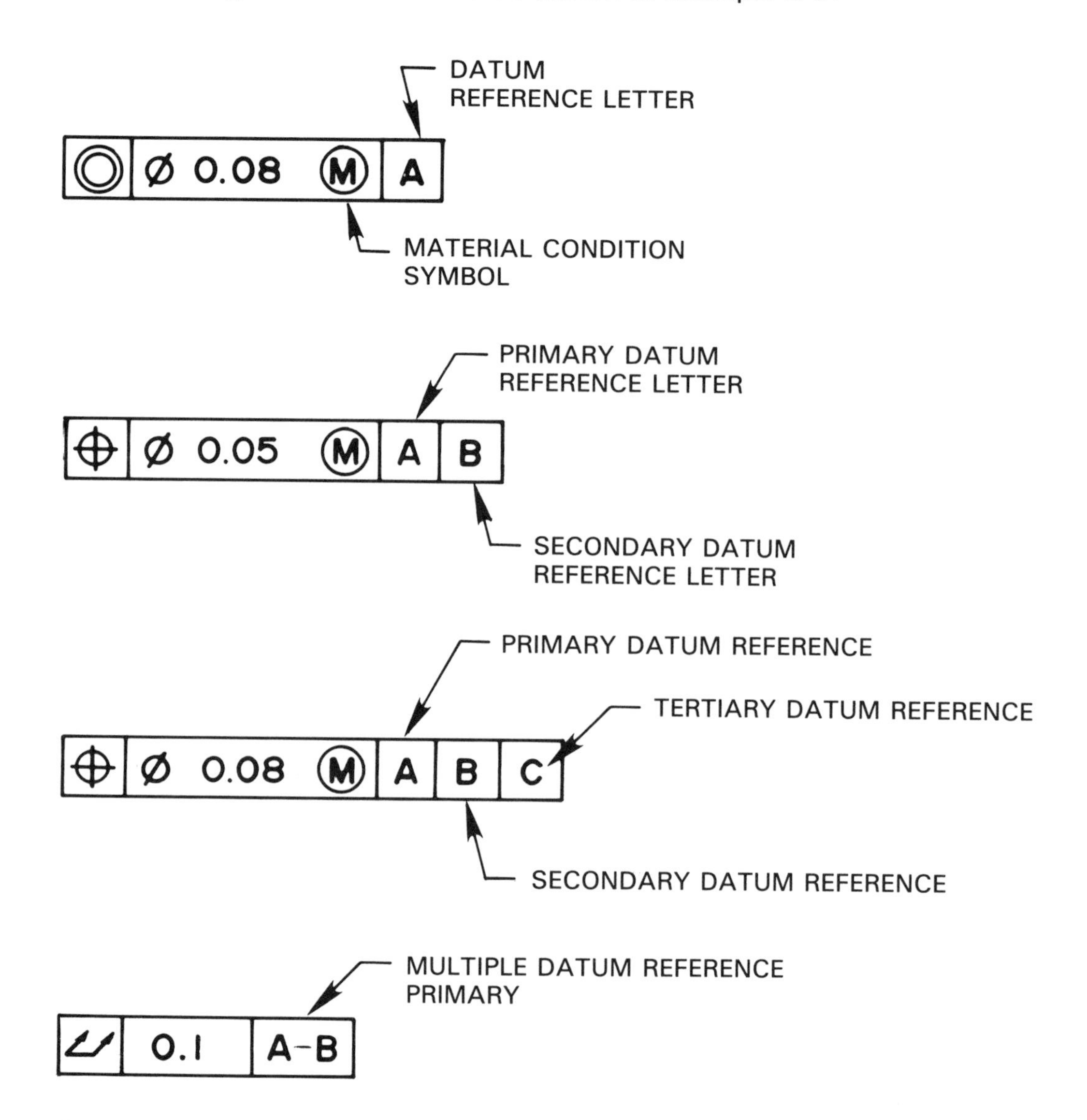

Example 2-8. Feature control frames with datum reference.

The order of elements in a feature control frame is shown in Example 2-9. Notice in Example 2-9 the datum reference letters may be followed by a material condition symbol where applicable. Draw each feature control frame compartment large enough to accommodate the symbols without crowding. Minimum compartment length is 2 × the lettering height.

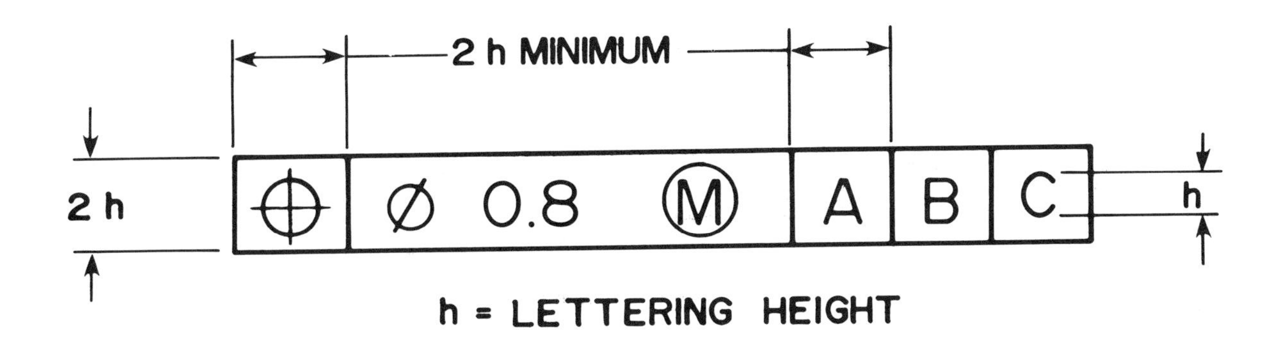

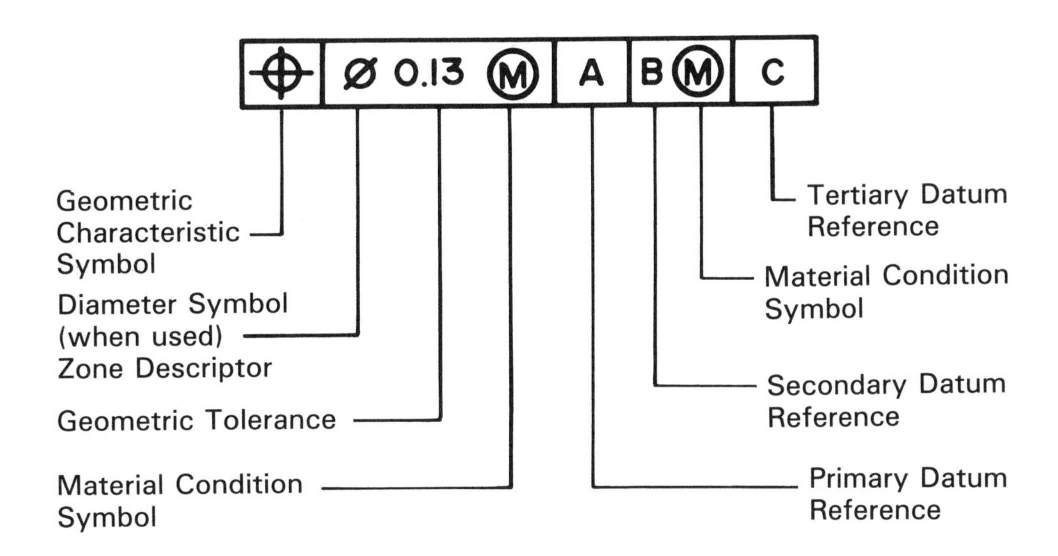

Example 2-9. Order of elements in a feature control frame.

BASIC DIMENSIONS

A BASIC DIMENSION is considered a theoretically perfect dimension. Basic dimensions are used to describe the theoretically exact size, profile, orientation, or location of a feature or datum target. These dimensions provide the basis from which permissible variations are established by tolerances on other dimensions, in notes, or in feature control frames. This text will show you specific situations where basic dimensions are optional or required. Basic dimensions are shown on a drawing by placing a rectangle around the dimension as shown in Example 2-10. A general note may also be used to identify basic dimensions in some applications, for example, UNTOLERANCED DIMENSIONS LOCATING TRUE POSITION ARE BASIC. This method is not recommended and should be avoided in most cases. The basic dimension symbol is clearer.

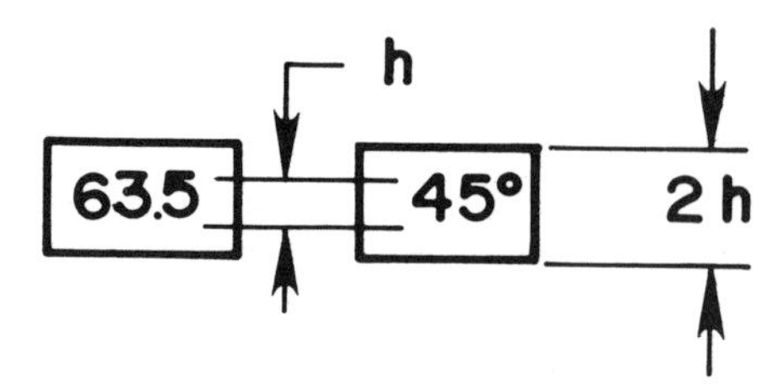

Example 2-10. Basic Dimensions.

The basic dimension symbol around a dimension is a signal to the reader to look for a geometric tolerance in a feature control frame elsewhere on the drawing.

ADDITIONAL SUPPLEMENTARY SYMBOLS

Symbols aid in clarity, ease of drawing presentation, and save time, especially when used in conjunction with Computer Aided Design and Drafting (CADD). Symbols should be drawn clearly using a template or CADD. Example 2-11 shows recommended symbols.

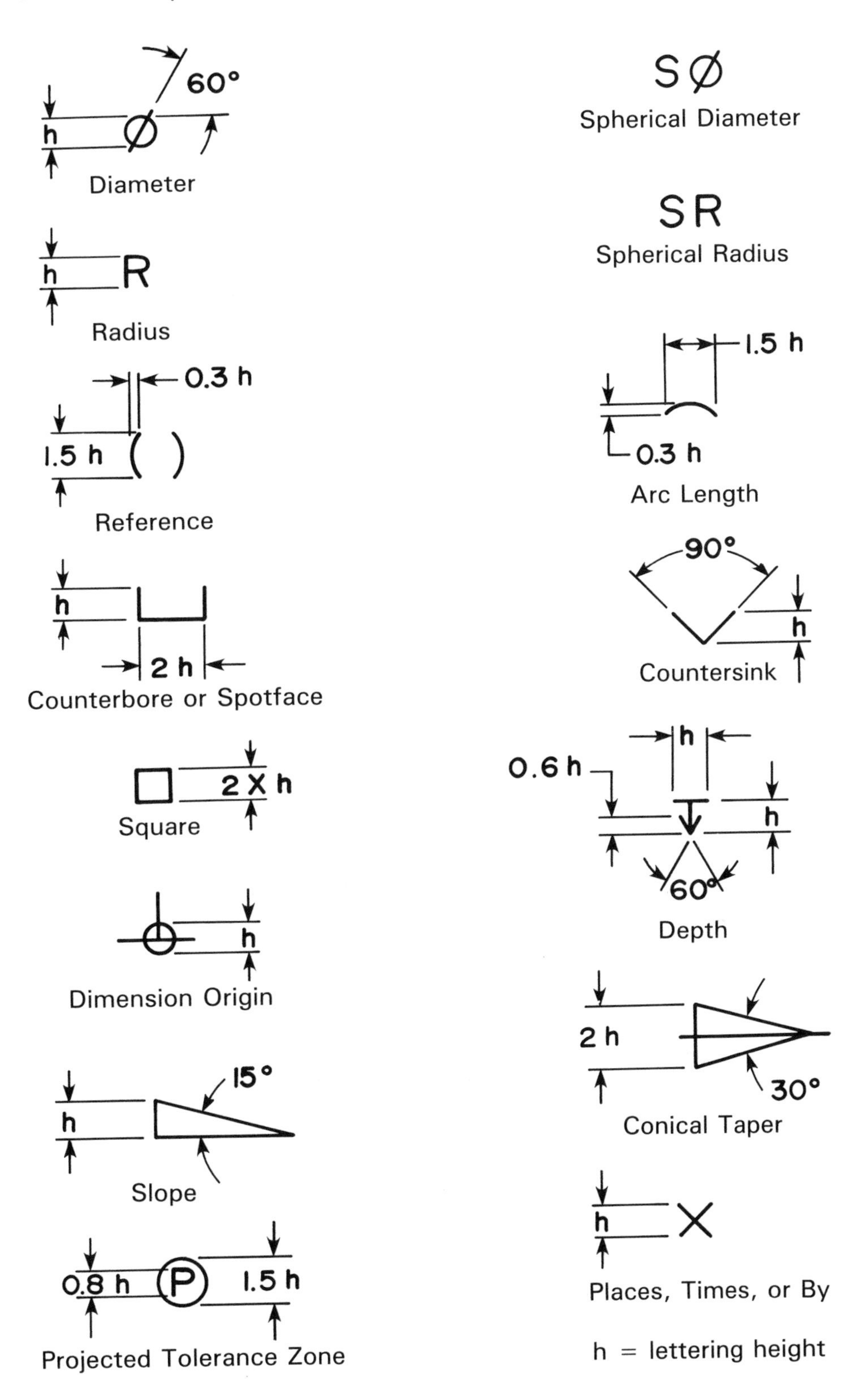

Example 2-11. Recommended Symbols.

COMPUTER AIDED DESIGN AND DRAFTING (CADD)

One of the advantages of using Computer Aided Design and Drafting (CADD) is the potential increase in productivity over manual drafting methods. This increase in productivity is achieved when the standard symbols used in drafting are created in a menu. The menu is a list of items such as computer commands, numbers, or symbols which the drafter may select as needed. There are several types of menus including keyboard menus, screen menus, tablet menus, button menus, and auxiliary menus.

The typewriter keyboard with additional function keys is commonly used with CADD workstations. Alphanumeric data (letters and numbers) may be entered into the computer to implement commands and drawing information.

The screen menu contains commands that may be selected using function keys, puck keys, a stylus, a mouse, or by touch depending on how the system is set up.

A tablet or digitizer menu is set up with commands and symbols that are customized for a specific purpose. This customized menu is often referred to as a symbol library, symbol directory, cell library, or CADD template.

Puck or mouse keys are used as button menus where screen or digitizer commands are selected by pressing a specific key (button). Additional buttons may be programmed to initiate a variety of commonly used commands.

Auxiliary menus include voice activated and auxiliary keyboard menus. Similar to the use of puck keys, computer functions and commands are activated by selecting the specific key or button on the auxiliary keyboard.

The advantage of using CADD exists because GT symbols may be placed on a drawing in much less time than it takes to manually draw the same symbol. The added advantage occurs when it is necessary to revise or change a drawing. The process of making engineering changes that often takes many hours manually may be done in a few minutes using CADD. It becomes a simple task to either remove or change symbols on the computer and then plot a new representation.

Pointing Devices

The most popular method of selecting items on a menu tablet is with a pointing device. Pointing devices refer to a variety of instruments that are attached to the digitizer or computer terminal. Some systems allow the drafter's finger to be used as the pointing device. These systems have function boards, buttons, sensitized menu boards, or sensitive computer screens.

The digitizer cursor is the most common pointing device. The digitizer cursor is an input device that is held in the hand and the screen cursor is a small box or lines crossing on the video display screen that indicate the current position. The puck and stylus shown in Example 2-12, are two forms of digitizer cursors.

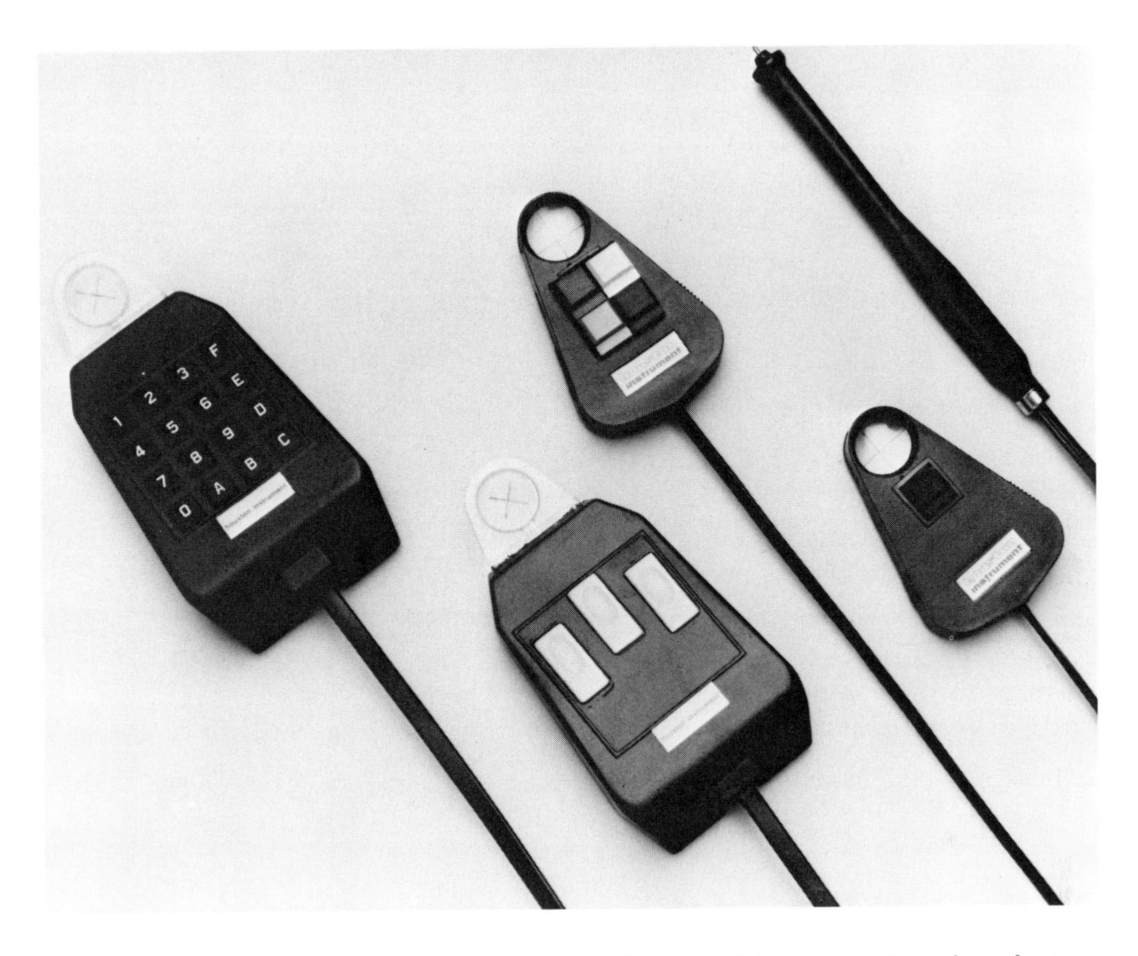

Example 2-12. Digitizer cursors. From left to right: a 16 button puck, a three button puck, a four button puck, a one button puck, and a stylus. (Houston Instrument)

A light pen is another type of pointing device that allows the drafter to digitize information into the computer by pressing a light pen on the screen at the desired location.

The mouse is a device that senses its position on a flat surface by movement of a ball or reflected light across a grid. When the mouse is moved across a flat surface, the screen cursor also moves. Buttons on the mouse activate specific functions or allow the user to choose from menu items displayed on the screen as shown in Example 2-13.

Example 2-13. Using a three button optical mouse to select items from a screen menu. (Summagraphics Corporation)

Digitizer

The digitizer, also referred to as a menu tablet or graphics tablet, is an electronic input device that allows data to be entered into the computer by pointing, using one of the pointing devices previously described. The puck or stylus senses movement through a magnetic field in the digitizer tablet. Several digitizer sizes are shown in Example 2-14.

Example 2-14. A variety of digitizer sizes. (Houston Instrument)

When the pointing device is placed on the digitizer cursor crosshairs or an aperture box, the video display screen registers the location. In most situations, a printed menu (referred to as a digitizer tablet menu) is placed on the digitizer tablet. Commands are displayed on the tablet menu which are used for specific applications such as mechanical or architectural drafting. Menus may also be customized with special symbols such as geometric tolerancing. A sample standard CADD menu, referred to as a template, is shown in Example 2-15. Notice that the top portion of this menu has been customized with GD&T symbols.

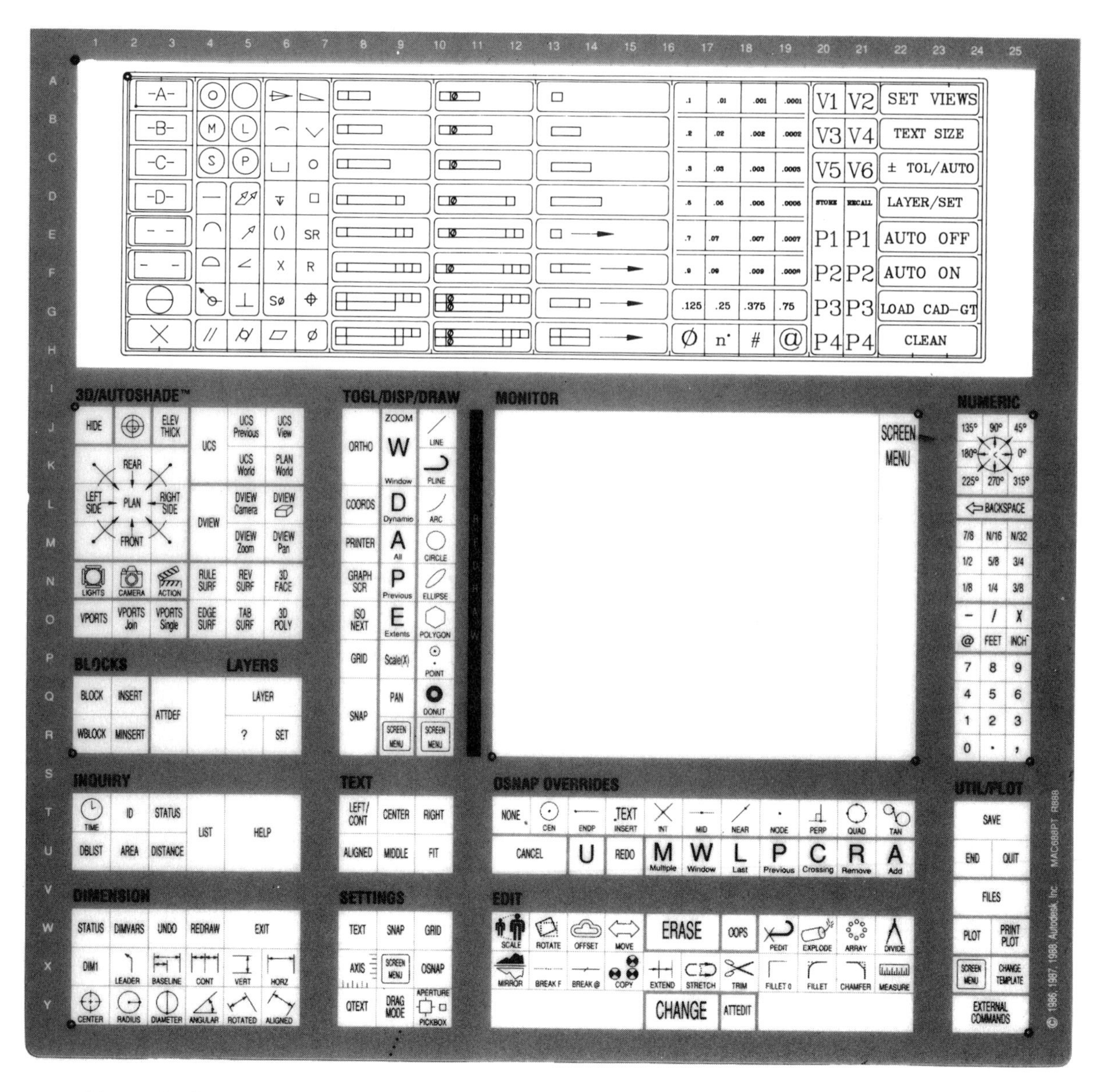

Example 2-15. A standard CADD menu. The top portion of this template may be customized with special symbol libraries. (AUTODESK, INC.)

The easiest way to use the menus is to place registration tabs on the digitizer that align with the prepunched holes on the preprinted menu sheet.

Menus are often made of polyester so they are durable and remain dimensionally stable. Customized overlays of specific symbols may be registered over the standard menu. The advantage is that several custom symbol libraries or templates may be designed and easily substituted by removing one and placing the next over the standard menu. An advantage of the tablet menu is the user can pick commands by association with a symbol on the menu.

CADD GENERATED GEOMETRIC TOLERANCING SYMBOLS

Before you begin creating a geometric tolerancing symbol refer to this text and the ANSI Y14.5M-1982 standard to determine the proper size and format for the symbols. Make a sketch of the desired symbol and decide how it will be placed on a drawing. Name the symbol, this will also be the CADD file name. Select a point on each symbol that will become a convenient point of origin when placing the symbol on a drawing. The point of origin is the location position for placing a symbol on a drawing. Example 2-16 shows the insertion point determined for a sample of geometric tolerancing symbols. The insertion point may be any corner of the symbol, but the user should consider the insertion point as the most common point of origin.

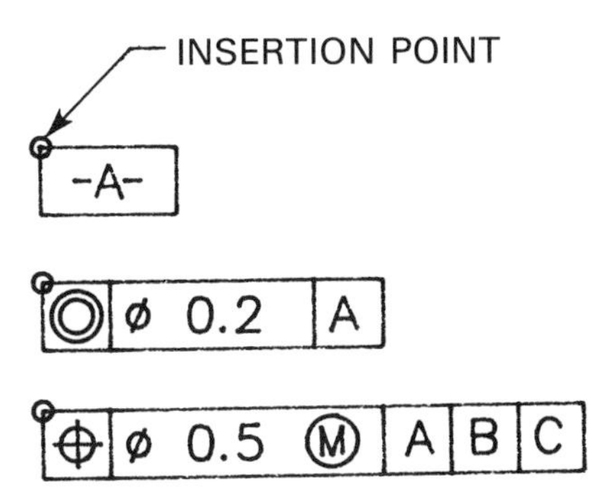

Example 2-16. Selecting the point of origin for geometric tolerancing symbols.

After you have drawn a specific symbol, it must be stored as a symbol by using such commands as SYMBOL TO DISK, STORE AS SYMBOL, or CREATE SYMBOL. When you have created a group of symbols it is time to organize them into a symbol library using a command such as CREATE MENU, or TEMPLATE. A common computer prompt will be: TYPE THE SYMBOL NAME. Then digitize the box on the template where you want the symbol located. Follow this same process until all of the symbols that you created are located on the template. When all of the symbols have been placed on the template, assign a name such as GEOMETRIC TOLERANCING TEMPLATE, or LIBRARY. Print out or plot the symbols on a sheet so they correspond to the locations in which they were placed on the template. This new symbol library may then be placed as an overlay on the menu tablet.

Symbol templates may be selected for use by entering a command such as SELECT, or ACTIVATE MENU; or by using the pointing device to select the template name from the menu tablet. When the symbol template is ready for use, individual symbols may be entered on a drawing by pressing the pointing device inside the specific box on the symbol library and then pressing the desired location on the drawing. Some systems automatically display the symbol at the crosshairs location on the screen. As the pointing device is moved, the symbol also moves on the screen. When the symbol is positioned in the desired location, a button is pushed that places the symbol. Some symbols may require informational prompts after placement such as: ZONE DESCRIPTOR, GEOMETRIC TOLERANCE, MATERIAL CONDITION, and/or DATUM REFERENCE.

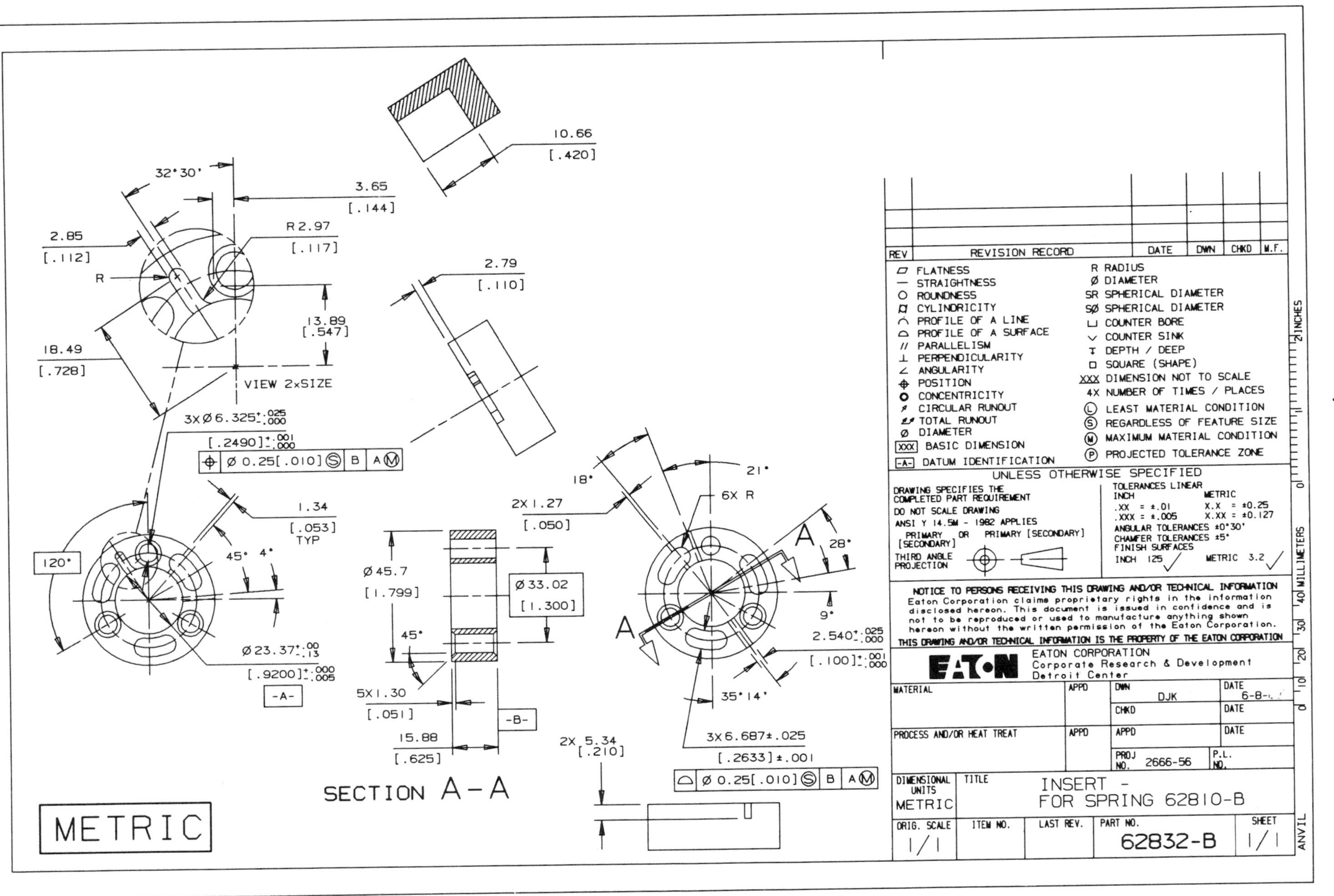

Example of typical industrial CADD drawing using GD&T symbols.

TEST 2, SYMBOLS AND TERMS

Name: ______________________

A dimensioning and tolerancing template is recommended for drawing proper symbols on this test and future tests.

1. List the five basic types of geometric tolerancing and dimensioning symbols.
   1. ______________________
   2. ______________________
   3. ______________________
   4. ______________________
   5. ______________________
2. Name the five types of geometric characteristic symbols.
   1. ______________________
   2. ______________________
   3. ______________________
   4. ______________________
   5. ______________________
3. Name each of the following geometric characteristic symbols.

— ______________	// ______________
⏥ ______________	⊥ ______________
○ ______________	∠ ______________
⌭ ______________	↗ or ↗ ______________
⌒ ______________	⌰ or ⌰ ______________
⌓ ______________	⌖ ______________
	◎ ______________

4. Any letter of the alphabet may be used to identify a datum except for ________ or ________.
5. When may datum feature symbols be repeated on a drawing?

   ____
6. What information is placed in the lower half of the datum target symbol?

   ____
7. What information is placed in the top half of the datum target symbol?

   ____

   ____

   ____

8. Label the parts of the following feature control frame.

⊕	Ø 0.25 Ⓜ	A	B	C

(A) ____________________

(B) ____________________

(C) ____________________

(D) ____________________

(E) ____________________

(F) ____________________

(G) ____________________

9. Completely define the term basic dimension.

10. How are basic dimensions shown on a drawing?

11. Name the following symbols.

Ø ____________	□ ____________
R ____________	( ) ____________
⌴ ____________	⏀ ____________
∨ ____________	SØ ____________
↧ ____________	SR ____________
X ____________	⌒ ____________

Chapter 3

DATUMS

Datums are requirements for referencing features of an object. You will see that DATUMS are considered to be theoretically perfect planes, surfaces, points, lines, or axes from which measurements are made. These datums are used by the machinist, toolmaker, or inspector of quality control to ensure that the part is in agreement with the drawing. Examples of manufacturing datums may be machine tables, surface tables, or specially designed rotation devices. The feature on the part that is called out as a datum feature may not be as perfect as the device that has been selected as the datum plane, because all measurements are made from the manufacturing and inspection tooling.

This chapter is intended to provide instruction on the interpretation and proper drafting of datum features and datum related symbols without regard to specific inspection and tooling techniques.

DATUMS

A DATUM is considered to be a plane, surface, surface point(s), line, or axis of an object or part. These items are referred to as datum features. A datum is assumed to be exact. Location and size dimensions are established from the datum. There are many concepts to keep in mind when datums are established, including the function of the part or feature, manufacturing processes, methods of inspection, the shape of the part, relationship to other features, assembly considerations, and design requirements. Datum features should be selected to match on mating parts, should be easily accessible, and should be of adequate size to permit control of the datum requirements.

DATUM FEATURE SYMBOL

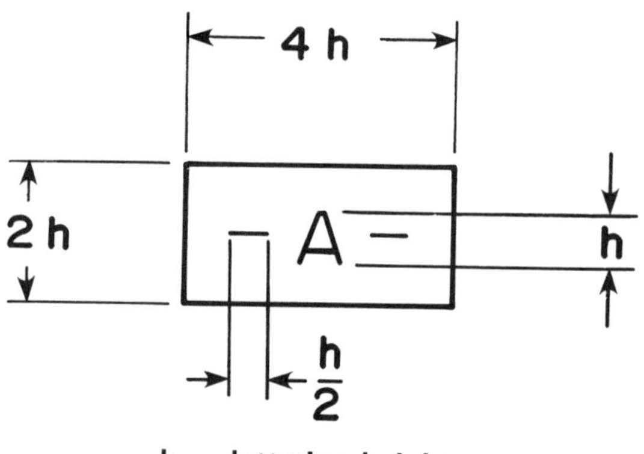

h = lettering height

Example 3-1. Datum feature symbol.

DATUM PLANE, DATUM SURFACE FEATURE, HIGH POINT CONTACT

A DATUM PLANE is a theoretically exact plane.

A DATUM SURFACE OR FEATURE is the actual surface of an object that is used to establish a datum plane. Look at Example 3-2. You will see that the datum surface touches the datum plane at the high points of the surface. This is known as high point contact. Keep in mind that Example 3-2 is a magnified view of the relationship between the datum plane and the datum surface.

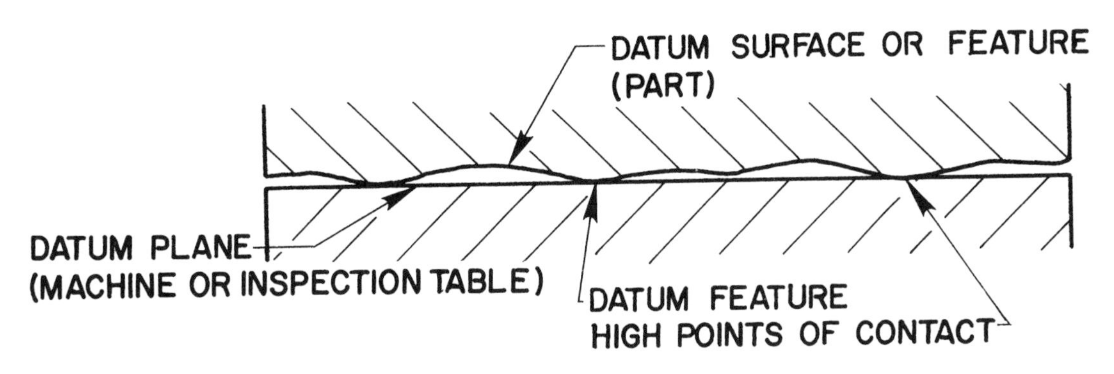

Example 3-2. Datum plane, datum surface, datum feature.

When a datum surface is used on a part, the datum feature symbol is placed on an extension line in the view where the surface appears as a line. See Example 3-3. A leader line may also be used to connect the datum feature symbol to the view in some applications.

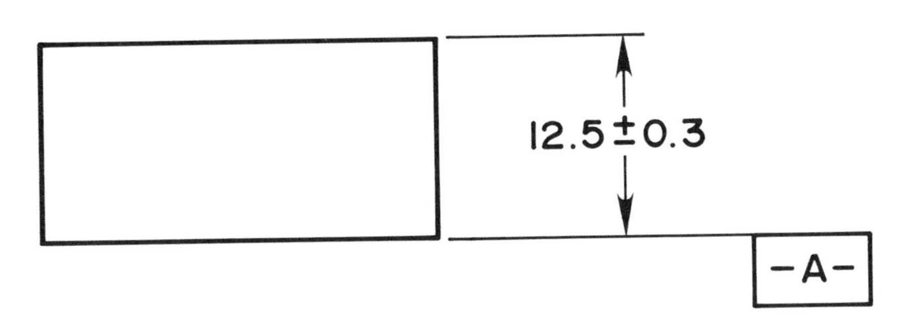

Example 3-3. The datum feature symbol is placed on an extension line in the view where the datum surface appears as a line.

When high point contact is used to establish a datum, the datum surface may be controlled by flatness. The flatness specification is not required in these cases. Measurements taken from a datum plane do not take into account any variations of the datum surface from the datum plane. Any geometric tolerance applied to a datum should only be specified if the design requires the control. Example 3-4 shows the feature control frame and datum feature symbol together. Notice that the feature control frame should be placed first.

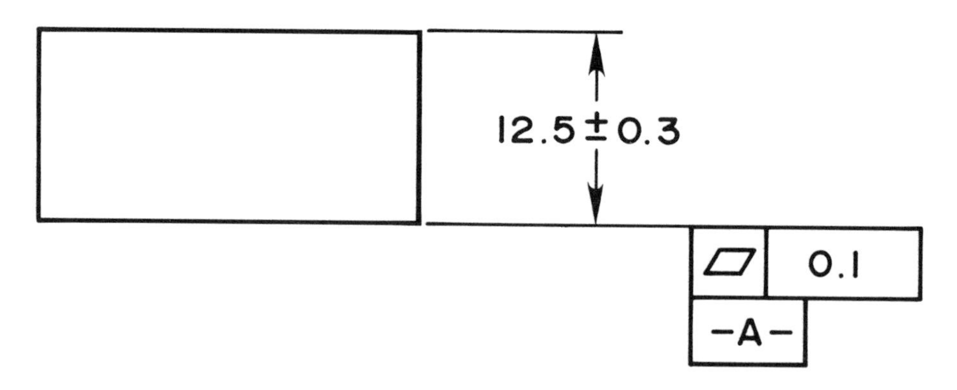

Example 3-4. A feature control frame and datum feature symbol combined.

Look at Example 3-5 and you will see a magnified representation showing the meaning of the drawing in Example 3-4.

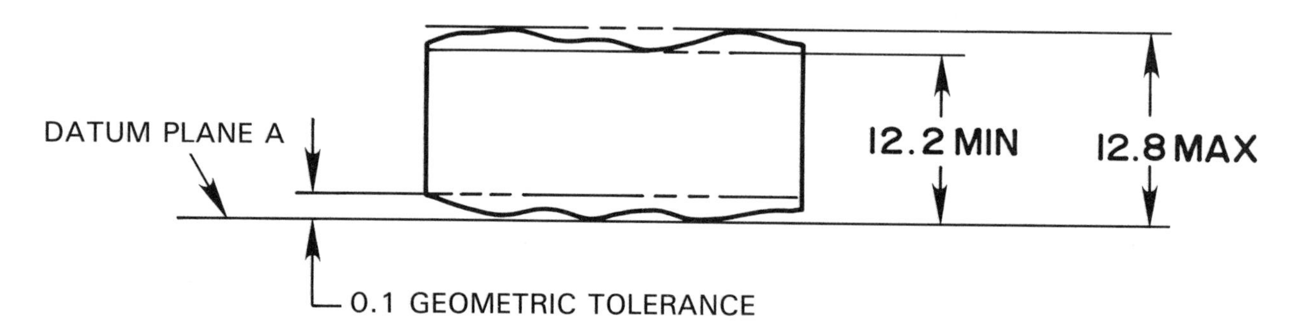

Example 3-5. The meaning of the drawing in Example 3-4.

The geometric tolerance of 0.1 is specified in the feature control frame.

The maximum size that the part can be produced is the upper limit of the dimensional tolerance or MMC = 12.5 + 0.3 = 12.8.

The minimum size that the part can be produced is the lower limit of the dimensional tolerance or LMC = 12.5 − 0.3 = 12.2.

DATUM REFERENCE FRAME CONCEPT

A complete datum frame is made up of three datum planes. They are identified as primary, secondary, and tertiary. See Examples 3-6.

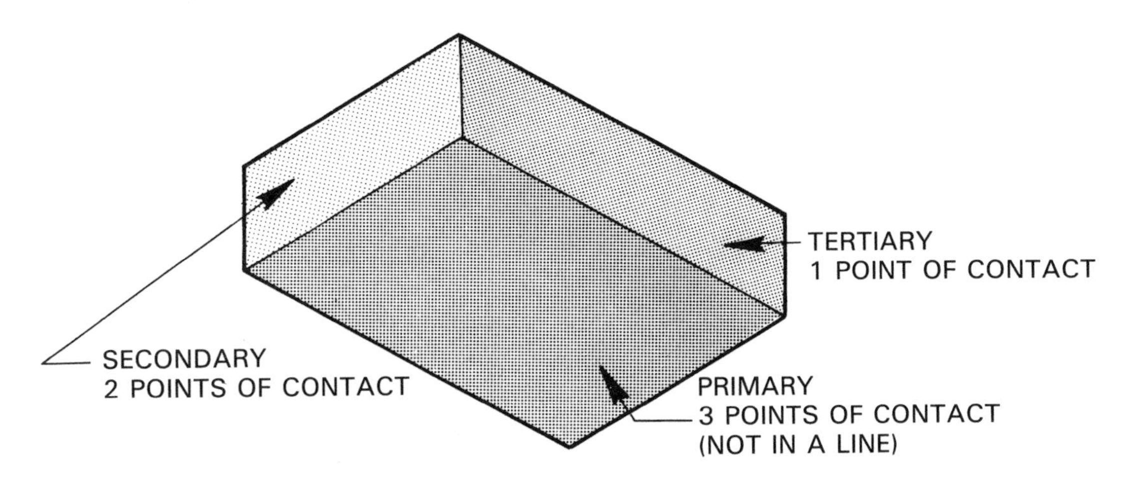

Example 3-6. Datum reference frame.

The datum reference frame exists in theory only, not on the part. Selection of the primary, secondary, and tertiary datums is made by the functional importance of the object. The secondary datum is established with perpendicularity to the primary datum, depending on design requirements. Often times parts may not always have surfaces 90° to each other such as castings, forgings, or sheet metal. Surfaces with draft or other characteristics may specify an angle other than 90°, for example 87°.

Sufficient datum features are chosen to position the part in relation to a set of three mutually perpendicular planes, as previously shown. This three plane concept has all datum planes on a part intersecting at right angles which are 90° basic by interpretation. It is necessary to establish a method for simulating the theoretical reference frame from the actual features of the part. This simulation is accomplished by positioning the part on appropriate datum features to relate the part to the reference frame and to restrict motion of the part in relation to the reference frame. Refer to Example 3-7.

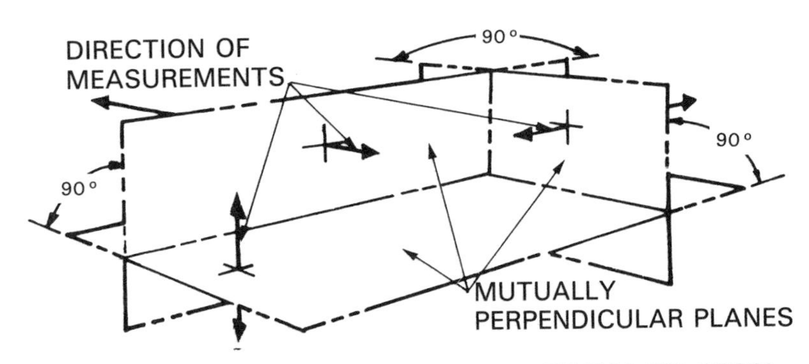

Example 3-7. Datum reference frame. (ANSI Y14.5M-1982)

When referring to the datum frame, the primary datum should be given first, followed by the secondary and tertiary datums. This is known as DATUM PRECEDENCE. For instructional purposes, this workbook labels datums conveniently as A, B, and C. In industry, other letters are also used to identify datums, such as D, E, F, or X, Y, Z. The letters O, Q, and I should be avoided as they may resemble numbers. Example 3-8 shows a part where three datum features are surfaces.

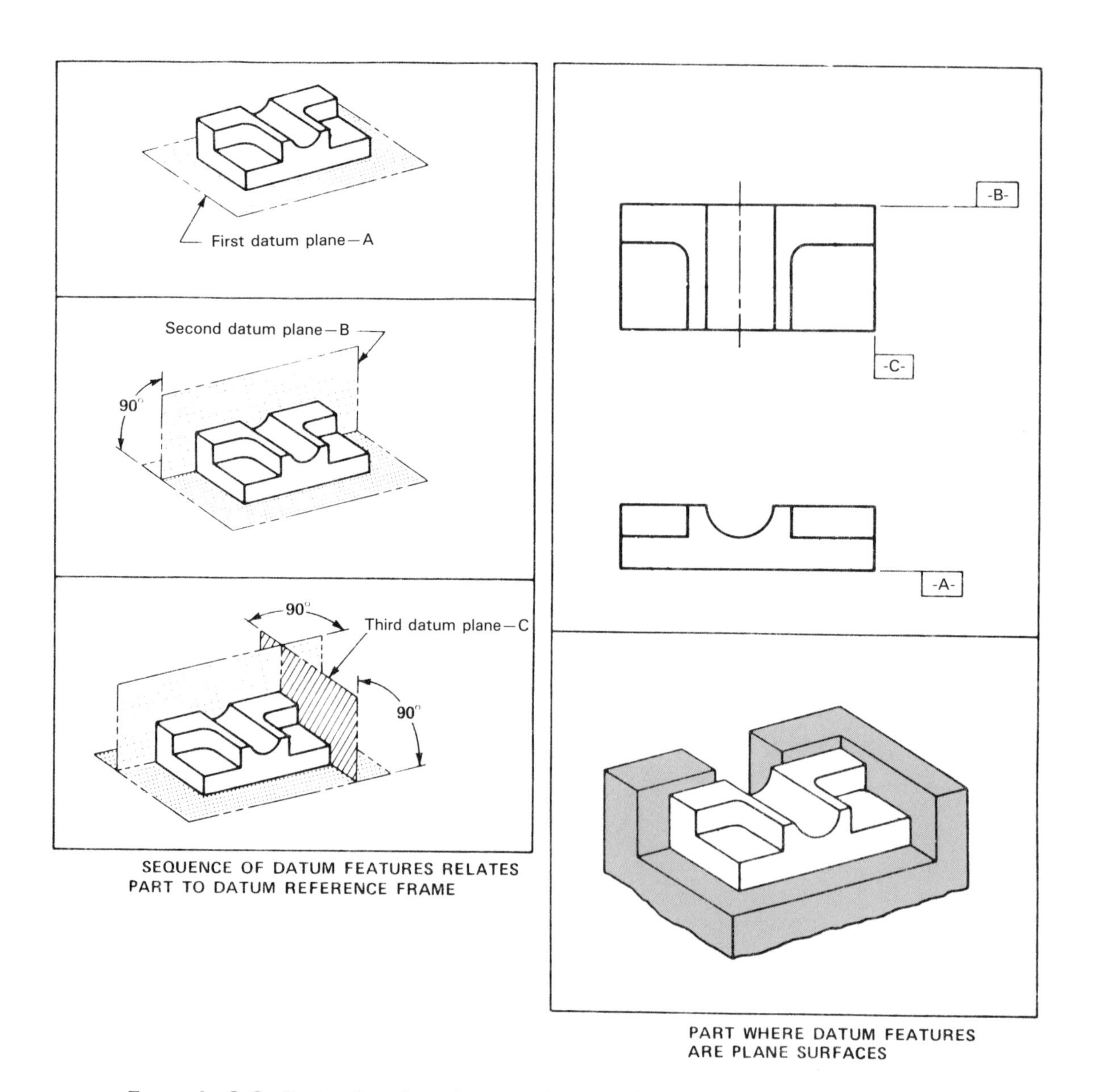

Example 3-8. Part related to datum reference frame. (ANSI Y14.5M-1982)

DATUM TARGET SYMBOLS

In many situations it is not possible to establish an entire surface or surfaces as datums. When this happens due to the size or shape of the part, then datum targets may be used to establish datum planes. This procedure is especially useful on parts with surface or contour irregularities such as some sheet metal, sand cast, or forged parts which are subject to bowing or warpage. This method may also be applied to weldments where heat may cause warpage. DATUM TARGETS are designated points, lines, or surface areas that are used to establish the datum reference frame. The datum target symbol is drawn as a circle using thin lines, and is connected with a leader that identifies a target point, line, or surface area. See Example 3-9, 3-10, and 3-11.

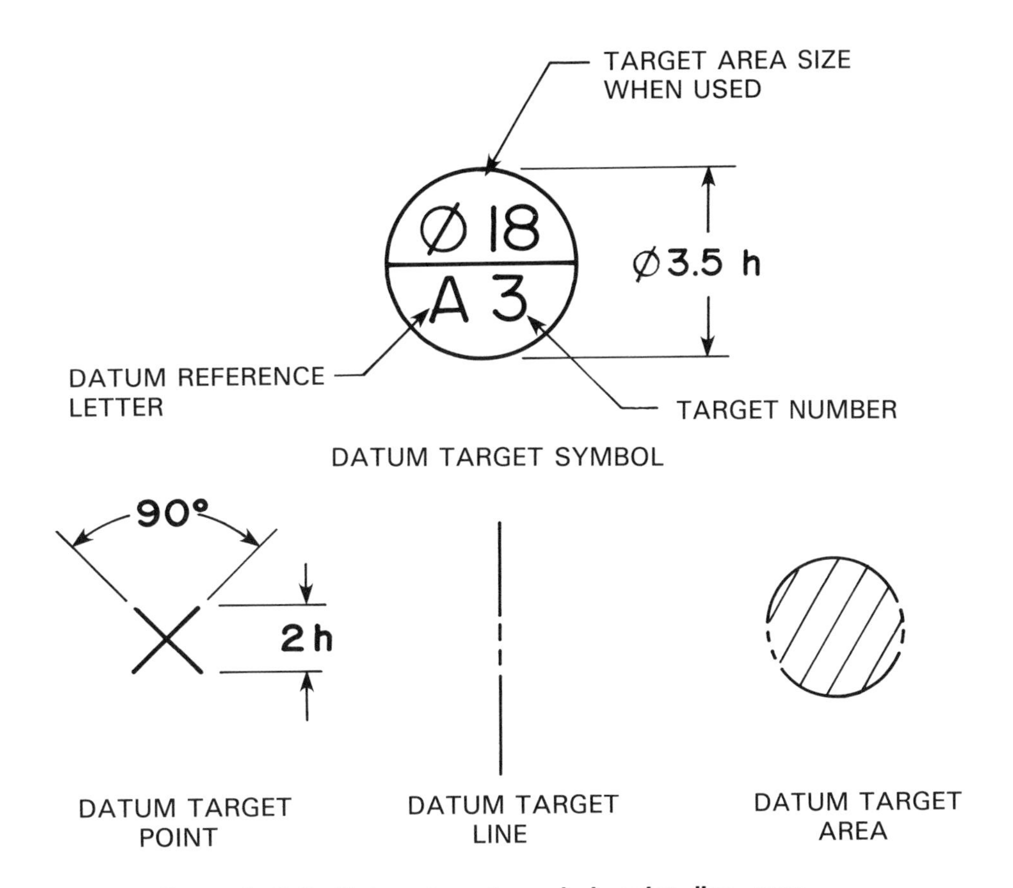

Example 3-9. Datum target symbol, point, line, area.

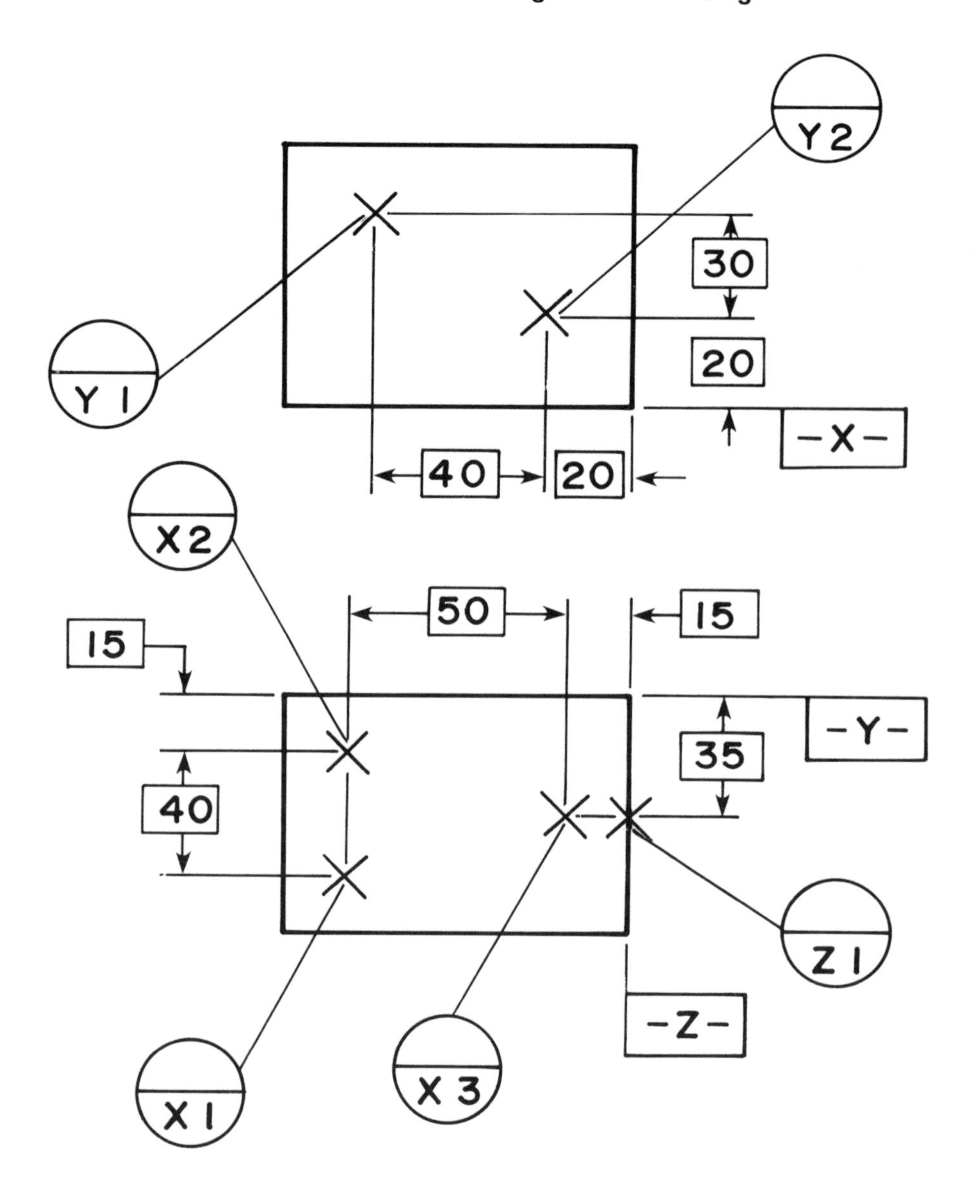

Example 3-10. Datum target symbol and point usage.

DATUM TARGET POINTS

Datum or chain dimensioning may be used to locate datum target points. The location dimensions must originate from datums. Datum target points are established on the drawing using basic or tolerance dimensions. Datum targets are established on the part with fixtures and with pins. These pins contact the part where the datum targets are specified. The datum planes are established by the datum points as follows:

The primary datum plane must be established by at least three points on the primary datum surface. These points are used to provide stability on the primary plane, similar to a three legged stool.

The secondary datum plane must be located by at least two points on the related secondary datum surface. Two points provide the required stability for the secondary plane.

The tertiary datum plane must be located by at least one point on the related tertiary datum surface. One point of contact at the tertiary datum plane is all that is required to complete the datum reference frame. Example 3-11 shows a pictorial drawing of the datum target points on the primary, secondary, and tertiary datums.

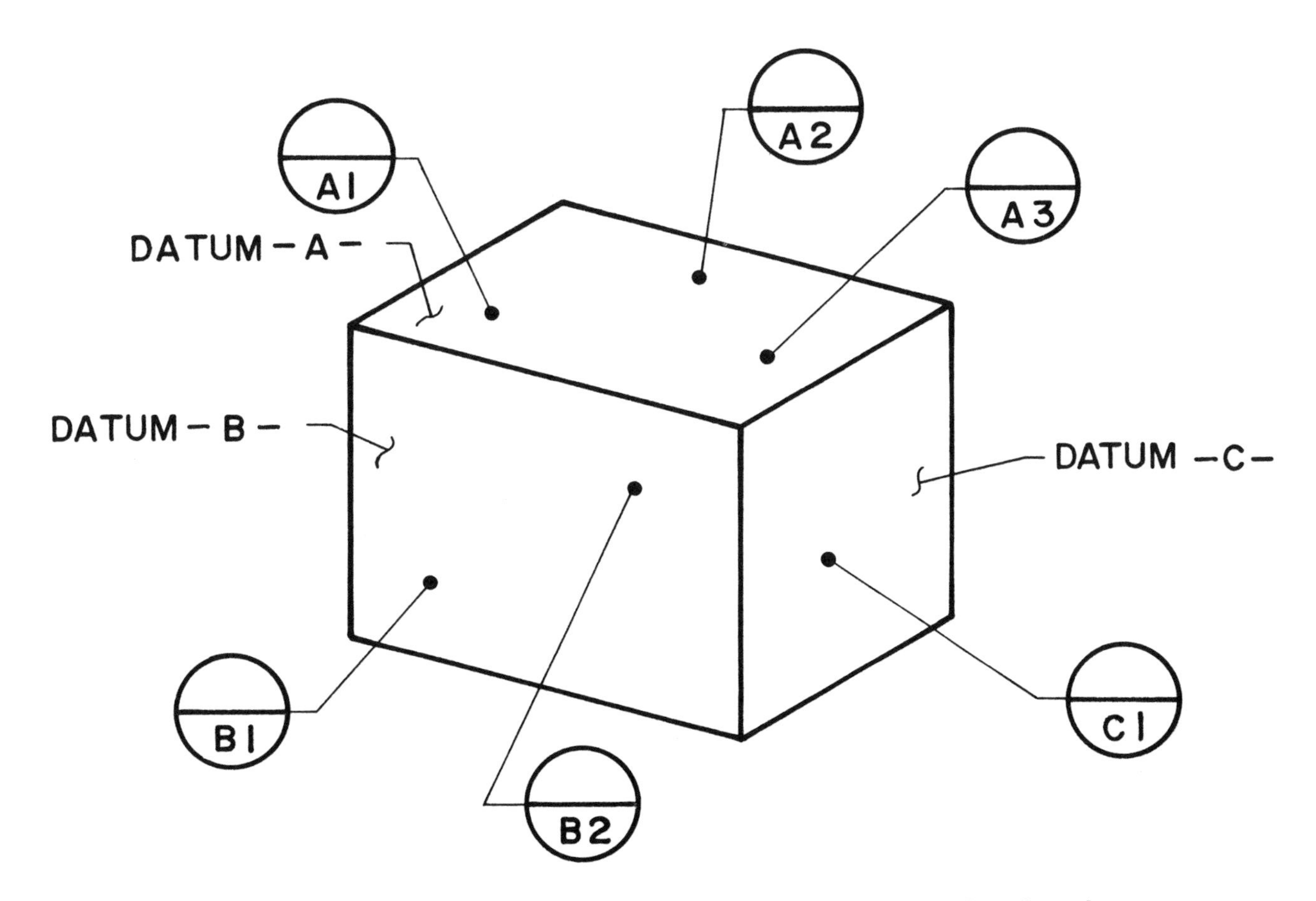

Example 3-11. Datum target points on the primary, secondary, and tertiary datums.

Note that each point is identified with a datum target symbol, for example

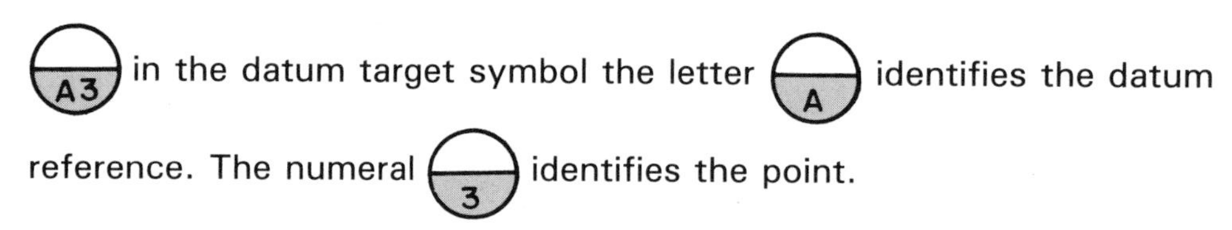

in the datum target symbol the letter identifies the datum reference. The numeral identifies the point.

The datum points may be located with basic dimensions or tolerance dimensions. Look at Example 3-12 for a multiview representation of the example in Example 3-11 using basic dimensions to locate the datum target points. Remember, the datum feature symbols appear on the view where the datum surface is a line and the datum points are located on the surface view of the related datum.

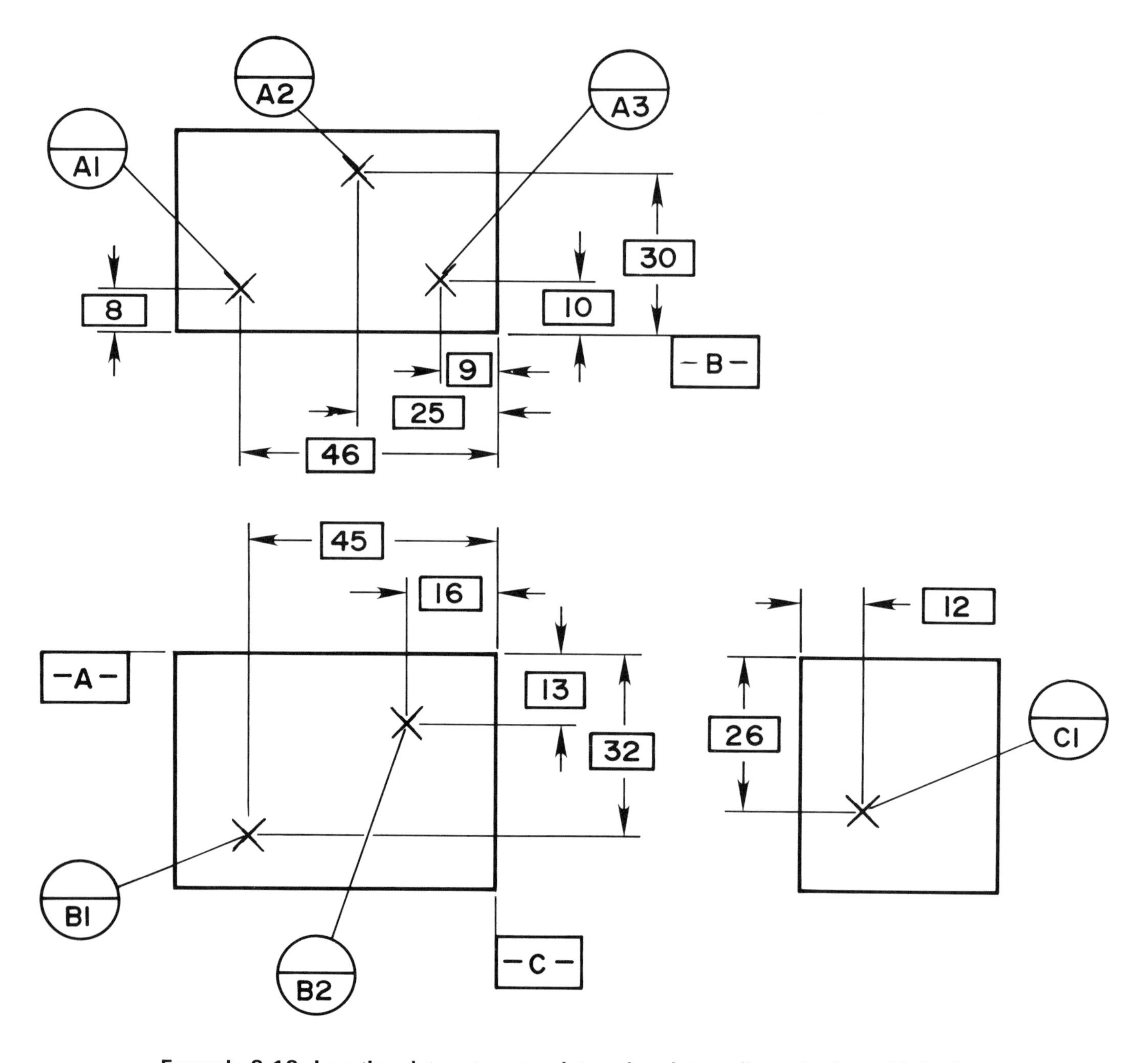

Example 3-12. Locating datum target points using datum dimensioning with basic dimensions, (chain dimensioning with tolerance dimensions may also be used).

When datum target points are used on a drawing to identify a datum plane, the actual datum feature is established by locating pins at the datum points as shown in the magnified representation in Example 3-13.

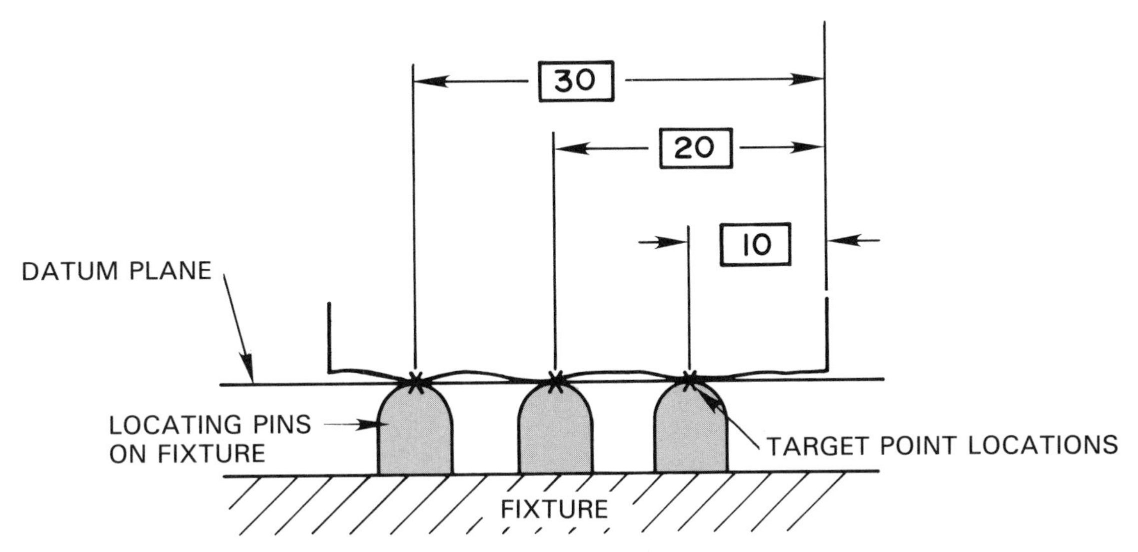

Example 3-13. Datum target points established with locating pins.

DATUM TARGET AREAS

Areas of contact may also be used to establish datums. When this is done, the shape of the datum target area is outlined by phantom lines with section or cross hatch lines through the area. Circular areas are dimensioned with basic dimensions to locate the center. The diameter of the target area is provided in the upper half of the datum target symbol as shown in Example 3-14.

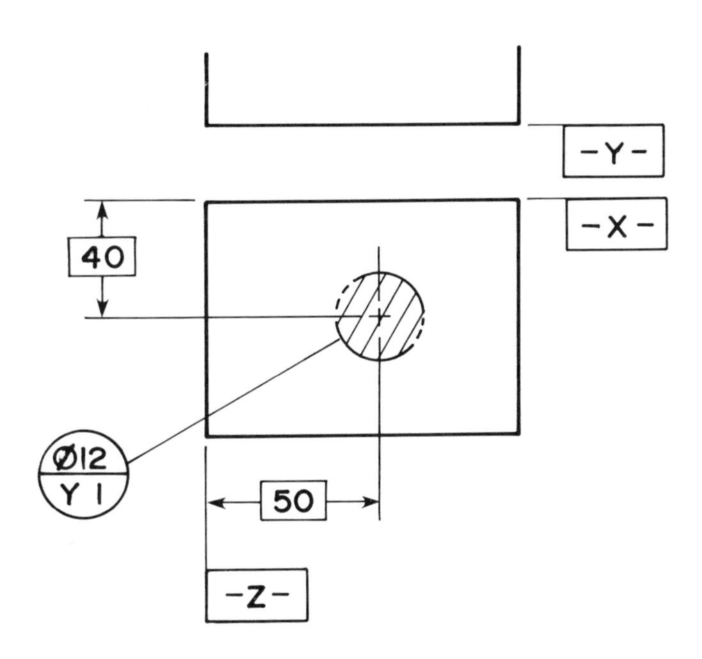

Example 3-14. Datum target area.

When the area is too small to draw or scale, then a datum target point is used at the center location. The datum target symbol identifies the diameter as shown in Example 3-15.

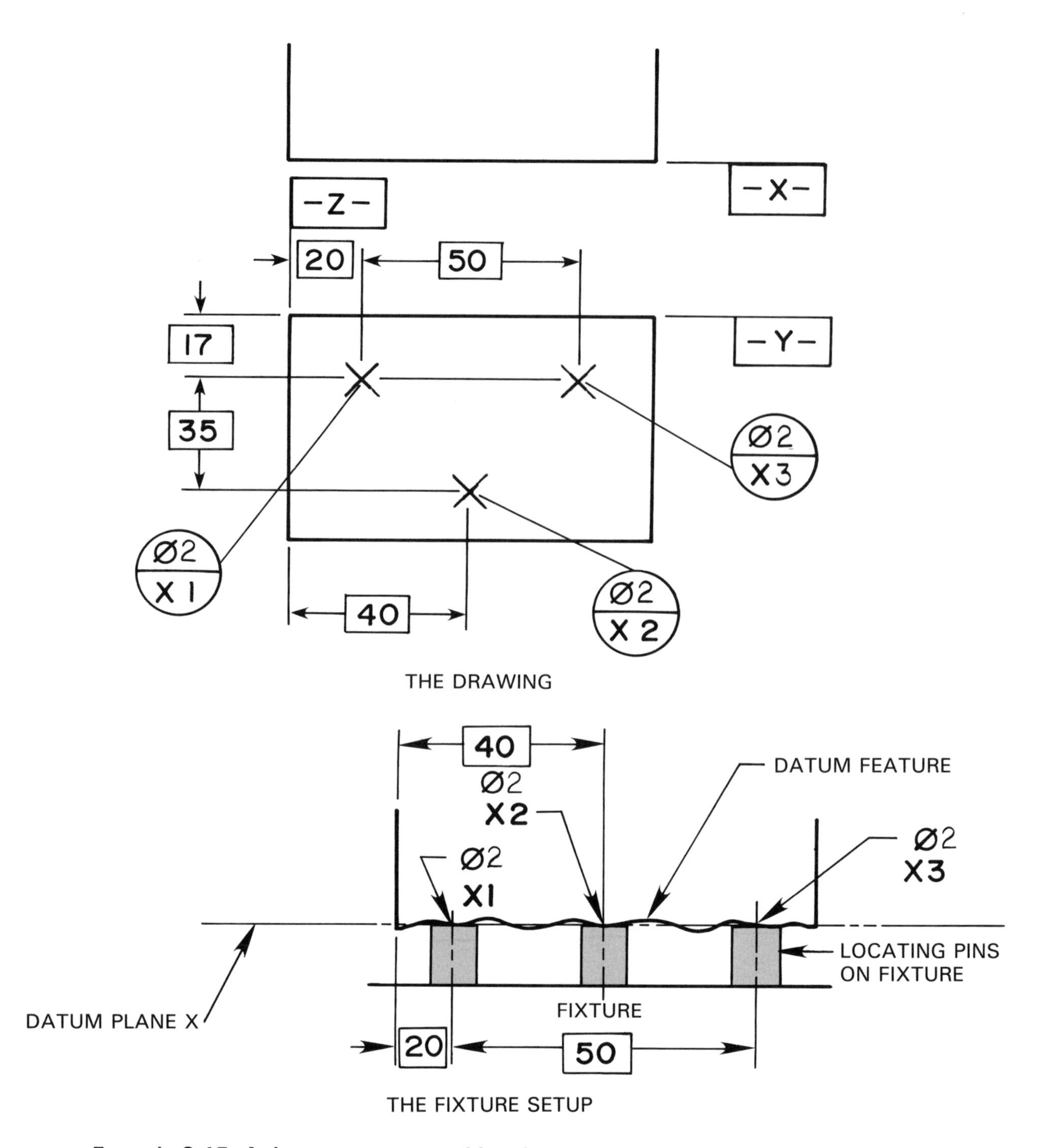

Example 3-15. A datum target area with a datum target point. The area size is identified in the top half of the datum target symbol.

Datum target areas are established by locating pins which contact areas established by pointed pins, a knife edge, rounded pins, or flat end pins as standard tooling hardware. The locating pins shown in Example 3-16 are the rounded type.

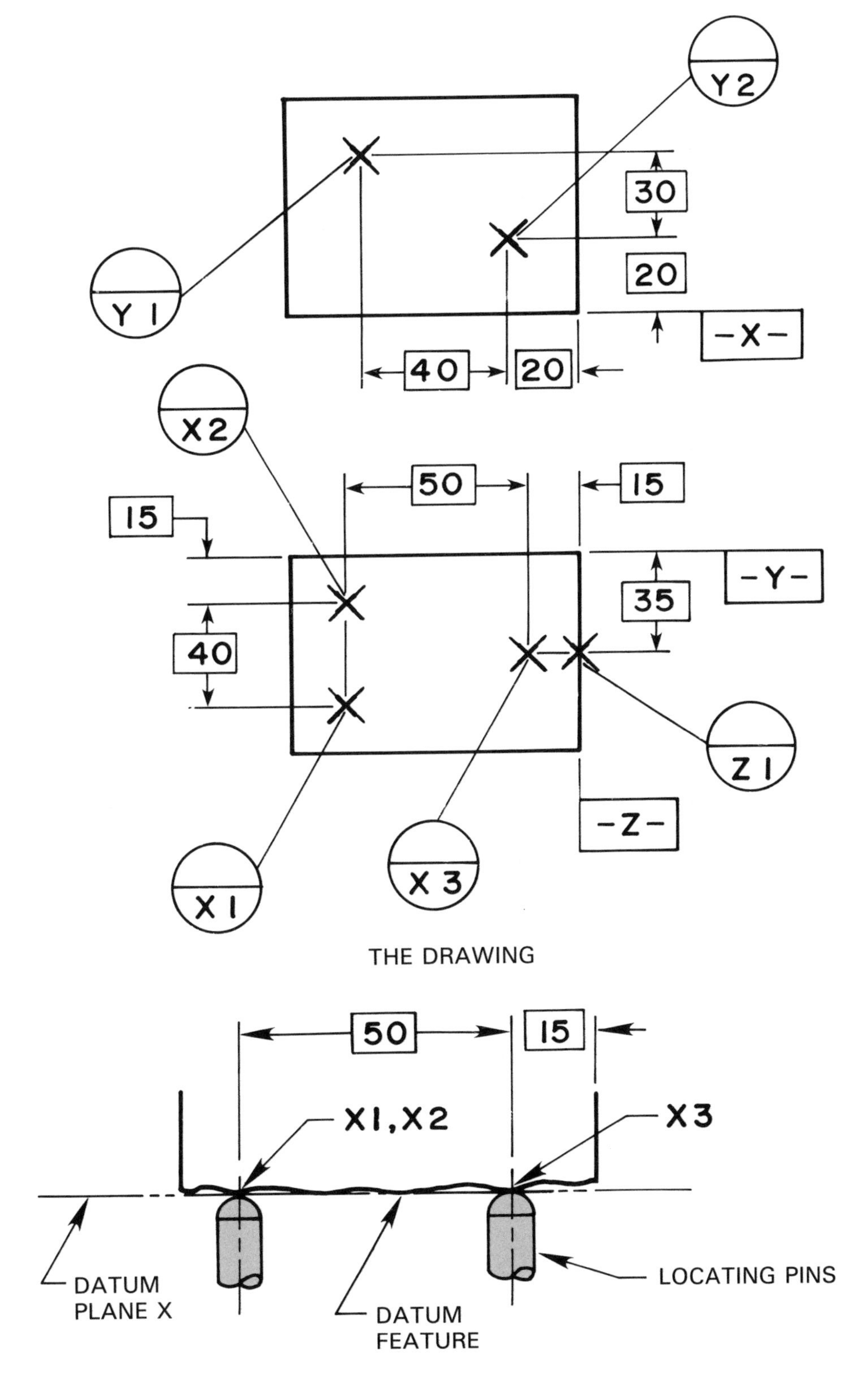

Example 3-16. Datum target points established with rounded locating pins.

DATUM TARGET LINES

A DATUM TARGET LINE is indicated by the target point symbol "X" on the edge view of the surface, or by a phantom line on the surface view, or by both methods. Both representations are clear, although not always possible, Example 3-17.

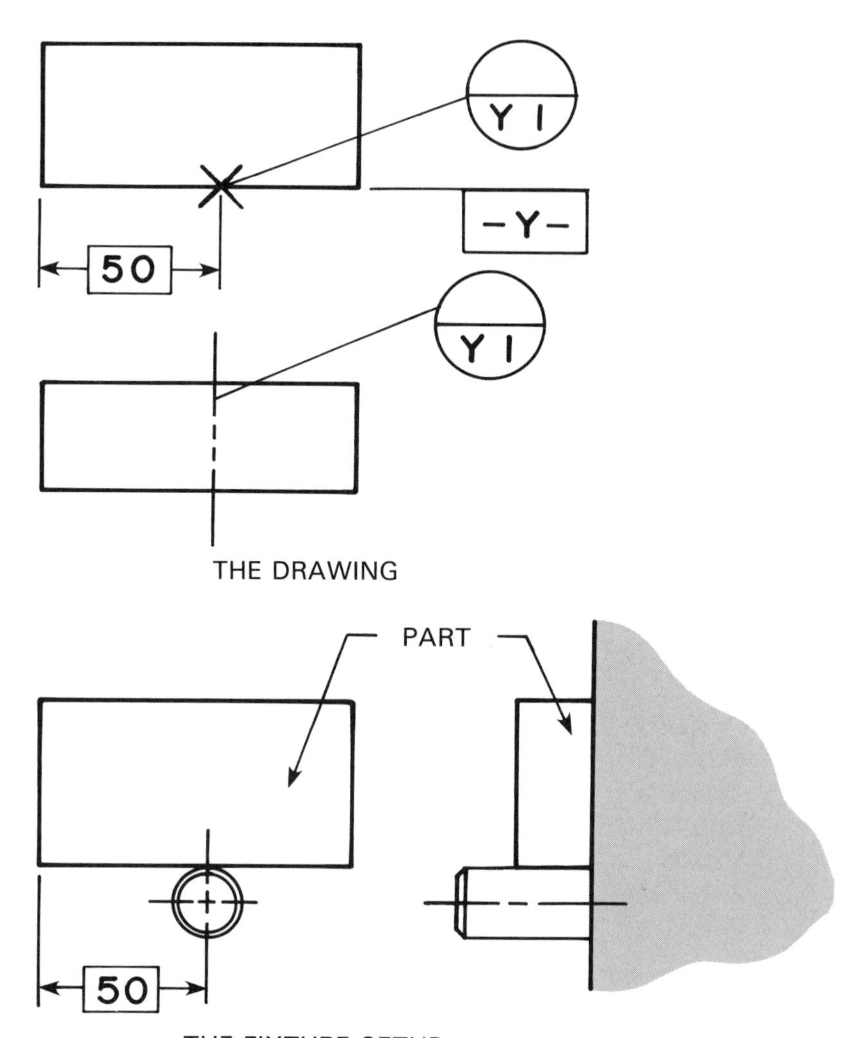

Example 3-17. Datum target line.

PARTIAL DATUM SURFACE

A portion of a surface may be used as a datum. For example, this may be done when a part has a hole or group of holes at one end where it may not be necessary to establish the entire surface as a datum to effectively locate the features. This may be accomplished on a drawing using a chain line dimensioned with a basic dimension to show the location and extent of the partial datum surface. The datum feature symbol has a connecting leader that points to the chain line as shown in Example 3-18.

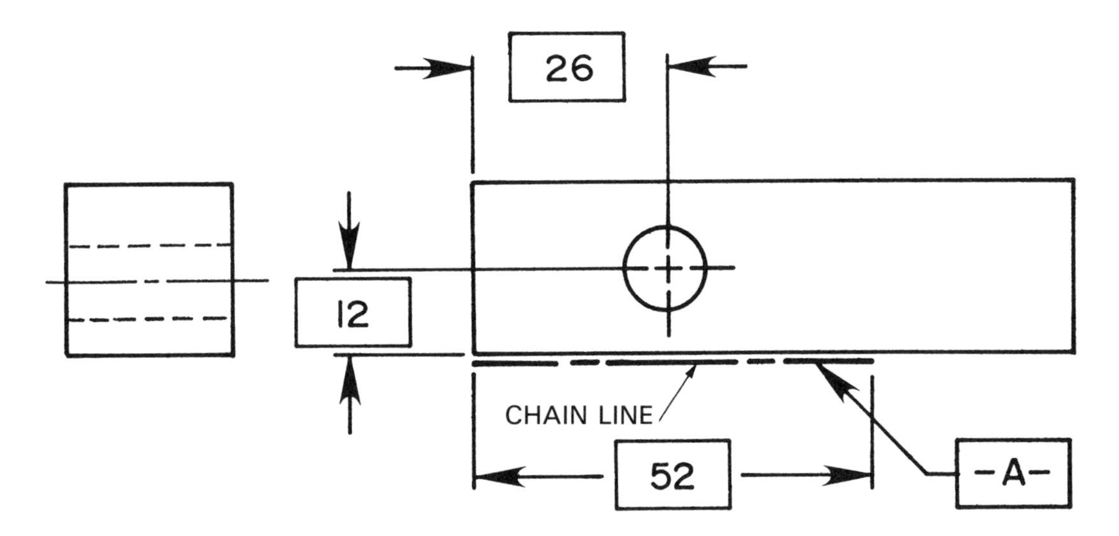

Example 3-18. Partial datum surface.

The datum plane is then established at the location of the chain line as shown in Example 3-19.

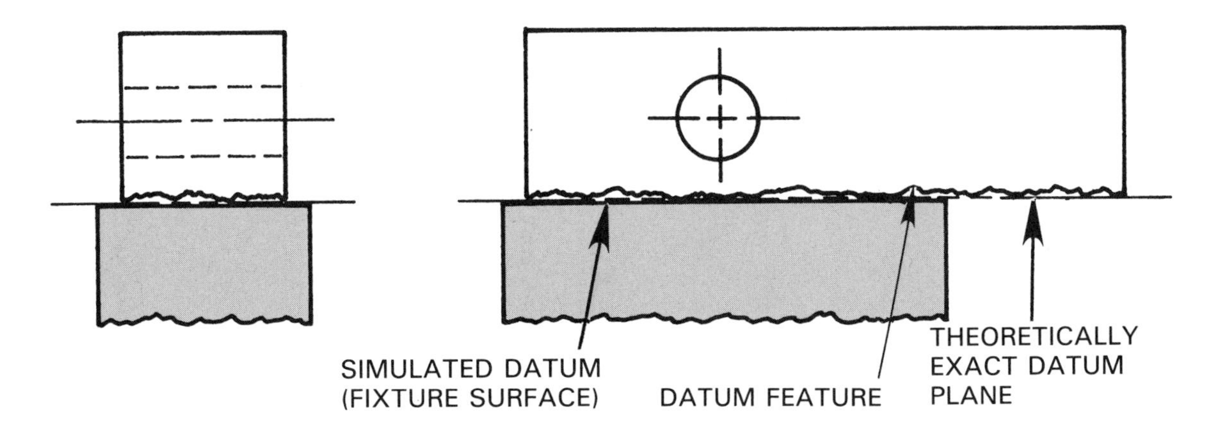

Example 3-19. The meaning of the partial datum surface.

COMPOUND DATUM FEATURES

When more than one datum feature is used to establish a single datum, the datum reference letters are separated by a dash and placed in one compartment of the feature control frame. These datum reference letters are of equal importance and may be placed in any order. See Example 3-20. This same application may be used for COAXIAL DATUM AREAS (datum features established on the same axis). A datum axis established by coaxial datum features is normally used as a primary datum used for controlling coaxiality of adjacent diameters.

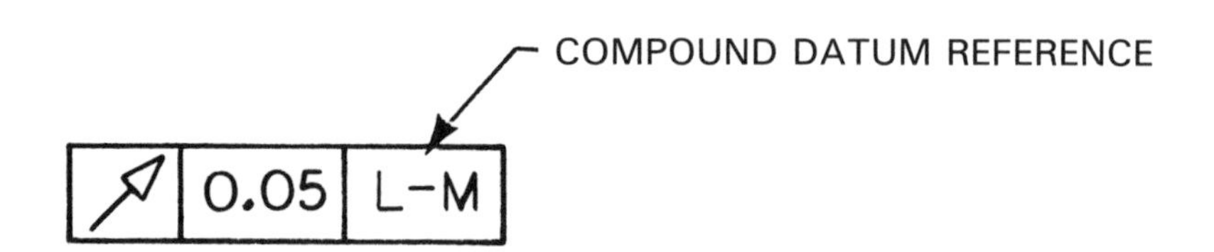

Example 3-20. Compound datum reference.

MULTIPLE DATUM REFERENCE FRAME

Depending on the functional requirements of a part, more than one datum reference frame may be established. In Example 3-21, datums X, Y, and Z constitute one datum reference frame, while datums L and M establish a second reference frame. The relationship between the two datum reference frames is controlled by the angularity tolerance on datum feature L.

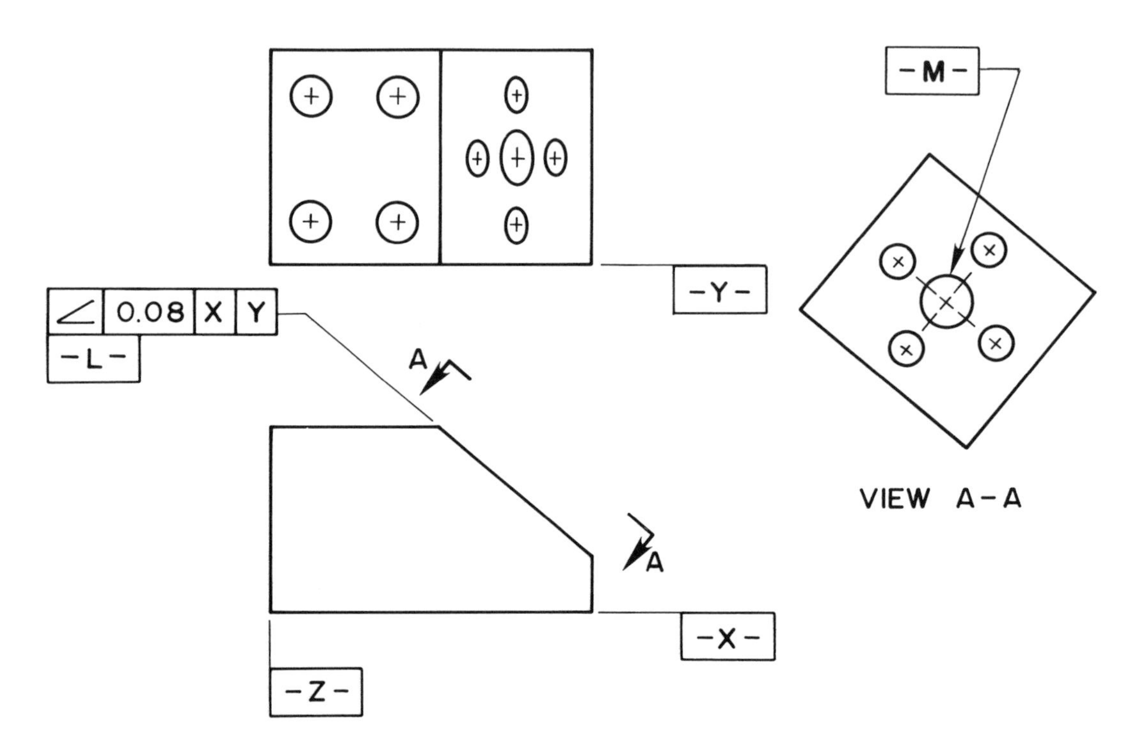

Example 3-21. Multiple datum reference frames.

COPLANAR SURFACE DATUMS

Coplanar surfaces are two or more surfaces that are in the same plane. The relationship of coplanar surface datums establishes the surfaces as one plane or datum in correlated feature control frame specifications. A phantom line is placed between the surfaces and the surfaces are treated as a single noncontinuous surface. The number of surfaces may be specified in a note such as, 2 (number of surfaces) SURFACES. See example 3-22. This concept will also be discussed in Chapter 5 with an application for profile tolerances of coplanar surfaces.

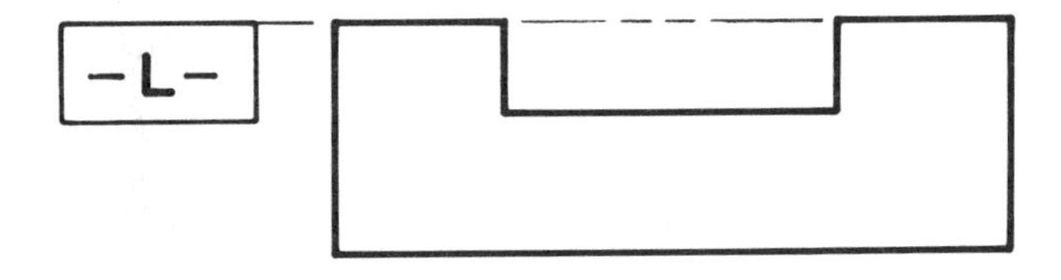

THE DRAWING

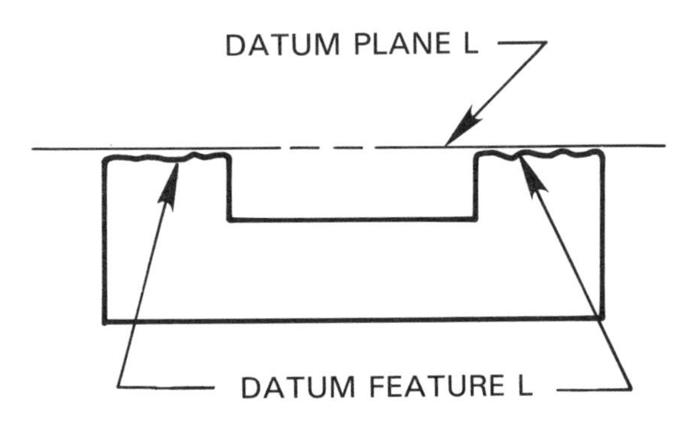

THE MEANING

Example 3-22. Coplanar surface datums.

DATUM AXIS

A cylindrical object may be a datum feature. When the cylindrical datum feature is used, the center axis is known as the datum axis. There are two theoretical planes intersecting at 90°, these planes are represented by the centerlines of the drawing. Where these planes intersect is referred to as the datum axis. The datum axis is the crigin for related dimensions while the X and Y planes indicate the direction of measurement as shown in Example 3-23.

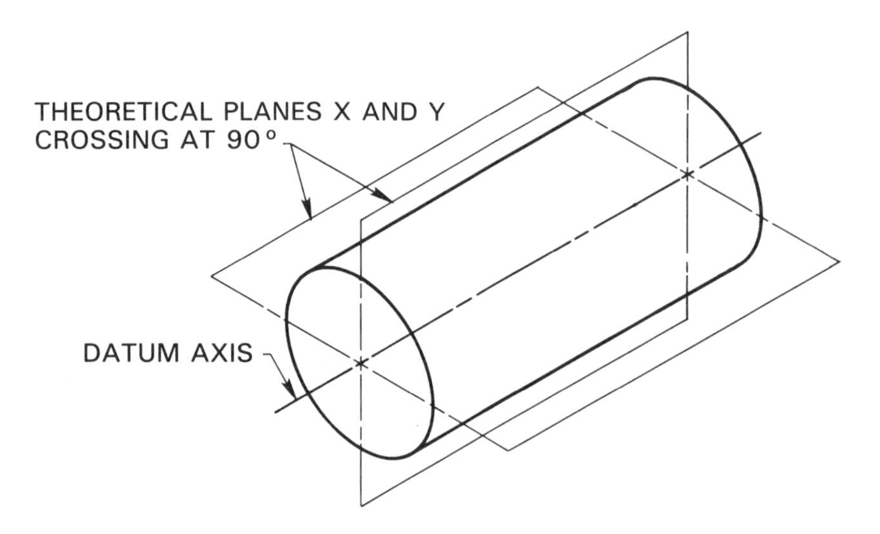

Example 3-23. Datum axis established by datum coordinates.

A third datum plane can be added to the end of the object to establish the datum frame as shown in Example 3-24.

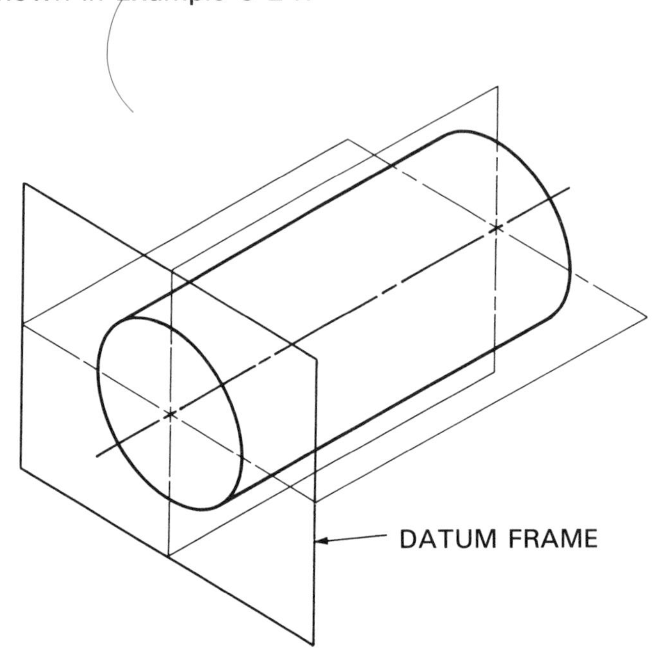

Example 3-24. Datum frame with datum axis.

The datum axis is represented on a drawing by placing the datum feature symbol below the diameter dimension or connected to the dimension line of the cylindrical feature when space is crowded as shown in Example 3-25. In this example, the datum axis is datum B. The datum feature symbol is not connected to the centerline.

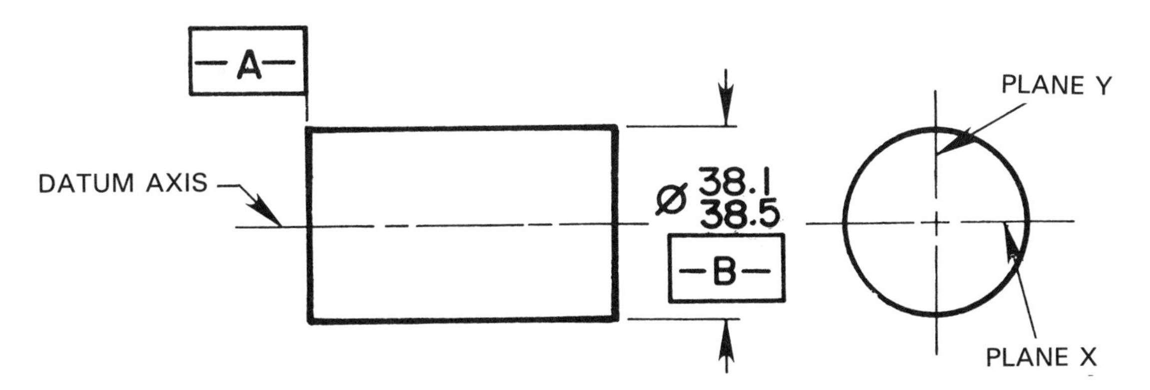

Example 3-25. Representing the datum axis. The X and Y center planes identified above are shown only for describing how the axis is established. The X and Y will not be placed on an actual drawing.

In Example 3-25, datum A is the primary datum reference because the relationship of all features to A will establish perpendicularity. The datum axis B is formed by the intersection of the center planes X and Y, and is the secondary datum feature. The datum axis B and the datum feature are established by a simulated datum as shown in Example 3-26.

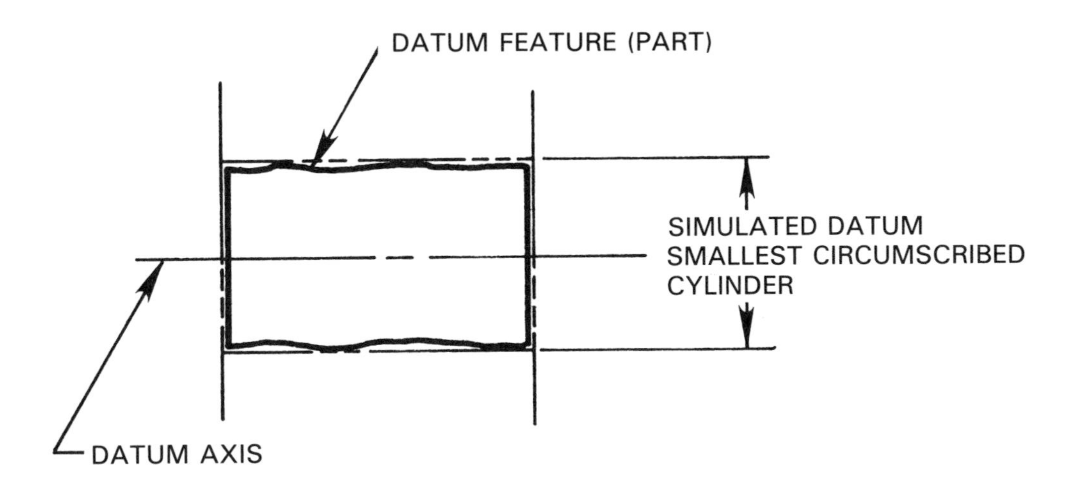

Example 3-26. Datum axis illustration.

Datum target points, lines, or surface areas may also be used to establish a datum axis. A primary datum axis may be established by two sets of three equally spaced targets; a set near one end of the cylinder and the other set near the other end as shown in Example 3-27.

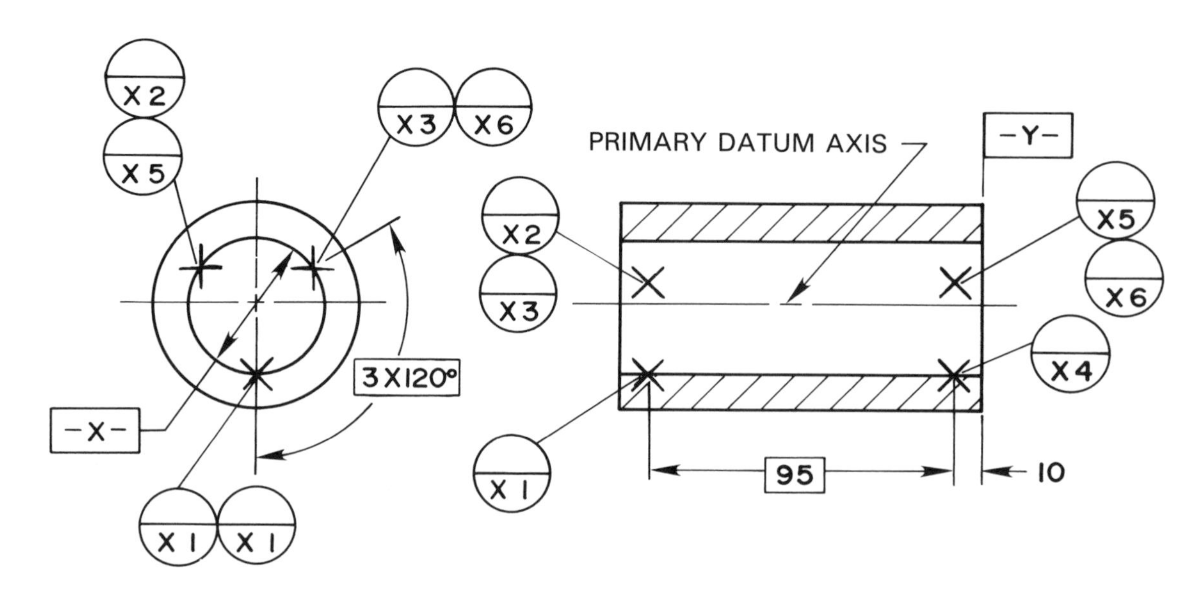

Example 3-27. Establishing a primary datum axis with target points.

When two cylindrical features of different diameters are used to establish a datum axis, then the datum target points are identified in correspondence to the adjacent cylindrical datum feature. See Example 3-28.

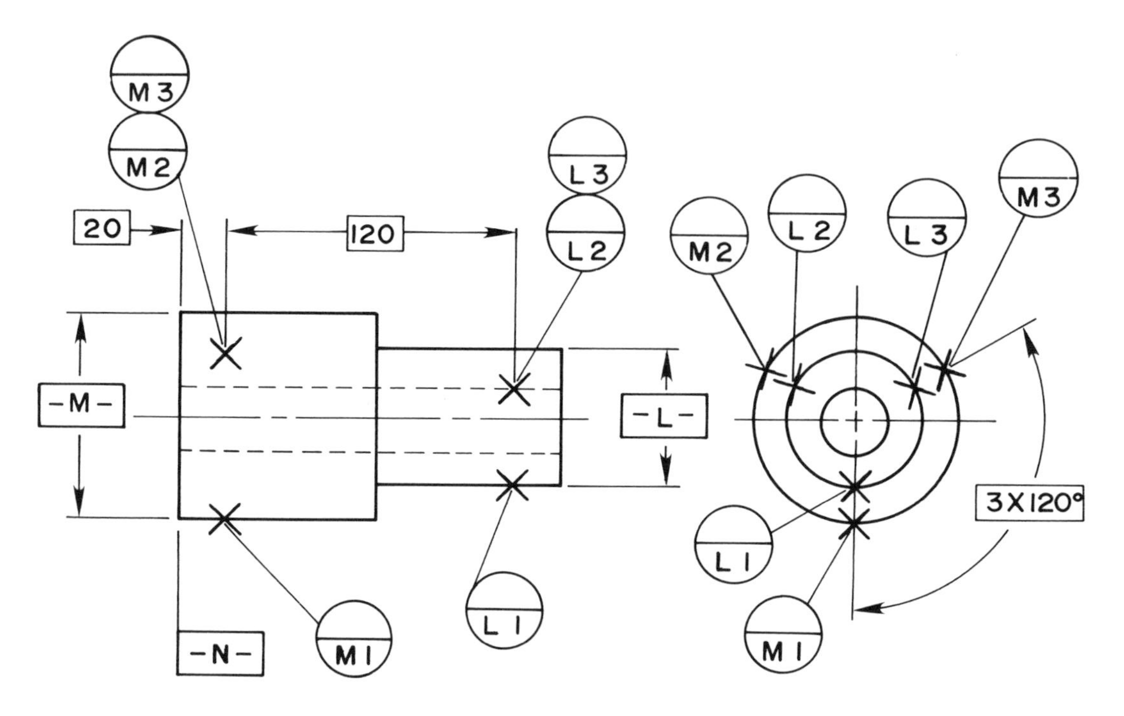

Example 3-28. Two cylindrical features of different diameters used to establish a datum axis.

Cylindrical datum target areas and circular datum target lines may also be used to establish the datum axis of cylindrical shaped parts as shown in Example 3-29.

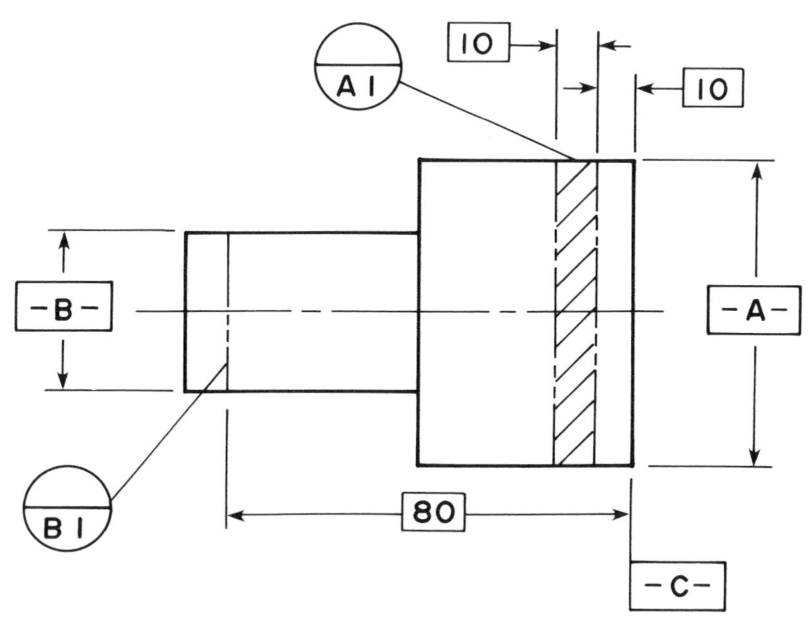

Example 3-29. Establishing datum axes with cylindrical datum target areas and circular datum target lines.

A secondary datum axis may be established by placing three equally spaced targets on the cylindrical surface. Example 3-30.

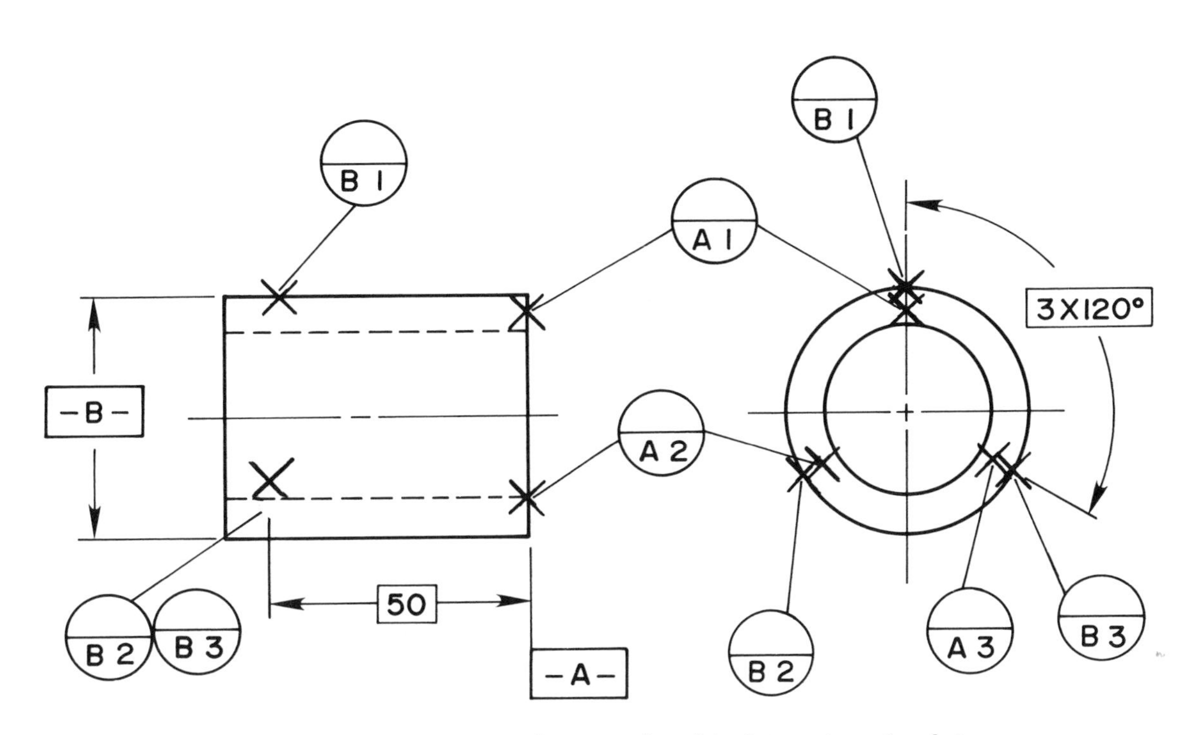

Example 3-30. Secondary datum axis with datum target points.

DATUM CENTER PLANE

Elements on a rectangular shaped symmetrical part or feature may be located and dimensioned in relationship to a datum center plane. The drawing representations and related meanings are shown in Example 3-31. The datum feature symbol is placed below the related feature dimension.

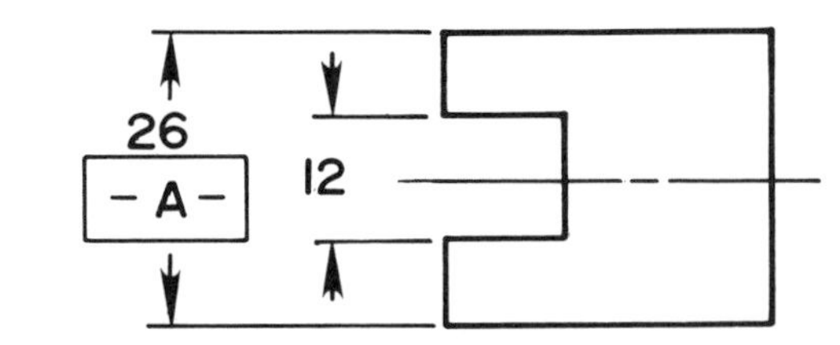

DATUM CENTERPLANE OF 26mm FEATURE

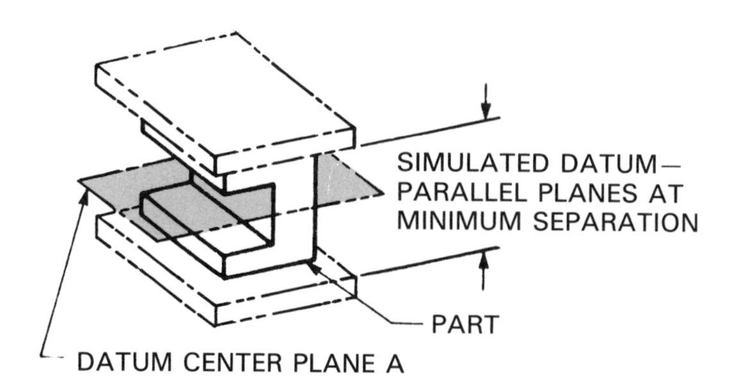

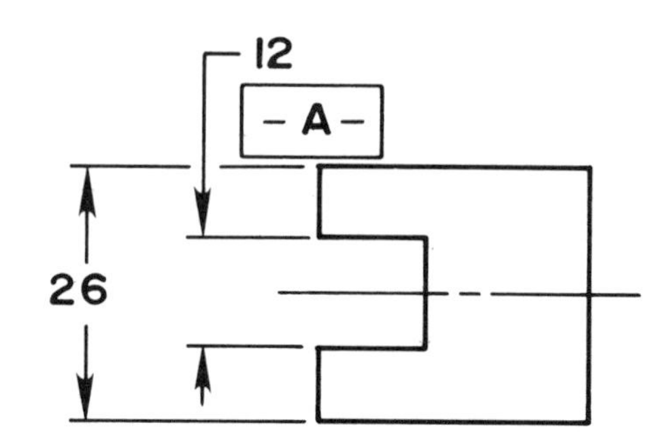

DATUM CENTERPLANE OF 12mm FEATURE

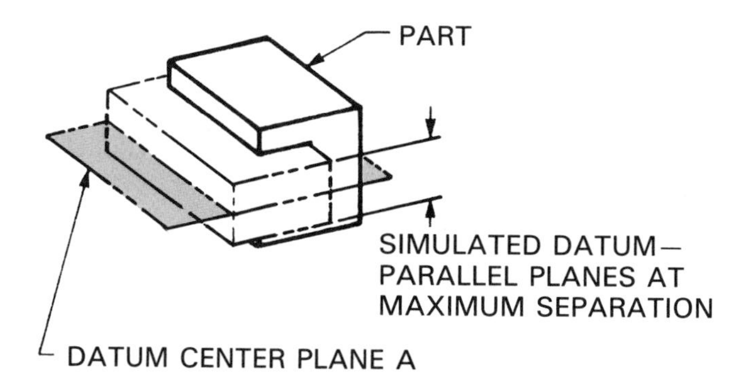

Example 3-31. Center plane datums.

AXIS AND CENTER PLANE DATUM EXAMPLES

Some correct and incorrect methods of representing axis and center plane datums are shown in Example 3-32.

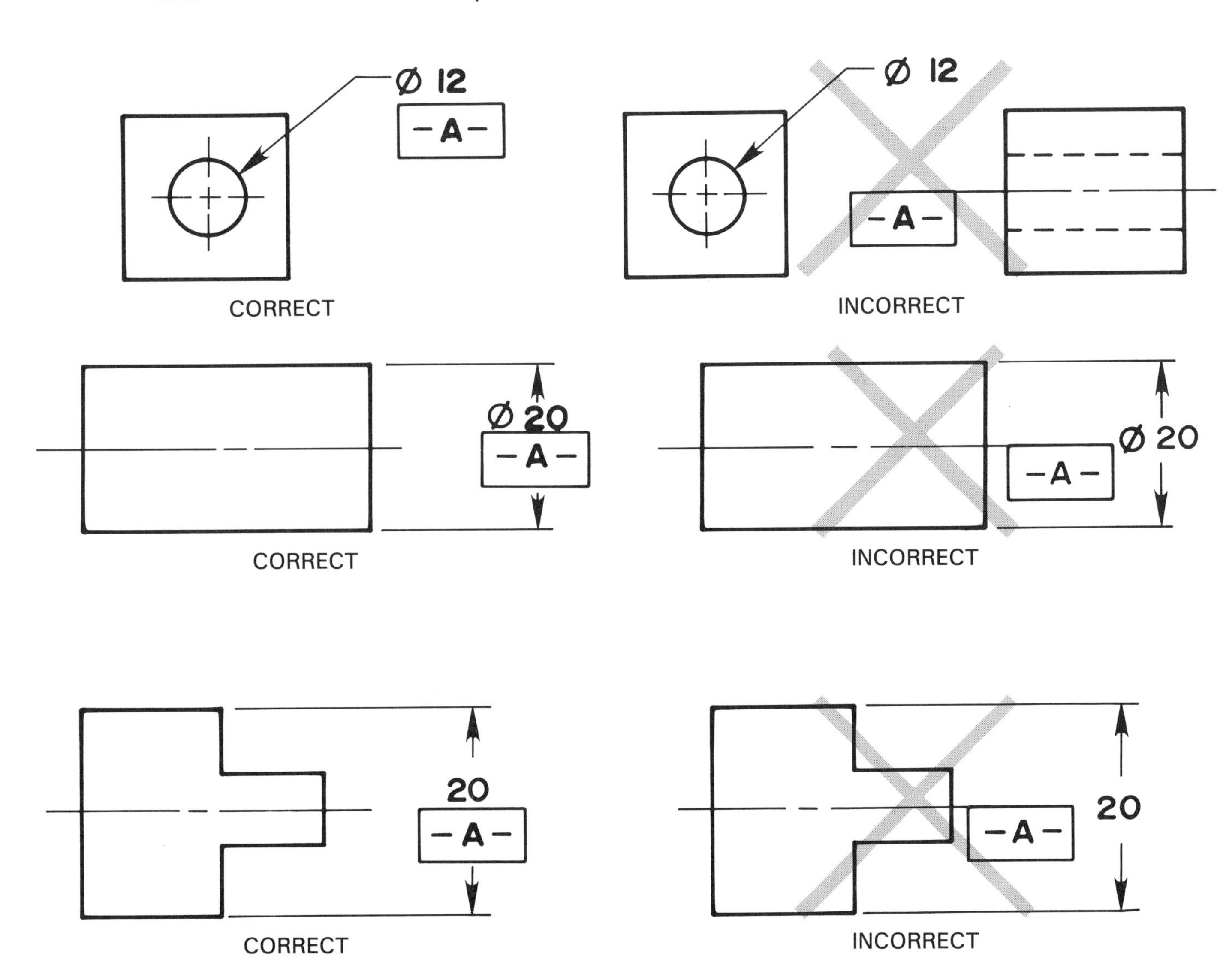

Example 3-32. Representing axis and center plane datums.

TEST 3, DATUMS

Name:____________________

1. List at least five items that may be considered as datums on an object or part.
   1. ____________________
   2. ____________________
   3. ____________________
   4. ____________________
   5. ____________________
2. A __________ plane is a theoretically exact plane.
3. A datum surface or feature is the __________ surface of an object that is used to establish a __________ plane.
4. Identify the datum feature or surface, and the datum plane on the following illustration.

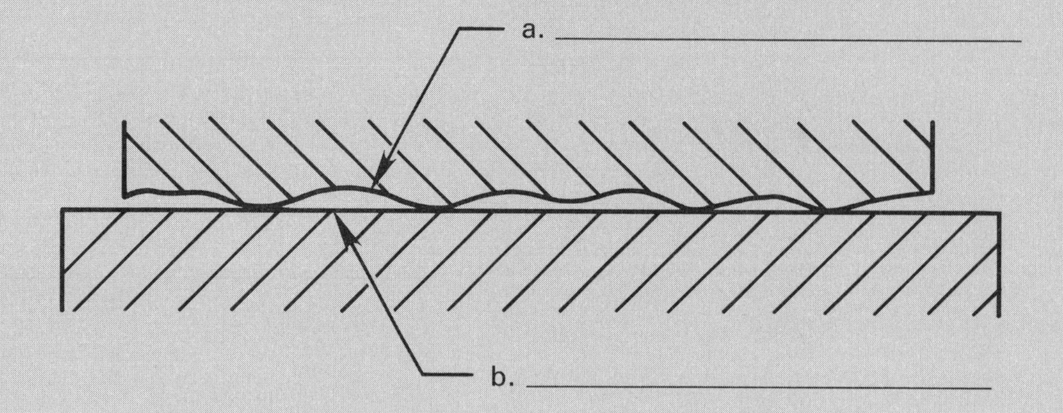

5. Complete this statement: When a datum surface is used on a part, the datum feature symbol is placed ____________________

   ____

6. Given the following drawing and related meaning, fill in the blanks at a, b, c, d, e, and f below. Provide the actual dimensions as related to the drawing at c, d, and e.

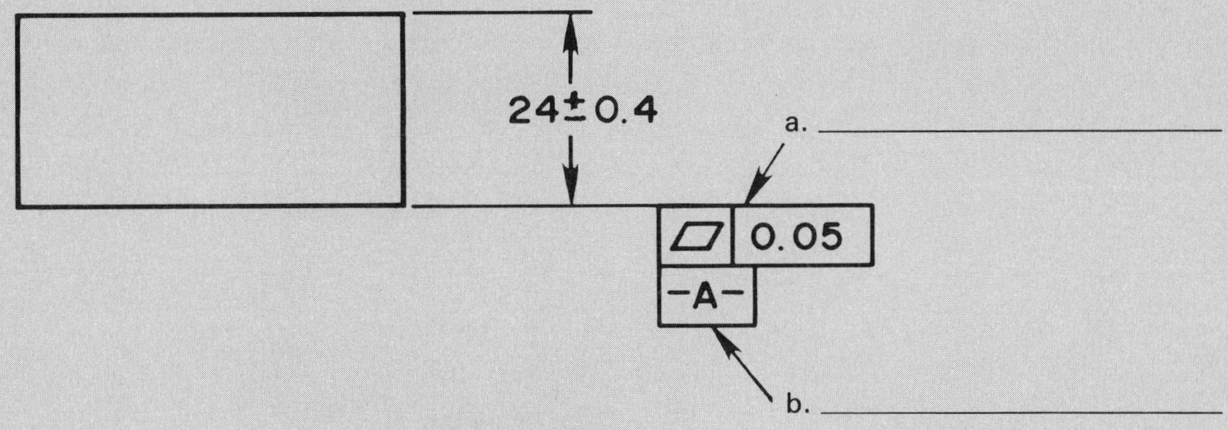

THE DRAWING

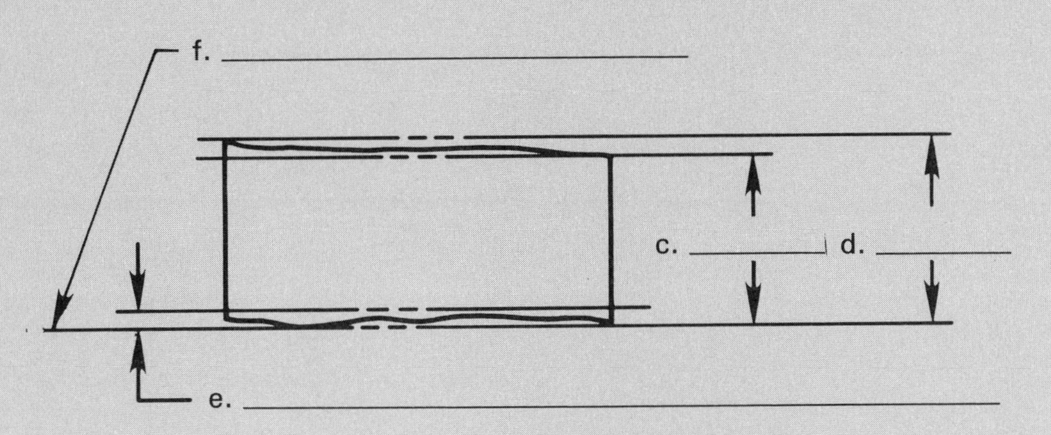

THE MEANING

7. Name the three planes of a complete datum reference frame.

____

____

____

8. When referring to the datum frame the ________________ datum should be given first followed by the ____________________ and ____________________ datums. This is known as datum ________________.

9. Define datum targets. ____
____
____

10. The primary datum plane must be established by at least __________ points on the primary datum surface.

11. The secondary datum plane must be established by at least _________ points on the related secondary datum surface.

12. The tertiary datum plane must be established by at least _________ point on the related tertiary datum surface.

13. How are datum target areas represented on a drawing? ____________
____
____
____

14. The circular datum target area is dimensioned with __________ dimensions to locate the __________ from datums and the diameter of the area is provided in the __________ half of the datum target symbol.

15. How are datum target areas treated on a drawing when the target area is too small to draw? ____
____
____
____

16. How should datum target lines be represented on a drawing? _______
____
____
____

17. When a portion of a surface is used as a datum this is referred to as a _____________ datum surface.

18. When more than one datum feature is used to establish a single datum this is referred to as _______________ datum features.

19. Depending on the functional requirements of a part, more than one datum reference frame may be established. This is referred to as a ____________________ datum reference frame.

20. Define coplanar surfaces. ______________________________

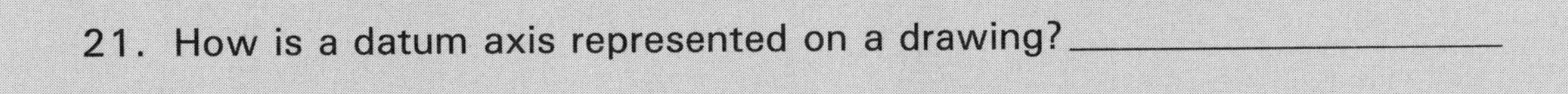

21. How is a datum axis represented on a drawing? ______________________________

22. Label the elements a, b, and c below that represent the meaning of a datum axis.

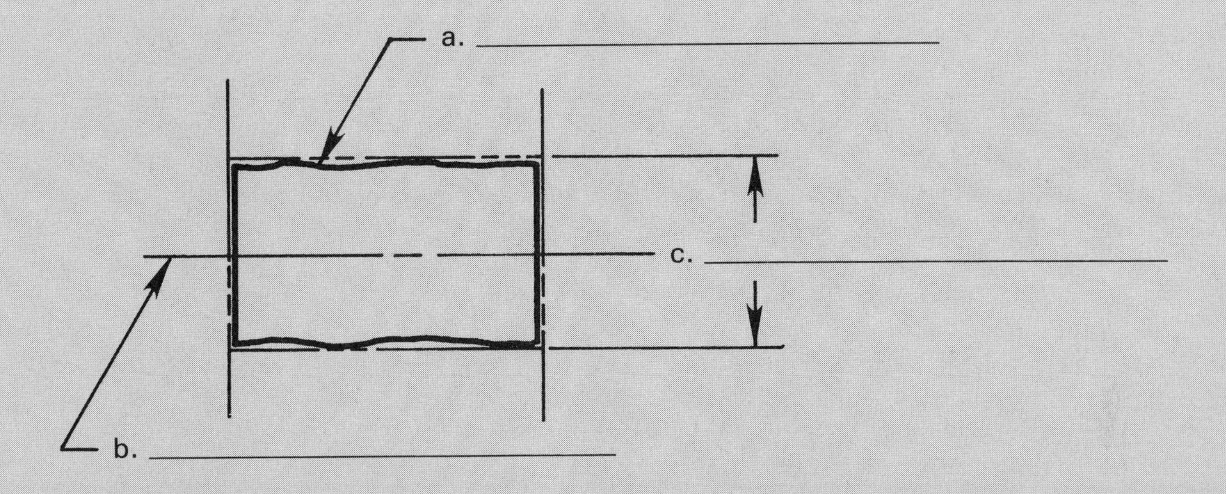

23. A primary datum axis may be established by two sets of three equally spaced target points. TRUE or FALSE?

24. Cylindrical datum target areas and circular datum target lines may be used to establish the datum axis of cylindrical shaped parts. TRUE or FALSE?

25. How is a datum centerplane shown on a drawing? ______________________________

PRINT READING EXERCISES FOR CHAPTER 3

Name:______________________

The following print reading exercise is designed for use in programs for machining, welding, tool and die, dimensional inspection, and other manufacturing curriculums where the objective is the reading and interpretation of prints rather than the development of drafting skills. An actual industrial print is used with related questions that require you to read and interpret specific dimensioning and geometric tolerancing representations. The answers and interpretations should be based on the previously learned content of this book. The prints used are based on ANSI standards; however, company standards may differ slightly. When reading these prints or any other industrial prints, a degree of flexibility may be required to determine how individual applications correlate with the ANSI standard.

PRINT READING EXERCISE

Refer to Hyster Company print of the PEDAL-ACCELERATOR found on page 216.

1. How many points of contact are used to establish the following datums? Datum A ______, Datum B ______, Datum C ______.
2. Interpret the datum target areas by giving the datum, specified number on the datum, and the area size and shape. ________________________

   ____
3. What is the distance between datum target point B-1 and B-2?

   ____
4. What does the box around the 74.00 dimension denote? ____________

   ____
5. What feature is datum D? ________________________

Refer to the Hyster Company print of the CASE-DIFF found on page 217.

6. Interpret datum feature A. ________________________
7. Interpret datum feature B. ________________________
8. Interpret datum feature C. ________________________
9. What does the feature control frame associated with datum A denote?

   ____

Refer to the Hyster Company print of the FLYWHEEL-DSL found on page 218.

10. How many datum features are identified on this part? ______.
11. What is the datum reference for the runout geometric tolerance at the ϕ376.81-376.76. ______________________________
12. Describe Datum J. ______________________________

Refer to DIAL INDUSTRIES print of the BODY, CONNECTOR SAMPLE & HOLD FIXTURE found on page 221.

13. Interpret Datum -A-. ______________________________
    ____
14. Interpret Datum -B-. ______________________________
    ____

Chapter 4

MATERIAL CONDITION SYMBOLS

Material Condition Symbols are used in conjunction with the feature tolerance or datum reference in the feature control frame. The MATERIAL CONDITION SYMBOLS establish the relationship between the size or location of the feature and the geometric tolerance. The use of different material condition symbols will alter the effect of this relationship. The material condition modifying elements are:

Maximum Material Condition, abbreviated MMC.
Regardless of Feature Size, abbreviated RFS.
Least Material Condition, abbreviated LMC.

The material conditions symbols are detailed in Example 4-1.

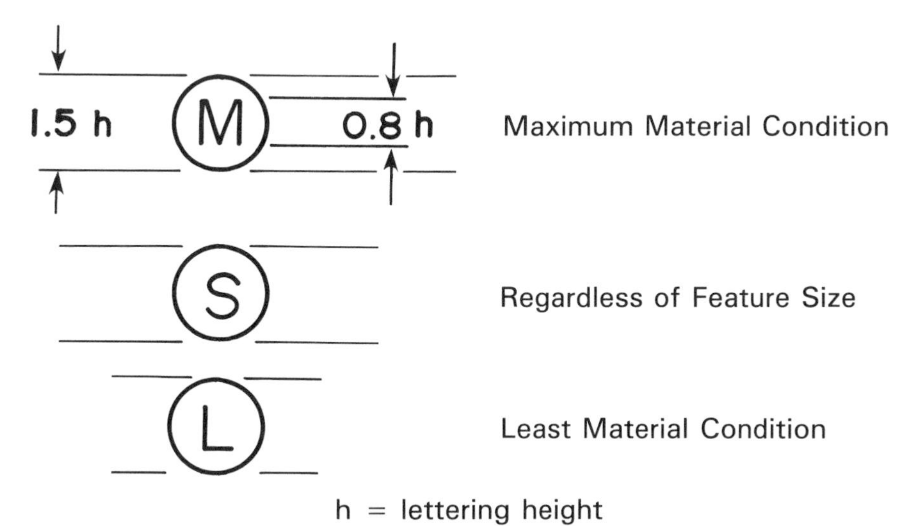

Example 4-1. Material condition symbols.

CONVENTIONAL TOLERANCE

Use of the term CONVENTIONAL TOLERANCING refers to tolerances related to conventional dimensioning practices without regard to geometric tolerancing practices. The limits of a size dimension determine the given variation allowed in the size of the feature or part. The part shown in Example 4-2 has a Maximum Material Condition of 6.5 and a Least Material Condition of 5.5. The MMC and LMC produced sizes represent the limits of the dimension. The actual part may be manufactured at any size between the limits. Some possible produced sizes in 0.1 mm increments are shown in the chart in Example 4-2.

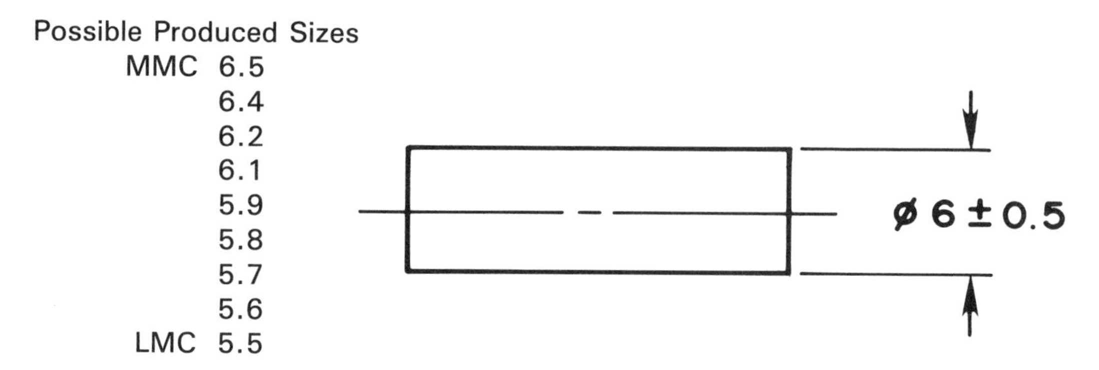

Example 4-2. Conventional tolerance.

Conventional tolerancing, without the addition of geometric tolerancing controls, permits a degree of variation in form, profile, or location because of the tolerance. The degree of form or location control can be increased or decreased by altering the tolerance. Conventional tolerancing does not necessarily take into account the geometric characteristics of the part's individual features or the relationship between mating elements unless expressly stated in note form. The amount of geometric control implied in a conventional tolerance is determined by the actual size of the feature or part which must be within the given tolerance at any cross section, as shown in Example 4-3.

LIMITS OF SIZE

The limits of size for individual features of size control the range within which the variation of size and geometric form are allowed. Also, the actual size of an individual feature shall be within the specified size tolerance at any cross section. The form of an individual feature is controlled by its size limits in the following ways:

- The surface of a feature may not extend beyond the MMC boundary. See PERFECT FORM BOUNDARY on page 82.
- When a features actual size departs from MMC a variation in form is allowed equal to the difference from MMC.
- If a feature is produced at LMC the geometric form may vary between the LMC and MMC boundaries.

See Example 4-3.

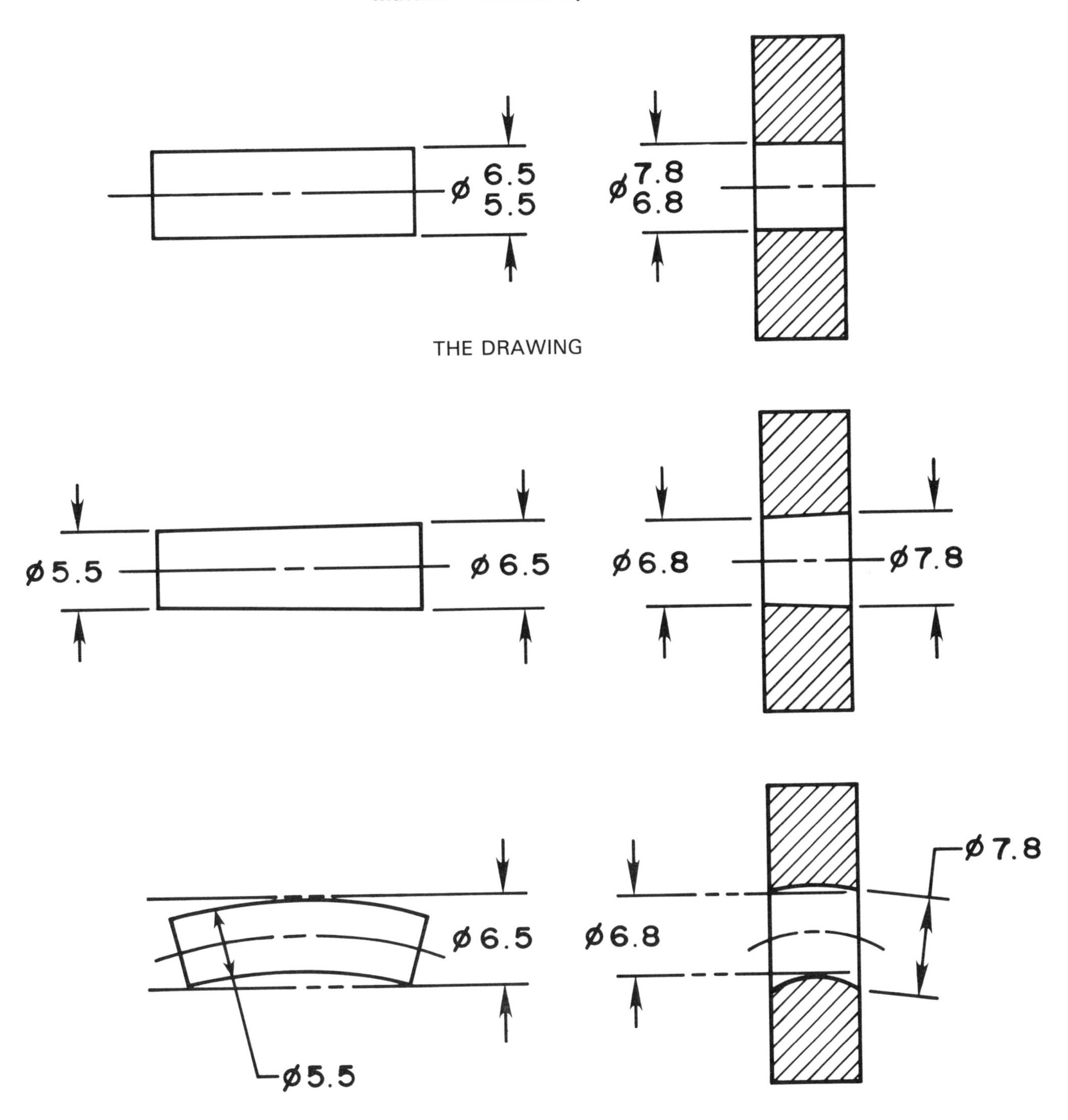

POSSIBLE EXTREME FORM VARIATIONS

Example 4-3. Extreme form variations of a given conventional tolerance.

PERFECT FORM BOUNDARY

The form of a feature is controlled by the size tolerance limits as demonstrated in Example 4-4. The boundary of these limits are established at MMC. The PERFECT FORM BOUNDARY is the true geometric form of the feature at MMC. Therefore, if the part is produced at MMC, it must be at perfect form. If a feature is produced at LMC, the form tolerance is allowed to vary within the geometric tolerance zone to the extent of the MMC boundary. In some applications it may be desirable to exceed the perfect form boundary at MMC. When this is done the note: PERFECT FORM AT MMC NOT REQUIRED, must accompany the size dimension. See Example 4-4.

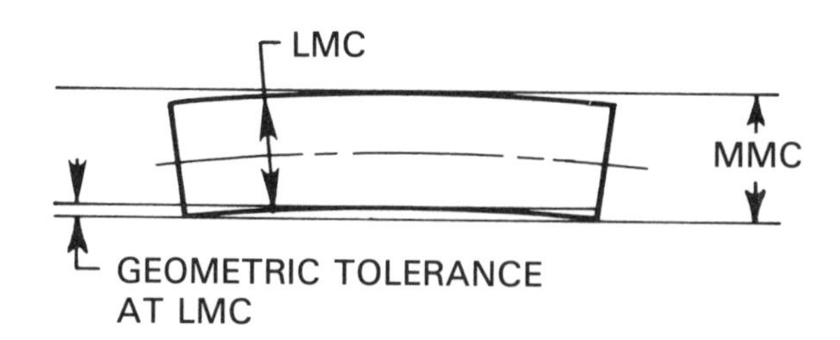

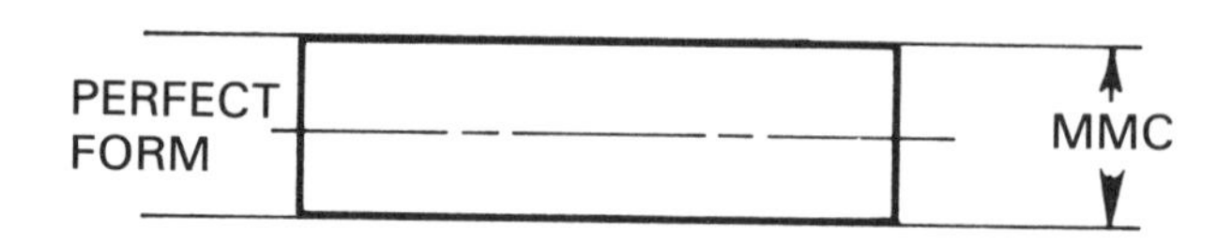

Example 4-4. Perfect Form at MMC.

Ⓢ, RFS, REGARDLESS OF FEATURE SIZE

REGARDLESS OF FEATURE SIZE is the term used to indicate that a geometric tolerance or datum reference applies at any increment of size of the feature within its size tolerance. Regardless of feature size is implied when any of the geometric characteristics shown in Example 4-5 are given in the feature control frame. Only the geometric characteristic symbol for position does not imply RFS.

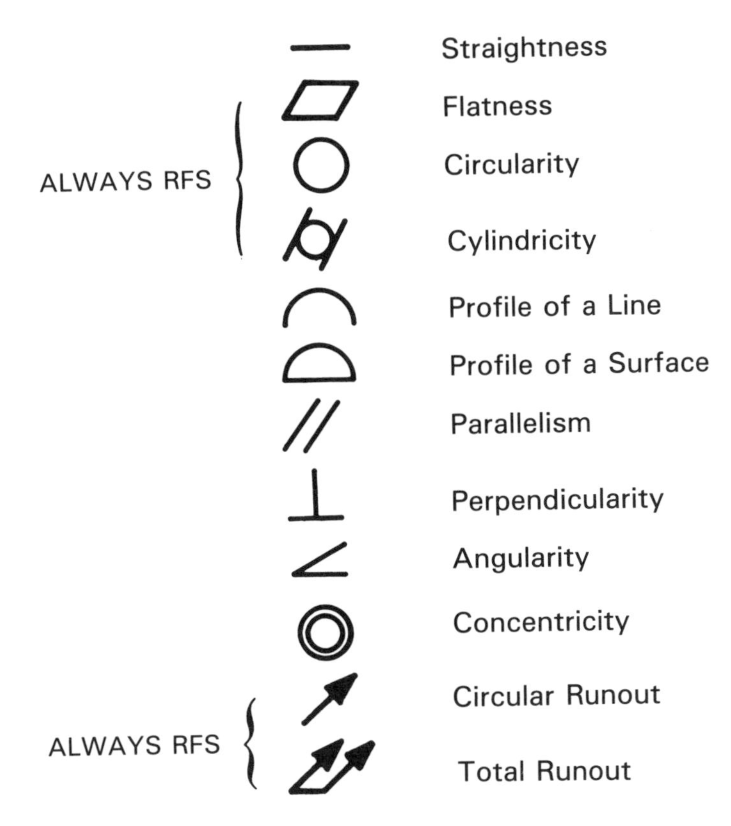

Example 4-5. Geometric characteristics that imply RFS unless otherwise specified.

SURFACE CONTROL, REGARDLESS OF FEATURE SIZE, RFS

SURFACE GEOMETRIC CONTROL is when the feature control frame is either connected with a leader to the surface of the object or feature, or extended from an extension line from the surface of the object or feature. The use of a leader connecting the feature control frame to the surface is shown in Example 4-6.

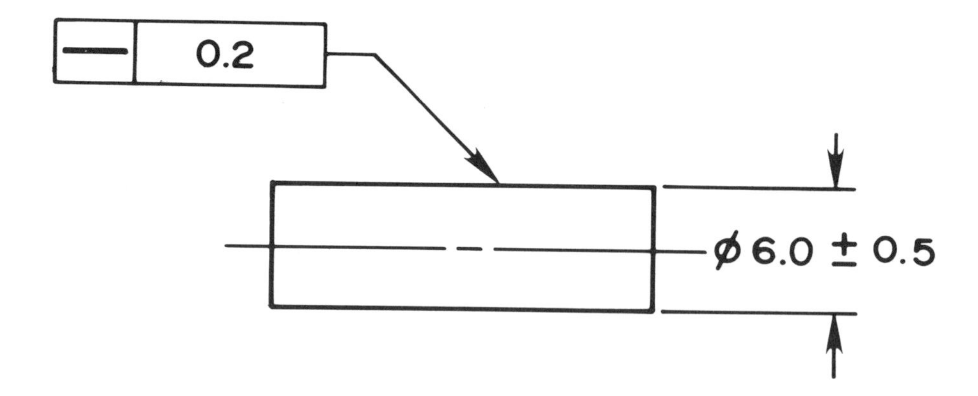

Example 4-6. Surface control showing straightness.

When any of the geometric characteristics shown in Example 4-5 are given in the feature control frame, RFS is implied. While some companies may select to show the RFS symbol for clarity, this is not an ANSI standard. The surface straightness specification, as shown in Example 4-6, means that each longitudinal element of the surface must lie between two parallel lines, 0.2 apart, where the two lines and the nominal axis of the part share a common plane. Also, the feature must be within the specified size limits and within the perfect form boundary at MMC. When the actual size of the feature is MMC, then zero geometric tolerance is required. When the actual produced size departs from MMC, then the geometric tolerance is equal to the amount specified in the feature control frame. Example 4-7 shows an analysis of regardless of feature size, RFS, surface control based on the drawing presented in Example 4-6.

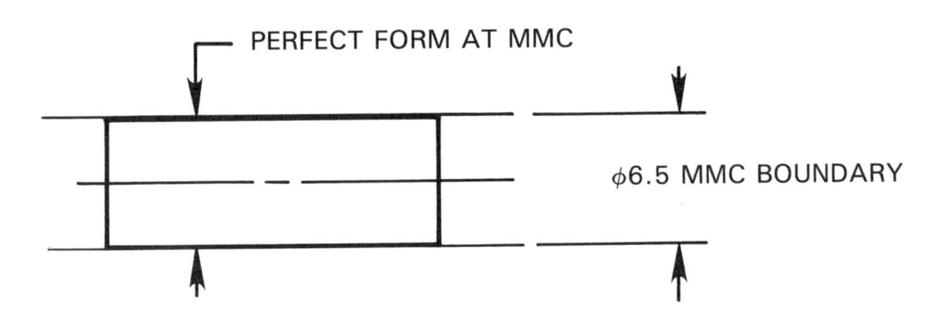

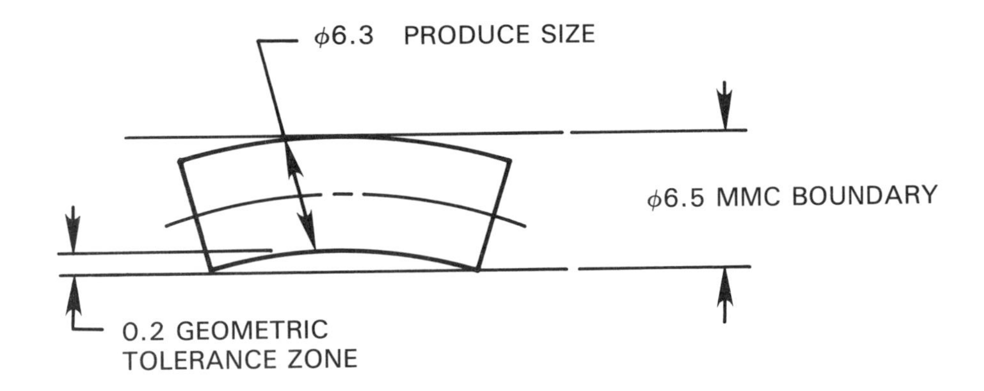

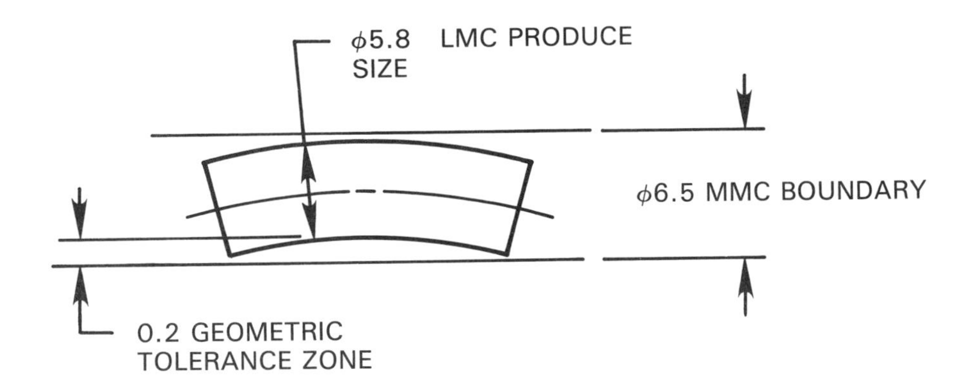

Possible Produced Sizes	Geometric Tolerances at Given Produced Sizes
MMC 6.5	Perfect Form 0 Geometric Tolerance
6.4	0.1
6.3	0.2
6.2	0.2
6.1	0.2
6.0	0.2
5.9	0.2
5.8	0.2
5.7	0.2
5.6	0.2
LMC 5.5	0.2

Example 4-7. Specifying straightness of surface elements.

AXIS CONTROL AT REGARDLESS OF FEATURE SIZE, RFS

AXIS GEOMETRIC CONTROL is implemented by placing the feature control frame with the diameter dimension of the related object or feature. A good way to remember the difference between surface and axis control is to recognize that surface control is when the feature control frame is connected to the surface, while axis control places the feature control frame with the diameter dimension which correlates with the axis. When axis control is used, a diameter tolerance zone must be specified by placing the diameter symbol in front of the tolerance in the feature control frame as shown in Example 4-8.

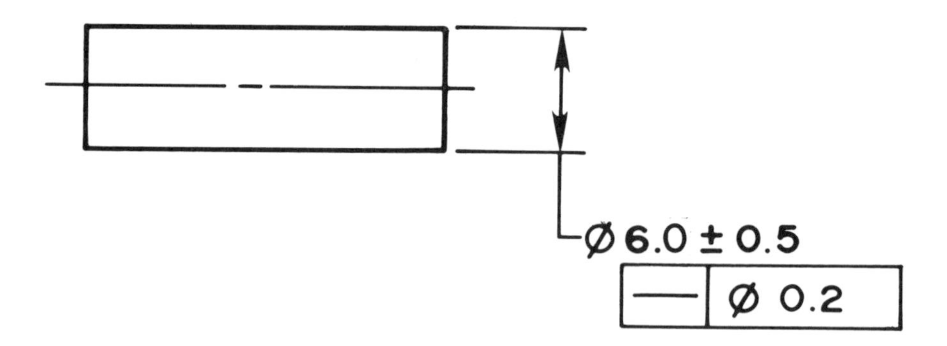

Example 4-8. RFS, axis control.

When axis control is specified, the perfect form boundary may be violated. This violation is permissable when the feature control frame is associated with the size dimension. When MMC is not specified, then RFS is implied. When this situation occurs, the geometric tolerance at various produced sizes remains the same as given in the feature control frame; even at MMC. Example 4-9 shows an analysis of axis control at regardless of feature size, RFS based on the drawing in Example 4-8.

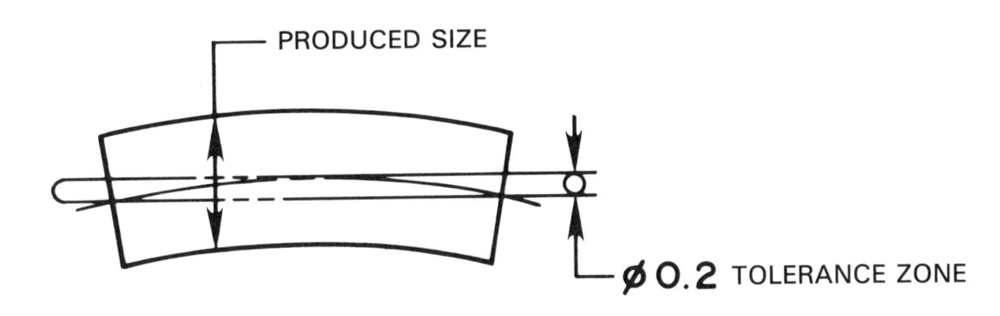

Possible Produced Sizes	Geometric Tolerances at Given Produced Sizes
MMC 6.5	0.2 Perfect form at MMC not required.
6.4	0.2
6.3	0.2
6.2	0.2
6.1	0.2
6.0	0.2
5.9	0.2
5.8	0.2
5.7	0.2
5.6	0.2
LMC 5.5	0.2

Example 4-9. Axis control at regardless of feature size.

Ⓜ, MAXIMUM MATERIAL CONDITION, MMC

The MAXIMUM MATERIAL CONDITION is the condition in which a feature contains the maximum amount of material within the stated limits of size. For example, maximum shaft diameter or minimum hole diameter. When MMC is used in the feature control frame, the given geometric tolerance is maintained when the feature is produced at MMC. Then as the actual produced size departs from MMC, the geometric tolerance is allowed to get larger equal to the amount of departure from MMC.

One of the following formulas may be used to calculate the geometric tolerance at any produced size when a MMC material condition symbol is used. The formula you should use is based on if you are working with an external feature such as a shaft or an internal feature such as a hole.

EXTERNAL FEATURE:
MMC − Produced Size + Given Geometric Tolerance = Applied Geometric Tolerance

INTERNAL FEATURE:
Produced Size − MMC + Given Geometric Tolerance = Applied Geometric Tolerance

AXIS CONTROL AT MAXIMUM MATERIAL CONDITION, MMC

When it is desirable to use MMC as the material condition symbol, then the MMC symbol must be placed in the feature control frame after the geometric tolerance. This axis control is also a diameter tolerance zone and the diameter symbol must preceed the geometric tolerance as shown in Example 4-10.

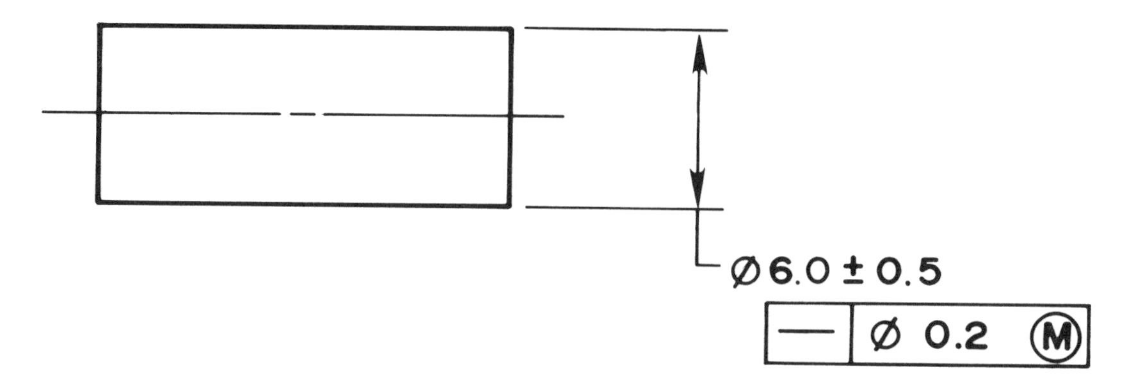

Example 4-10. Axis control at MMC.

When a MMC material condition symbol is used, the geometric tolerance is the same as specified in the feature control frame at the MMC produced size. Then, as the produced sizes depart from Maximum Material Condition, the geometric tolerance is allowed to increase equal to the amount of departure from MMC. For example, the geometric tolerance at MMC in Example 4-10 is 0.2. If the part were produced at MMC, the geometric tolerance would be 0.2. If the part were produced at 6.1 the allowed geometric tolerance would be MMC (6.5) − PRODUCED SIZE (6.1) + GIVEN GEOMETRIC TOLERANCE (0.2) = APPLIED GEOMETRIC TOLERANCE 0.6. The maximum geometric tolerance is at the LMC produced size. LMC is the condition in which a feature of size contains the least amount of material within the limits. Refer to the drawing of the part in Example 4-10 for the complete interpretation shown in Example 4-11.

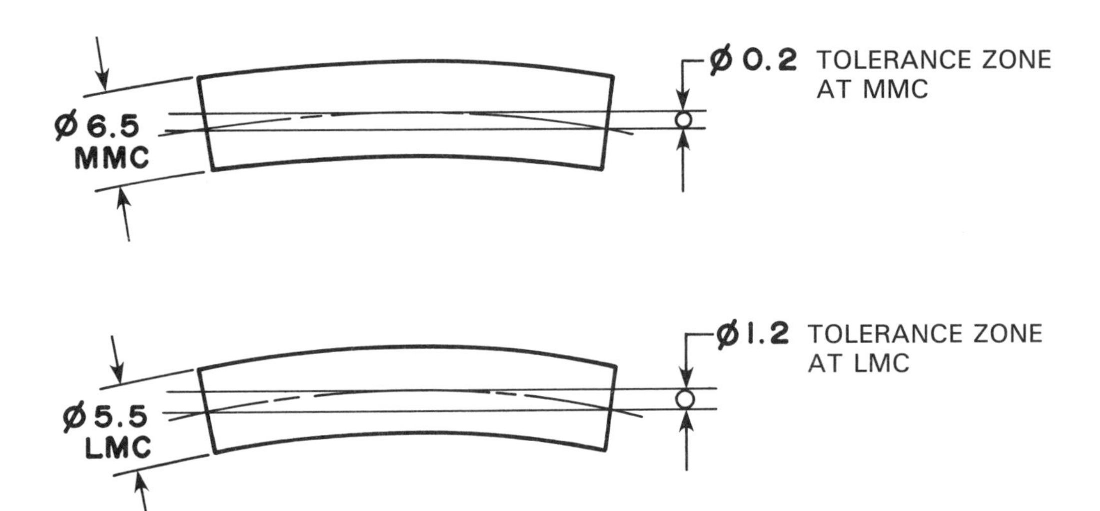

Possible Produced Sizes	Geometric Tolerances at Given Produced Sizes
MMC 6.5	0.2
6.4	0.3
6.3	0.4
6.2	0.5
6.1	0.6
6.0	0.7
5.9	0.8
5.8	0.9
5.7	1.0
5.6	1.1
LMC 5.5	1.2

Example 4-11. Interpretation of axis control with MMC.

The concepts of surface and axis straightness previously discussed may also be applied on an RFS or MMC basis to noncylindrical features of size. When this is done, the derived center plane must lie in a tolerance zone between two parallel planes separated by the amount of the tolerance. Otherwise, the feature control frame placement is the same as previously discussed. The diameter tolerance zone symbol is not used in front of the geometric tolerance because the tolerance zone is noncylindrical, established by two parallel planes as shown in Example 4-12.

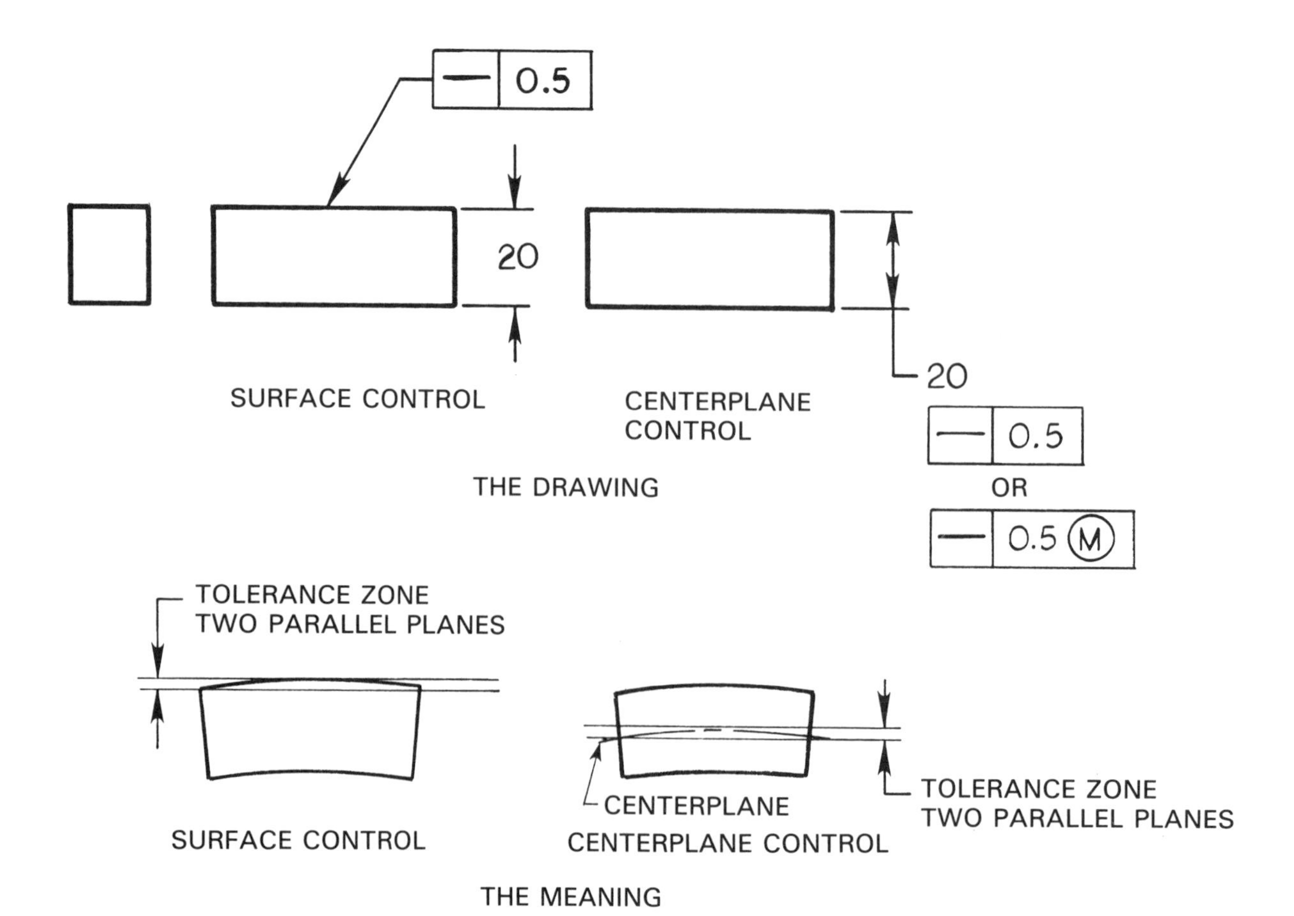

Example 4-12. Surface and center plane controls for noncylindrical features.

(L), LEAST MATERIAL CONDITION, LMC

The LEAST MATERIAL CONDITION is the condition where the feature of size contains the least amount of material. For example, minimum shaft diameter or maximum hole diameter. LMC is the opposite of MMC. When an LMC material condition symbol is used in the feature control frame, the given geometric tolerance is held at the LMC produced size. When the actual produced size departs from LMC toward MMC, the geometric tolerance is allowed to increase equal to the amount of departure. The maximum geometric tolerance is at the MMC produced size. The formula for an internal feature: Actual Size − LMC + Given Geometric Tolerance = Applied Geometric Tolerance may be used to calculate the geometric tolerance at any produced size when an LMC material condition symbol is used.

Example 4-13 shows an application of LMC in the feature control frame where the axis perpendicularity of a hole must be within a 0.2 diameter geometric tolerance zone, at LMC, to datum A. When the feature size is at LMC (12.5) the geometric tolerance is held as specified in the feature control frame. As the actual produced size decreases toward MMC, the geometric tolerance increases equal to the amount of change from LMC to the maximum change at MMC. The analysis in Example 4-13 shows the possible geometric tolerance variation from LMC to MMC. This specification may be used to control the minimum wall thickness of the part.

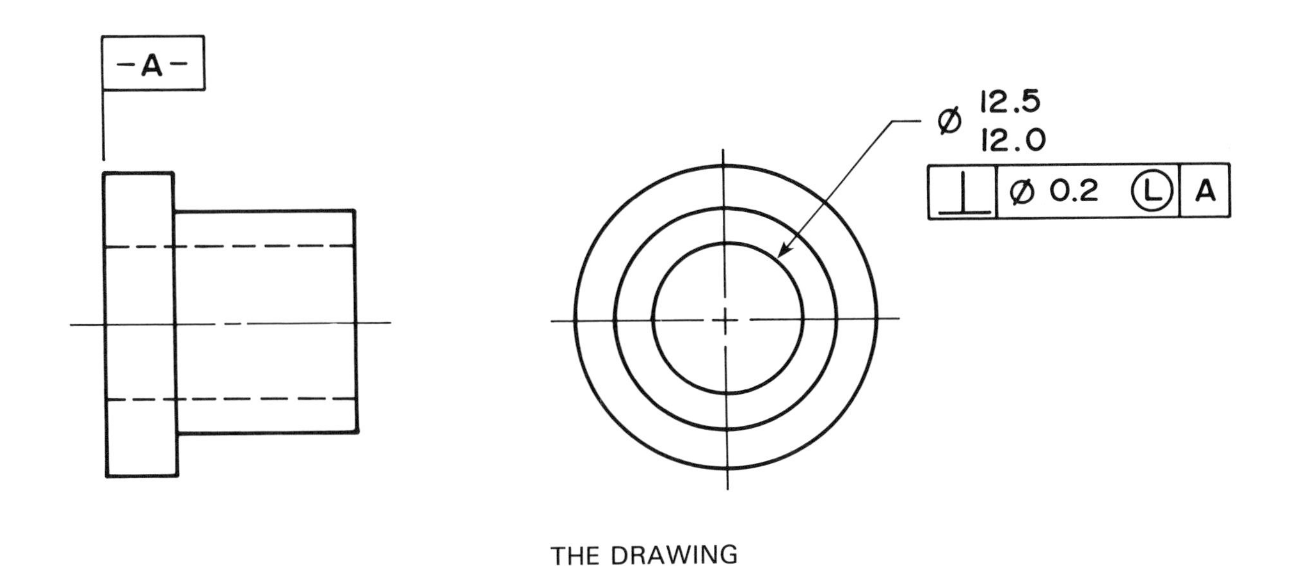

THE DRAWING

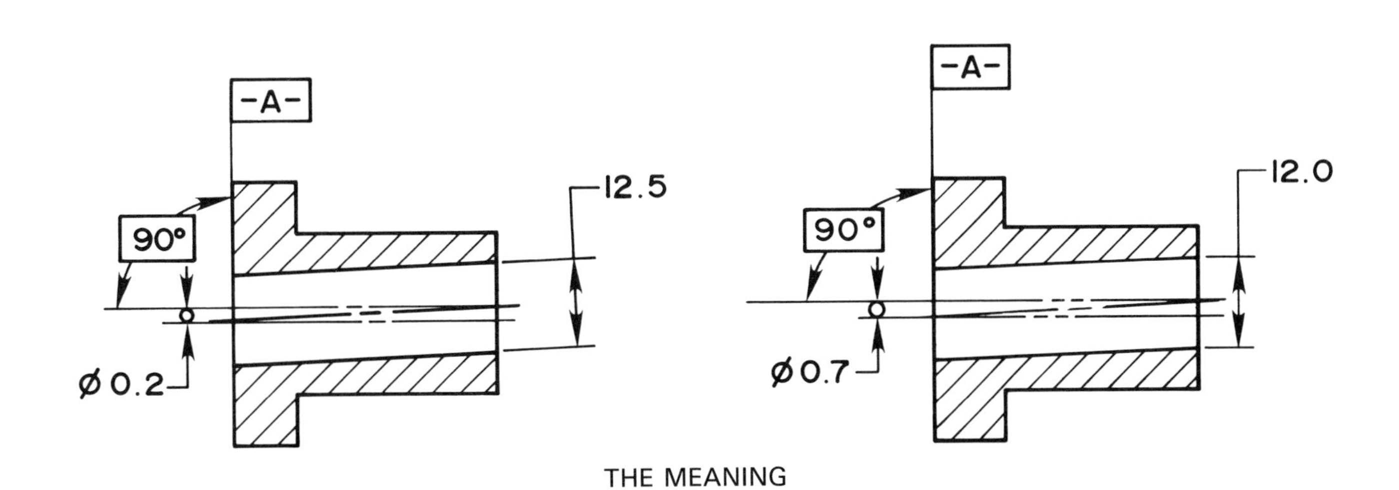

THE MEANING

Possible Produced Sizes	Geometric Tolerances at Given Produced Sizes
MMC 12.0	0.7
12.1	0.6
12.2	0.5
12.3	0.4
12.4	0.3
LMC 12.5	0.2

Example 4-13. Application of LMC.

DATUM FEATURE RFS

Datum features such as diameters and widths that are influenced by size variations, are also subject to variations in form. RFS is implied in these cases unless otherwise specified. When a datum feature has a size dimension and a geometric form tolerance, the size of the simulated datum is the MMC size limit. This applies, except for axis straightness where the boundary is allowed to exceed MMC. The effect of RFS on the primary datum feature with axis and center plane datums is shown in Example 4-14.

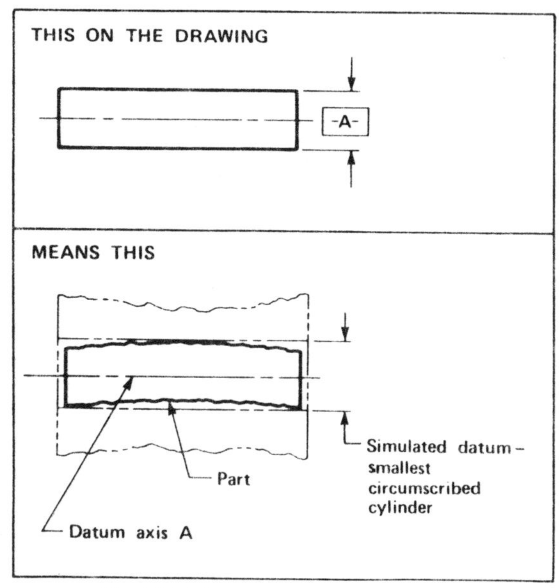

PRIMARY EXTERNAL DATUM DIAMETER – RFS

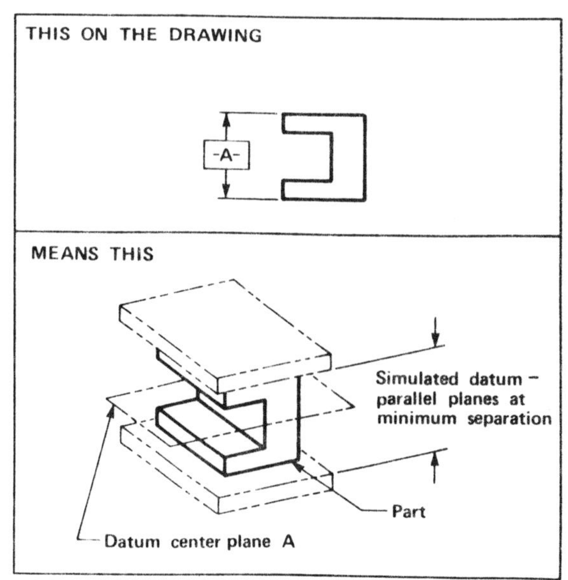

PRIMARY EXTERNAL DATUM WIDTH – RFS

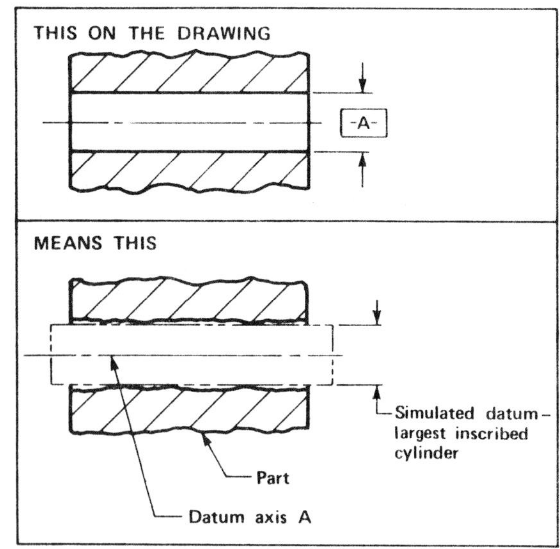

PRIMARY INTERNAL DATUM DIAMETER – RFS

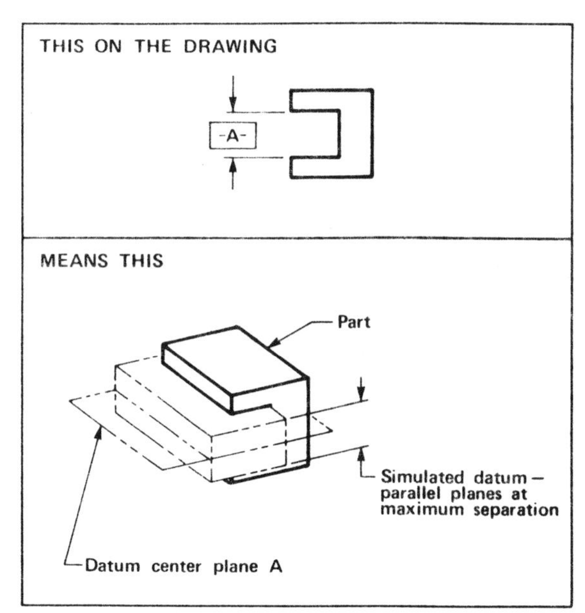

PRIMARY INTERNAL DATUM WIDTH – RFS

Example 4-14. Effect of RFS on primary datum features. (ANSI Y14.5M-1982)

When the datum features are secondary or tertiary, then the axis or centerplane shall also have an angular relationship to the primary datum as shown in Example 4-15.

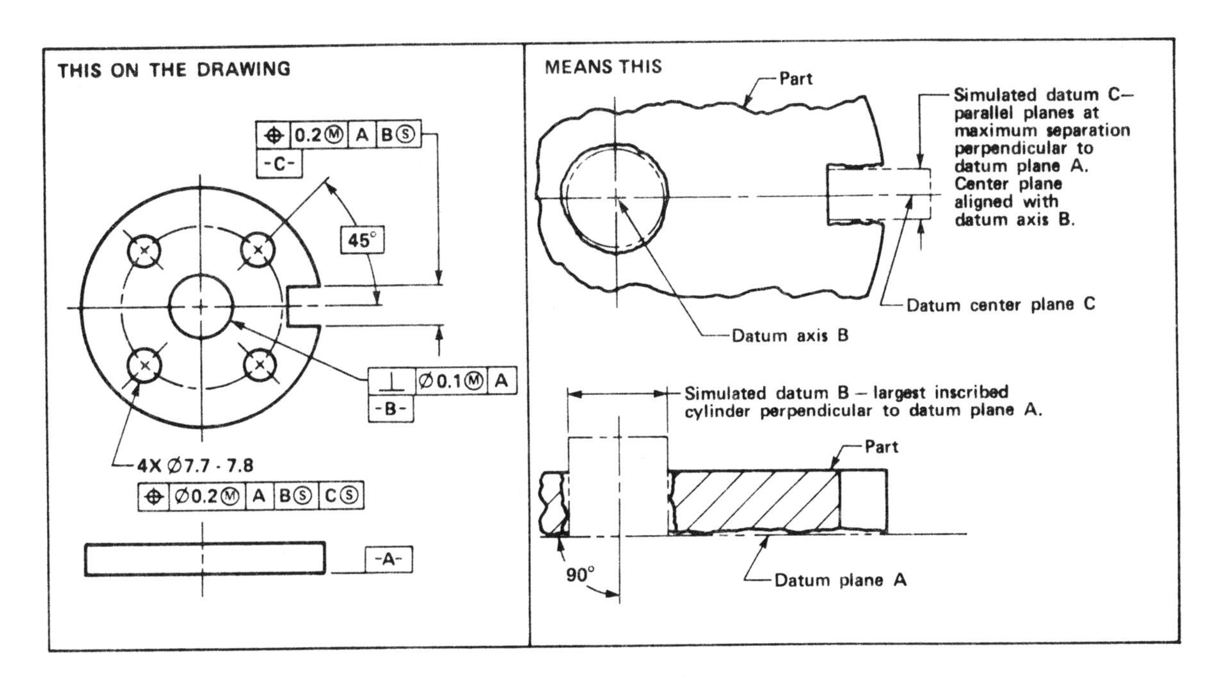

Example 4-15. Secondary and tertiary datum features RFS. (ANSI Y14.5M-1982)

DATUM PRECEDENCE AND MATERIAL CONDITION

The effect of material condition on the datum and related feature may be altered by changing the datum precedence and the applied material condition symbol. The DATUM PRECEDENCE is established by the order of placement in the feature control frame. The first datum listed is the primary datum, followed by the secondary and tertiary datums. See Example 4-16.

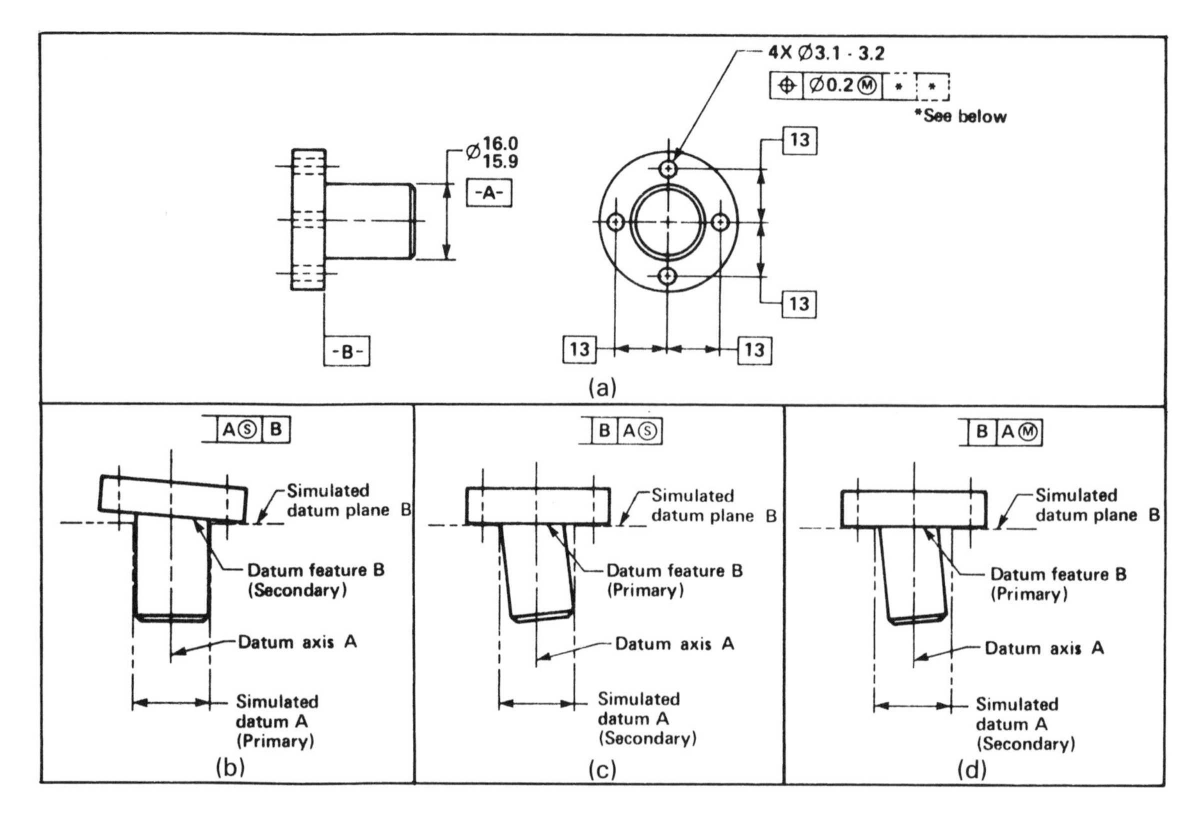

Example 4-16. Effect of altering datum precedence and material condition. (ANSI Y14.5M-1982)

The drawing in Example 4-16a shows a part with a pattern of holes located in relation to ϕ16.0-15.9 Datum A, and the surface Datum B. The datum requirements for the position tolerance associated with the location of the 4 holes may be specified in three different ways as shown in Example 4-16 b-d.

The illustration at (b) shows that the diameter Datum A is the primary datum feature applied at RFS, and surface Datum B is the secondary datum.

The drawing at (c) shows the surface Datum B is primary and the diameter Datum A is secondary with RFS applied.

The illustration at (d) shows the surface Datum B is primary and the diameter Datum A is secondary with MMC applied. Also, with MMC applied, as the diameter Datum A goes away from its maximum material condition size the axis is allowed to depart from the datum axis equal to the difference between the produced size and MMC.

TEST 4, MATERIAL CONDITION SYMBOLS Name:________________

1. Define perfect form boundary. ______________________________
____
____
____

2. Define regardless of feature size (RFS). ______________________________
____
____
____

3. There is only one geometric characteristic symbol that, when used, does not imply RFS. Name that geometric characteristic.________________

4. How is a feature control frame connected to a related feature when surface control is intended. ______________________________
____
____
____

5. May perfect form at MMC be violated for surface straightness? YES or NO.

6. Given the following drawing and a list of possible produced sizes specify the geometric tolerance at each possible produced size.

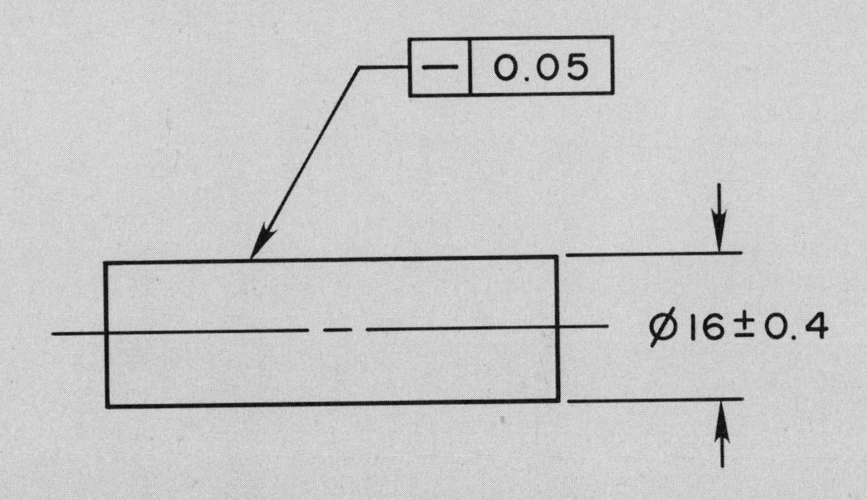

Possible Produced Sizes	Geometric Tolerance at Given Produced Sizes
16.4 MMC	________________
16.2	________________
16.0	________________
15.8	________________
15.6 LMC	________________

7. How is an axis geometric control specified? ______________________

____

____

8. When axis straightness is specified, may the perfect form boundary be violated? YES or NO.

9. When axis straightness control is used, a ______________ tolerance zone must be specified by placing the ______________ symbol in front of the geometric tolerance in the feature control frame.

10. Given the following drawing and a list of possible produced sizes, specify the geometric tolerance at each possible produced size.

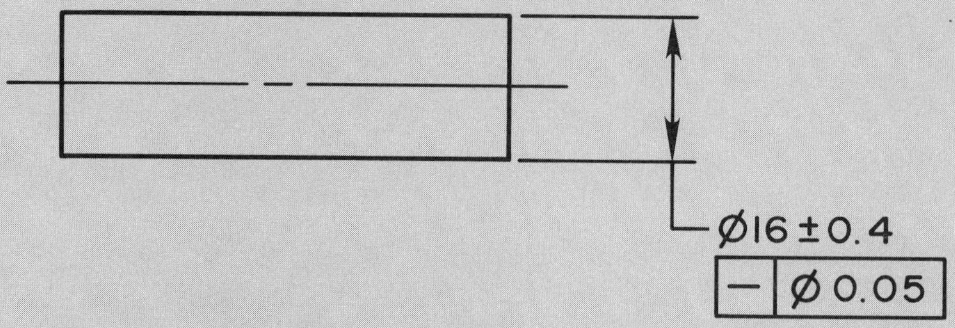

Ø16 ±0.4

—	Ø0.05

Possible Produced Sizes	ϕ Geometric Tolerance at Given Produced Sizes
16.4 MMC	______________
16.2	______________
16.0	______________
15.8	______________
15.6 LMC	______________

11. Give the formula that may be used for calculating the geometric tolerance at a given produced size when MMC is specified with the geometric tolerance in the feature control frame.

____

____

____

12. Given the following drawing and a list of possible produced sizes, specify the geometric tolerance at each possible produced size.

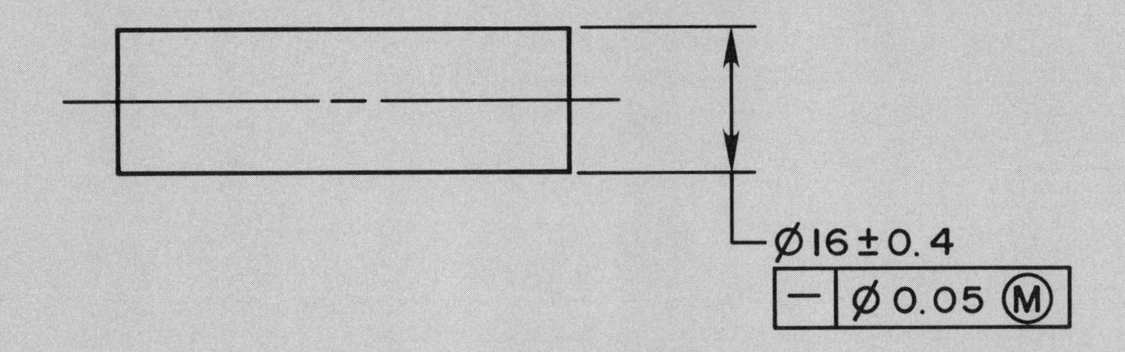

Possible Produced Sizes	ϕ Geometric Tolerance at Given Produced Sizes
16.4 MMC	__________
16.2	__________
16.0	__________
15.8	__________
15.6 LMC	__________

13. Give the formula that may be used for calculating the geometric tolerance at a given produced size when LMC is specified with the geometric tolerance in the feature control frame.

____

____

14. The use of the LMC material condition symbol after the geometric tolerance in a feature control is often used to control minimum wall thickness. TRUE or FALSE?

15. When a datum feature has a size dimension and a geometric form tolerance, the size of the simulated datum is the __________ size limit, except for __________ straightness applications where the boundary is allowed to exceed MMC.

16. The effect of material condition on the datum and related feature may be altered by changing the datum precedence and the applied material condition symbol. TRUE or FALSE?

17. Given the following drawing and a list of possible produced sizes, specify the geometric tolerance at each possible produced size.

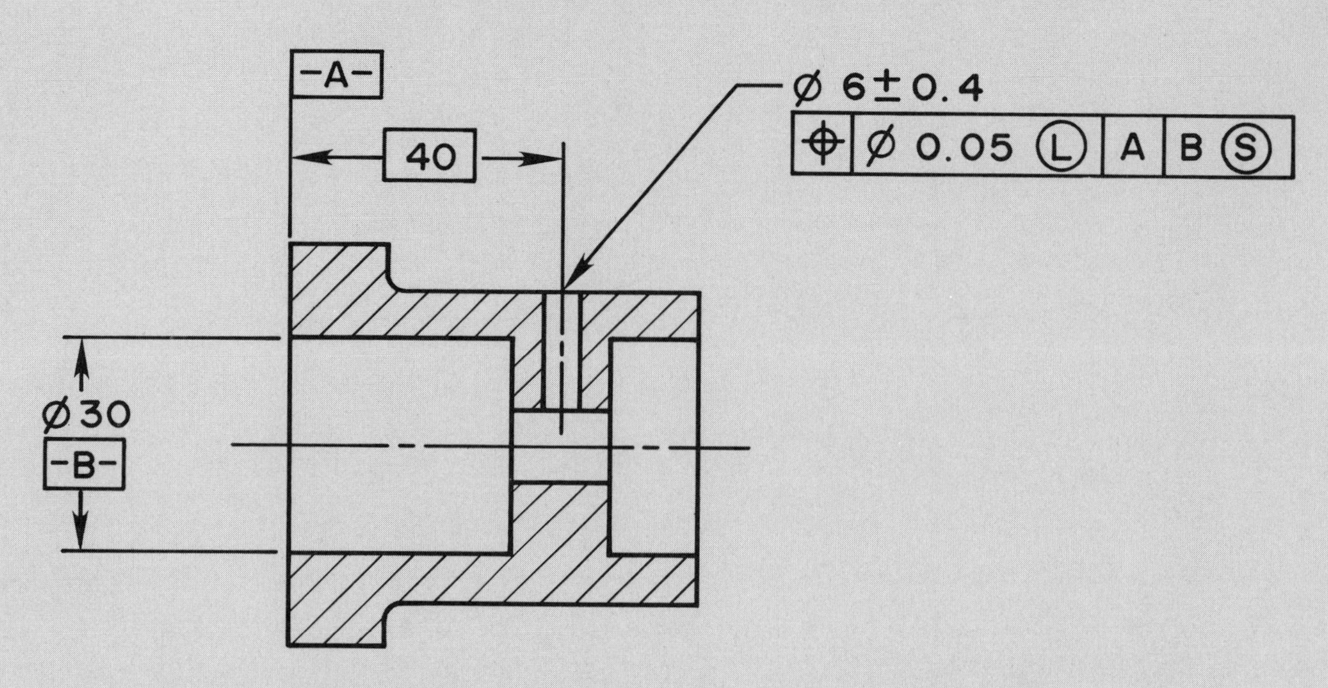

Possible Produced Sizes	ϕ Geometric Tolerance at Given Produced Sizes
6.4 LMC	________
6.2	________
6.0	________
5.8	________
5.6 MMC	________

18. Which of the following statements are true in regards to datum precedence? (More than one may be selected).
    a. Datum precedence is established by the order of placement in the feature control frame.
    b. Datum precedence is established by alphabetical order of datum reference letters.
    c. The first datum listed in the feature control frame is the primary datum.

PRINT READING EXERCISES FOR CHAPTER 4

Name:______________________

The following print reading exercise is designed for use in programs for machining, welding, tool and die, dimensional inspection, and other manufacturing curriculums where the objective is the reading and interpretation of prints rather than the development of drafting skills. An actual industrial print is used with related questions that require you to read and interpret specific dimensioning and geometric tolerancing representations. The answers and interpretations should be based on the previously learned content of this book. The prints used are based on ANSI standards; however, company standards may differ slightly. When reading these prints or any other industrial prints, a degree of flexibility may be required to determine how individual applications correlate with the ANSI standard.

PRINT READING EXERCISE

Refer to the Hyster Company print of PEDAL-ACCELERATOR found on page 216.

1. Refer to the 19.10-19.08 diameter dimension at datum D and answer the following related questions:

   a. What is the MMC? __________

   b. What is the LMC? __________

   c. The position tolerance for this feature is referenced to which datums?

   ____

   d. Why is there no material condition symbol after the perpendicularity geometric tolerance? ______________________________________

   ____

   ____

   e. Given the following list of possible produced sizes, determine the geometric tolerance at each produced size for perpendicularity and position:

Produced sizes	Geometric Tolerance	
	Perpendicularity	Position
19.10	______	______
19.095	______	______
19.09	______	______
19.085	______	______
19.08	______	______

Refer to the Hyster Company print of the FLYWHEEL-DSL found on page 218.

2. Refer to the ϕ133.37-133.35 dimension and the associated feature control frame:
   a. What is the material condition symbol associated with the perpendicularity geometric tolerance? ___________
   b. What is the MMC of this feature? ___________
   c. What is the LMC of this feature? ___________
   d. Given the following list of possible produced sizes determine the geometric tolerance at each produced size:

Produced sizes	Geometric tolerance
133.370	___________
133.365	___________
133.360	___________
133.355	___________
133.350	___________

Refer to the Hyster Company print on the CASE-DIFF found on page 217.

3. Refer to the ϕ42.70 dimension and the associated feature control frame:
   a. Where is the tolerance for this dimension specified? ___________
   ____
   b. What is the tolerance for this dimension? ___________
   c. Name the geometric characteristic symbol.___________
   d. What is the material condition symbol reference for the geometric tolerance? ___________
   e. Why is the material condition symbol not shown? ___________
   ____
   ____
   ____
   f. What is the MMC of the dimension? ___________
   g. What is the LMC of the dimension? ___________

h. Given the following list of possible produced sizes determine the geometric tolerance at each produced size:

Produced sizes	Geometric tolerance
42.58	________
42.64	________
42.68	________
42.72	________
42.76	________
42.82	________

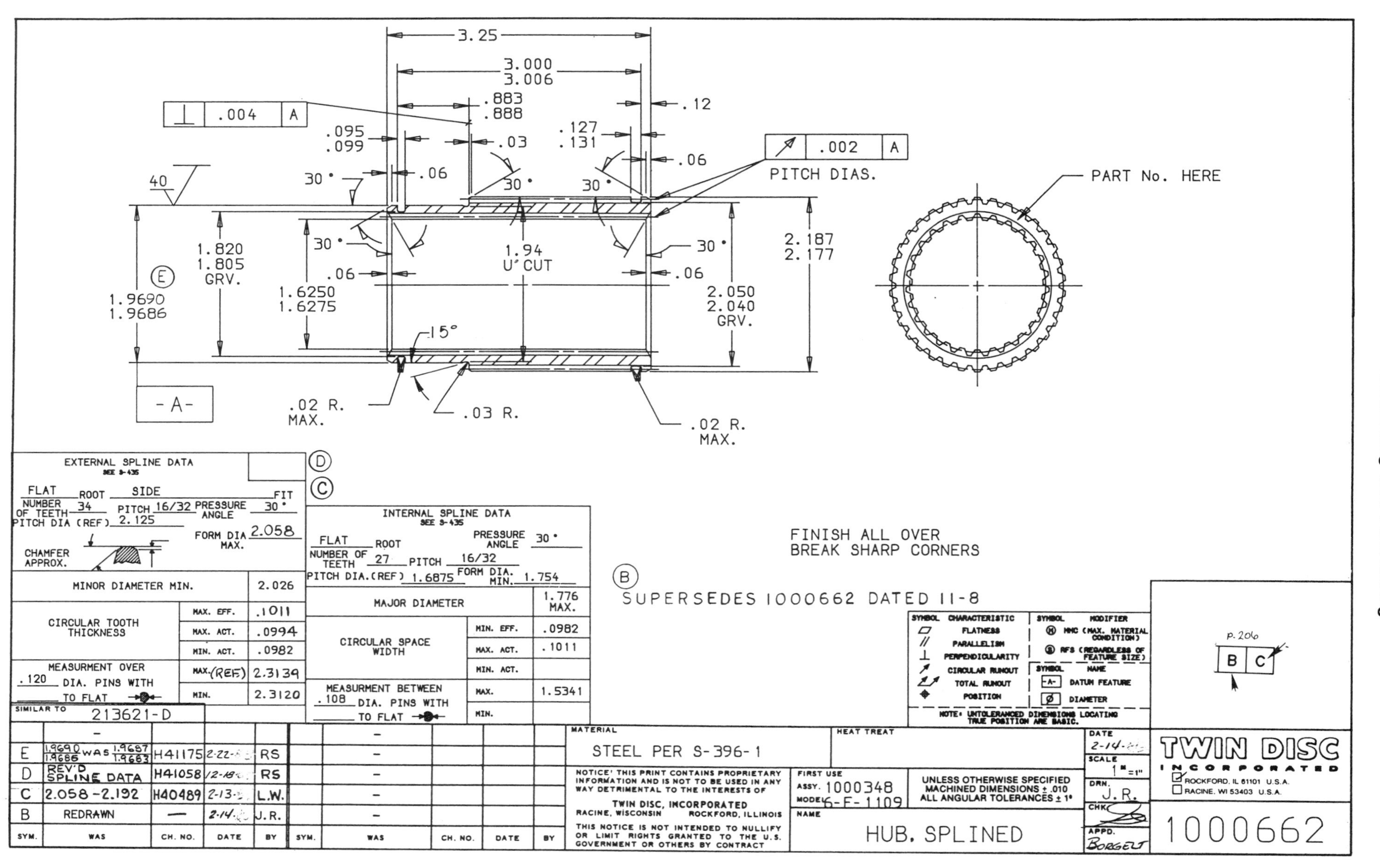

Typical industry print showing GD&T symbols.

Chapter 5

TOLERANCES OF FORM AND PROFILE

This chapter explains the concepts and techniques of dimensioning and tolerancing to control the form and profile of geometric shapes. FORM TOLERANCES control:

1. Straightness.
2. Flatness.
3. Circularity.
4. Cylindricity.

PROFILE TOLERANCES control the form and/or orientation of straight lines or surfaces, arcs, and irregular curves.

When size tolerances provided in conventional dimensioning do not provide sufficient control for the functional design and interchangeability of a product, then form and/or profile tolerances should be specified. As discussed in Chapter 4, size limits control a degree of form. The extent of this control should be evaluated before specifying geometric tolerances of form or profile.

STRAIGHTNESS TOLERANCE

STRAIGHTNESS is a condition where an element of a surface or an axis is in a straight line. Straightness is a form tolerance. The STRAIGHTNESS TOLERANCE specifies a zone within which the required surface element or axis must lie. Example 5-1 shows a detailed example of the straightness geometric characteristic symbol used in a feature control frame.

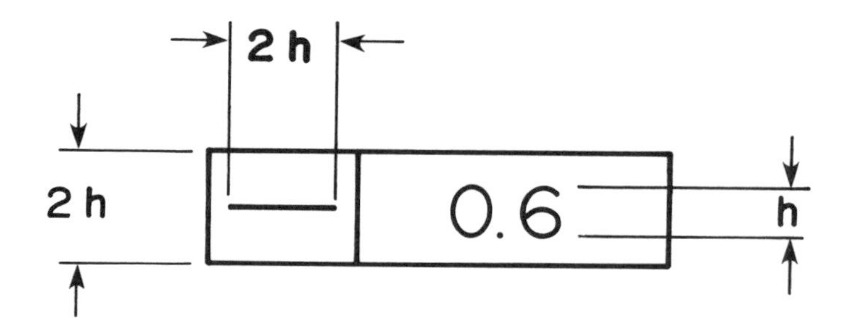

Example 5-1. Feature control frame with straightness geometric characteristic symbol.

The SURFACE STRAIGHTNESS TOLERANCE is represented by connecting the feature control frame to the surface with a leader. Or, the SURFACE STRAIGHTNESS TOLERANCE may be represented by connecting the feature control frame by an extension line in the view where the element to be controlled is shown as a line. The feature may not exceed the MMC envelope and perfect form must be maintained if the actual size is produced at MMC, otherwise RFS is implied and the geometric tolerance remains the same at any produced size. Example 5-2 shows a drawing and an exaggerated representation of what happens when a surface straightness tolerance is applied. Remember, straightness implies RFS as does the other form characteristics. The chart shows the maximum out of straightness at several possible produced sizes in Example 5-2. With surface straightness, the straightness tolerance must be less than the size tolerance.

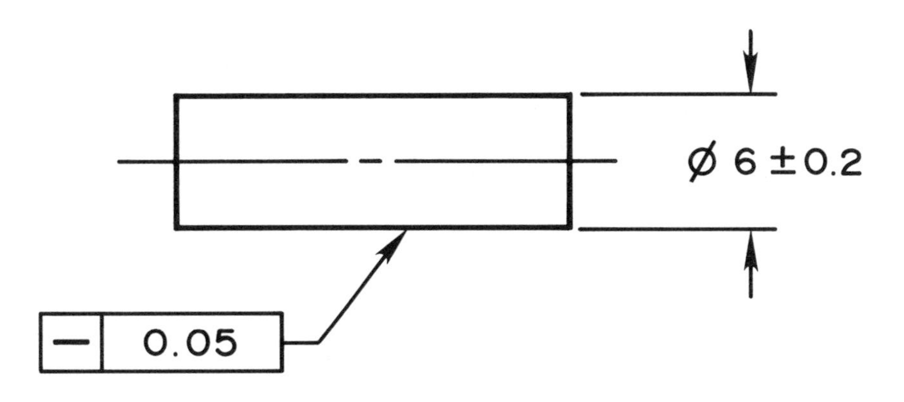

THE DRAWING

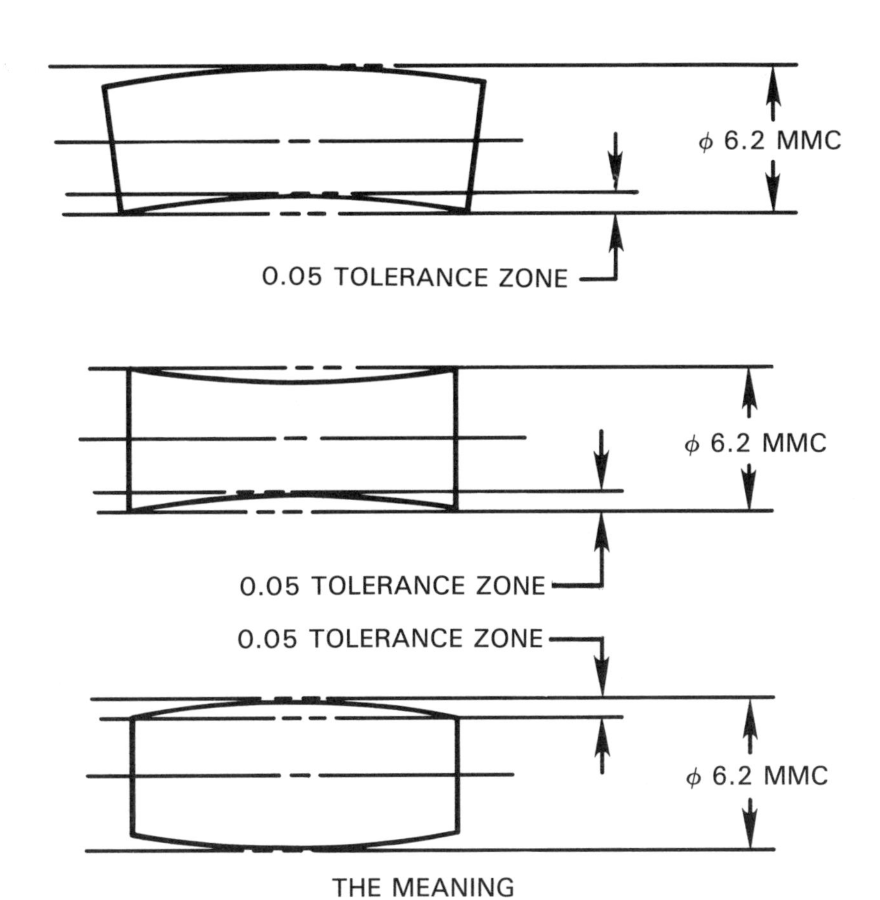

THE MEANING

Possible Produced Sizes	Maximum Out-of-Straightness
6.2 MMC	0 Perfect form
6.1	0.05
6.0	0.05
5.9	0.05
5.8 LMC	0.05

Example 5-2. Surface straightness.

AXIS STRAIGHTNESS is specified on a drawing by placing the feature control frame below the diameter dimension, and a diameter symbol is placed in front of the geometric tolerance to denote a cylindrical tolerance zone. RFS is implied, as shown in Example 5-3. Notice in the chart with Example 5-3 that this application allows a violation of perfect form at MMC.

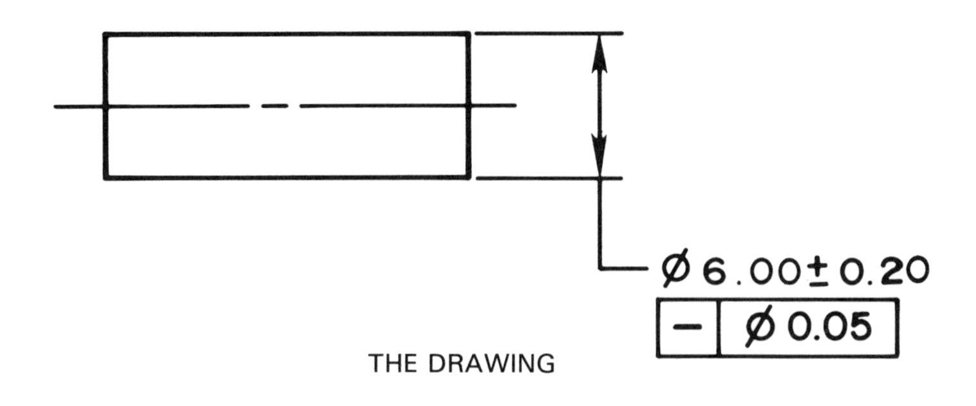

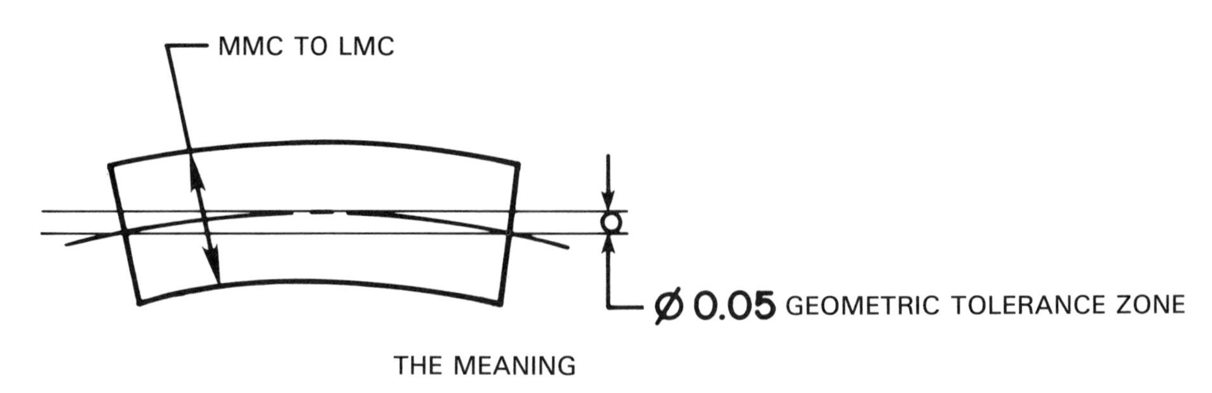

Possible Produced Sizes	Maximum Out-of-Straightness
6.20 MMC	0.05
6.10	0.05
6.00	0.05
5.90	0.05
5.80 LMC	0.05

Example 5-3. Axis straightness, RFS implied.

Axis straightness may also be specified on an MMC basis by placing the MMC material condition symbol after the geometric tolerance. The specified geometric tolerance is then held at MMC and allowed to increase as the actual size departs from MMC. The geometric tolerance is at MMC as shown in Example 5-4. In some instances the straightness tolerance may be greater than the size tolerance where necessary.

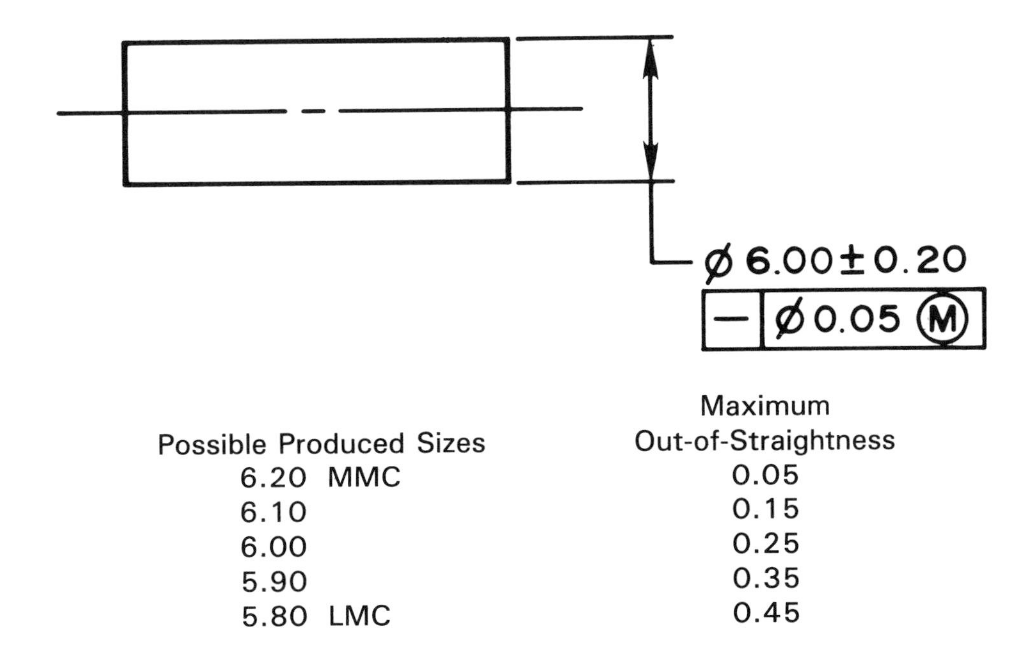

Possible Produced Sizes	Maximum Out-of-Straightness
6.20 MMC	0.05
6.10	0.15
6.00	0.25
5.90	0.35
5.80 LMC	0.45

Example 5-4. Axis straightness, MMC specified.

STRAIGHTNESS PER UNIT of measure may be applied to a part or feature in conjunction with a straightness specification over the total length. This may be done as a means of preventing an abrupt surface variation within a relatively short length of the feature. The specified tolerance over the total length is greater than the unit tolerance and is normally given to keep the unit tolerance from getting out of control when applied to the length of the feature. The per unit specification may be given as a tolerance per inch or per 25 millimeters of length. When this technique is used, the feature control frame is doubled in height and split so that the tolerance over the total length may be specified in the top half and the per unit control placed in the bottom half as shown in Example 5-5. Caution should be exercised when using unit straightness as it could cause waviness in the feature or part.

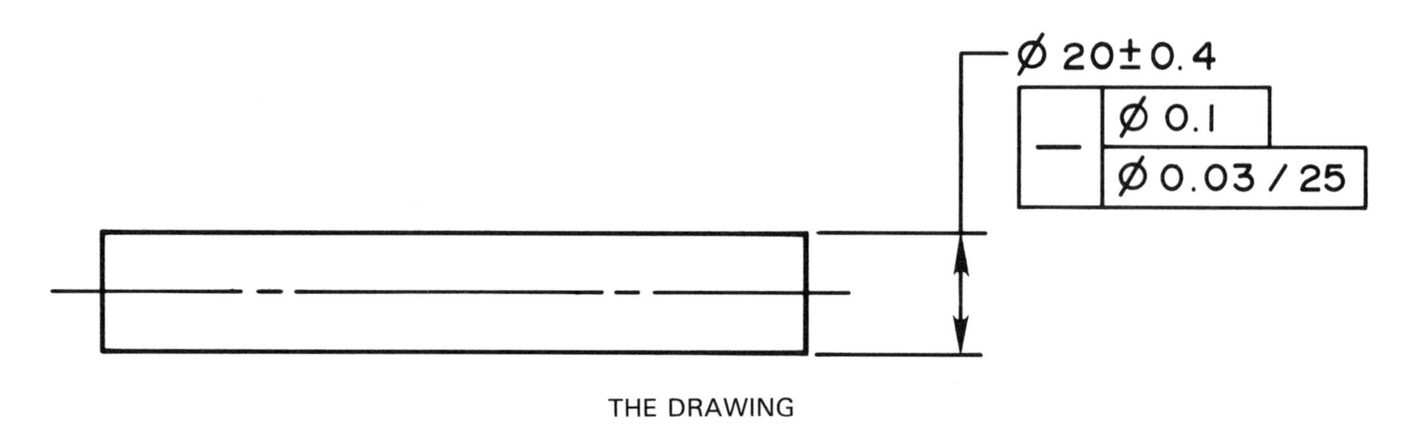

THE DRAWING

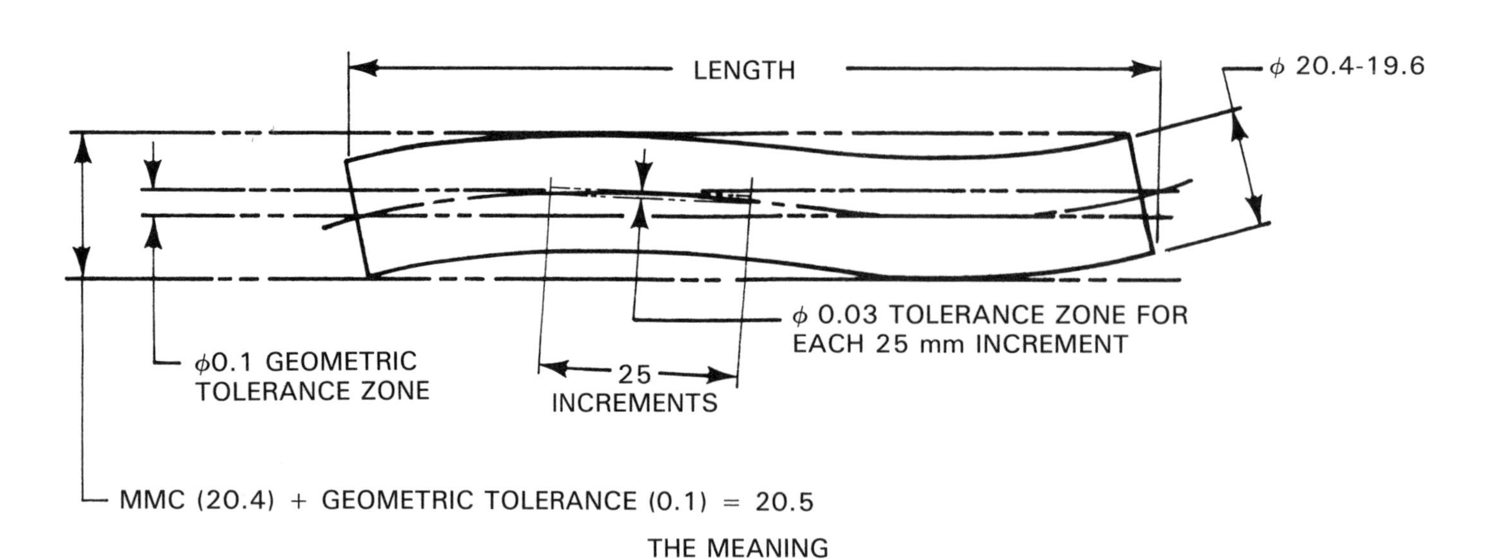

THE MEANING

Example 5-5. Unit straightness specified on a drawing.

The derived axis or center line of the actual feature must lie within a cylindrical tolerance zone of 0.1 diameter for the total length and within a 0.03 cylindrical tolerance zone for any 25 mm length, regardless of feature size. Additionally, each circular element of the surface must be within the specified limits of size.

Straightness may also be applied on an RFS or MMC basis to noncylindrical features of size. When this is done, the associated center plane must lie within two parallel planes separated by a distance equal to the specified geometric tolerance zone. The feature control frame may be attached to the view where the surface appears as a line by using a leader or an extension line. In this situation, the diameter symbol is not placed in front of the geometric tolerance.

Straightness may also be applied to a flat surface. When this is done, the straightness geometric tolerance may control single line elements on the surface in one or two directions. The direction of the tolerance zone is determined by the placement of the feature control frame as shown in Example 5-6.

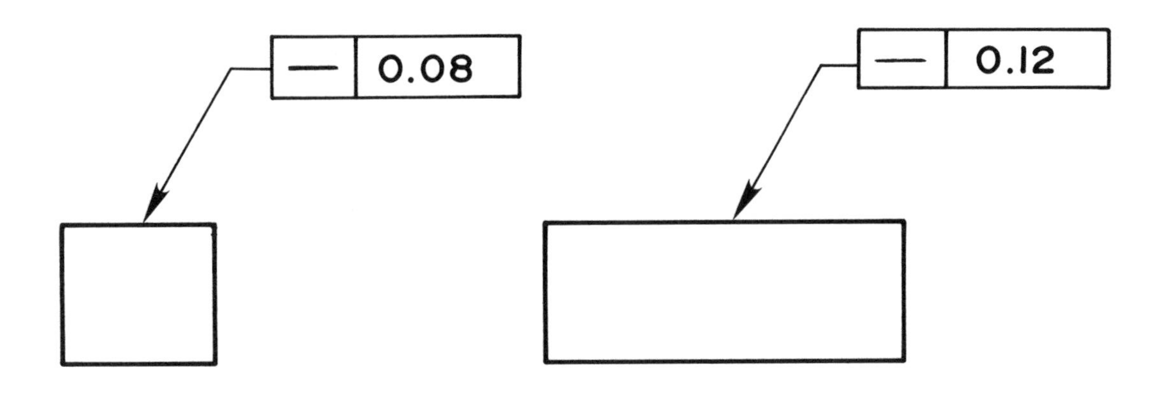

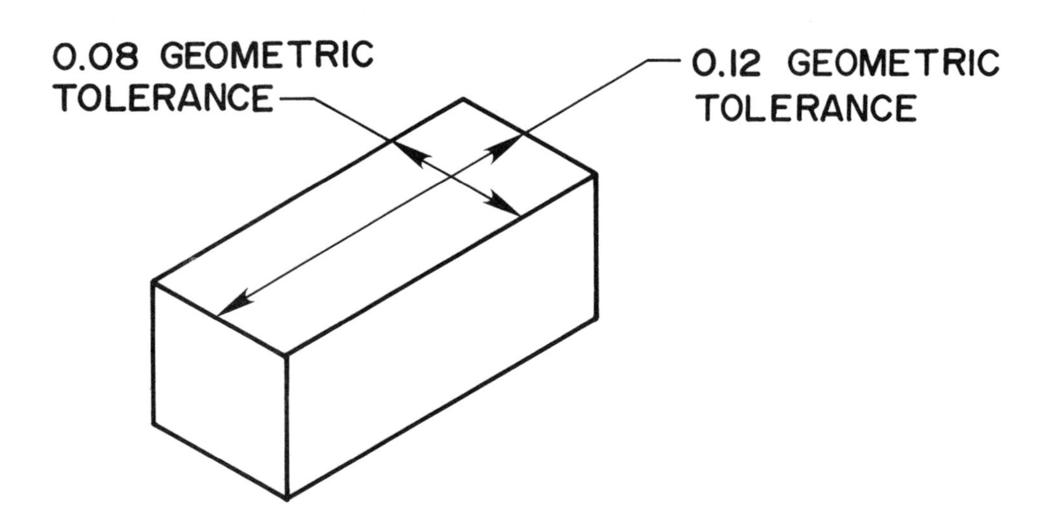

Example 5-6. Straightness applied to a flat surface.

FLATNESS TOLERANCE

PERFECT FLATNESS is the condition of a surface where all of the elements are in one plane. Flatness is a form tolerance. A FLATNESS TOLERANCE ZONE establishes the distance between two parallel planes within which the surface must lie. The flatness feature control frame is detailed in Example 5-7.

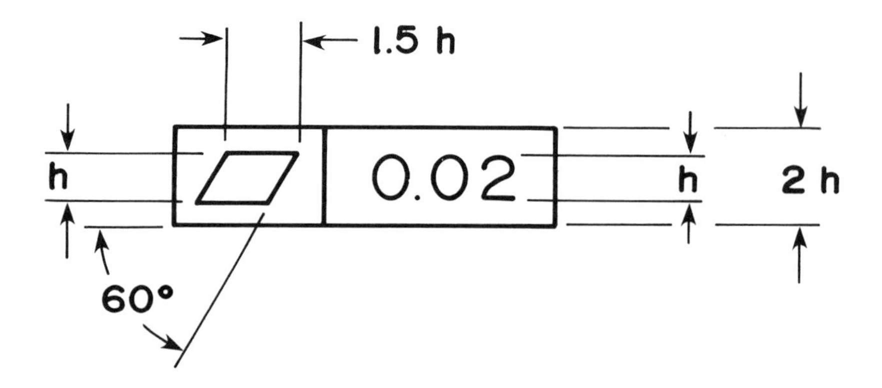

Example 5-7. Flatness feature control frame.

When a flatness geometric tolerance is specified, the feature control frame is connected by a leader or an extension line in the view where the surface appears as a line. Refer to Example 5-8. All of the points of the surface must be within the limits of the specified tolerance zone. The smaller the tolerance zone, the flatter the surface. The flatness tolerance must be less than the size tolerance.

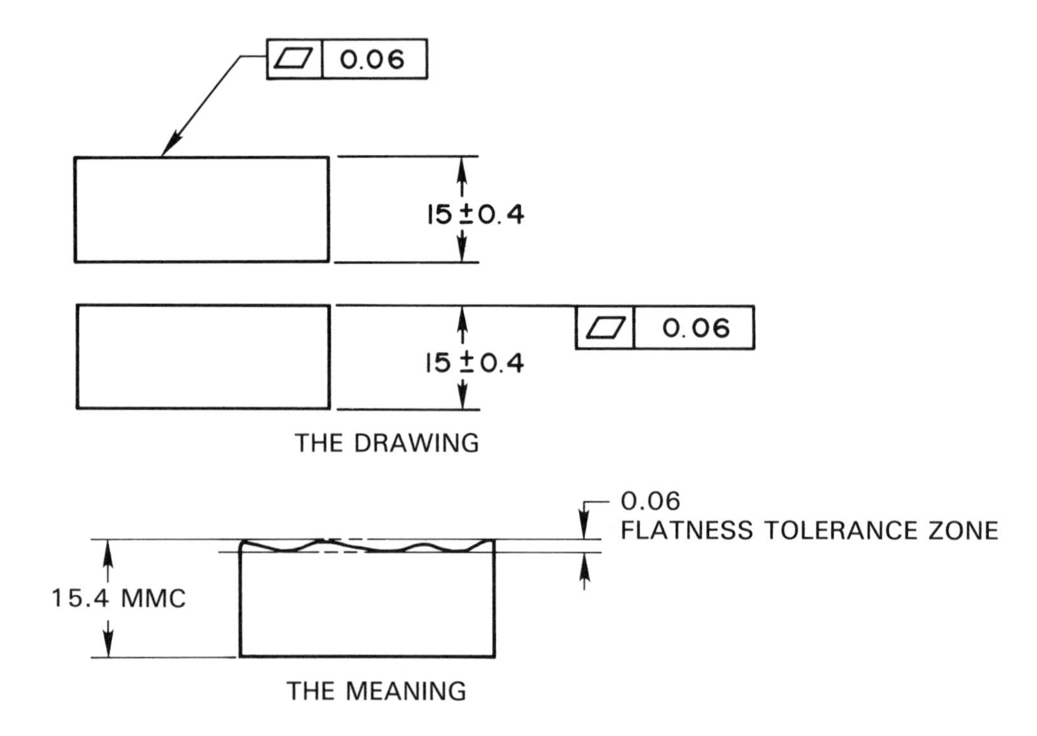

Example 5-8. Flatness tolerance.

At times it may be necessary to provide a flatness callout for only a specific area of a surface. This procedure is known as SPECIFIC AREA FLATNESS. Specific area flatness may be used when a large cast surface must be flat, but it may be possible to finish a relatively small area rather than an expensive operation of machining the entire surface as shown in Example 5-9.

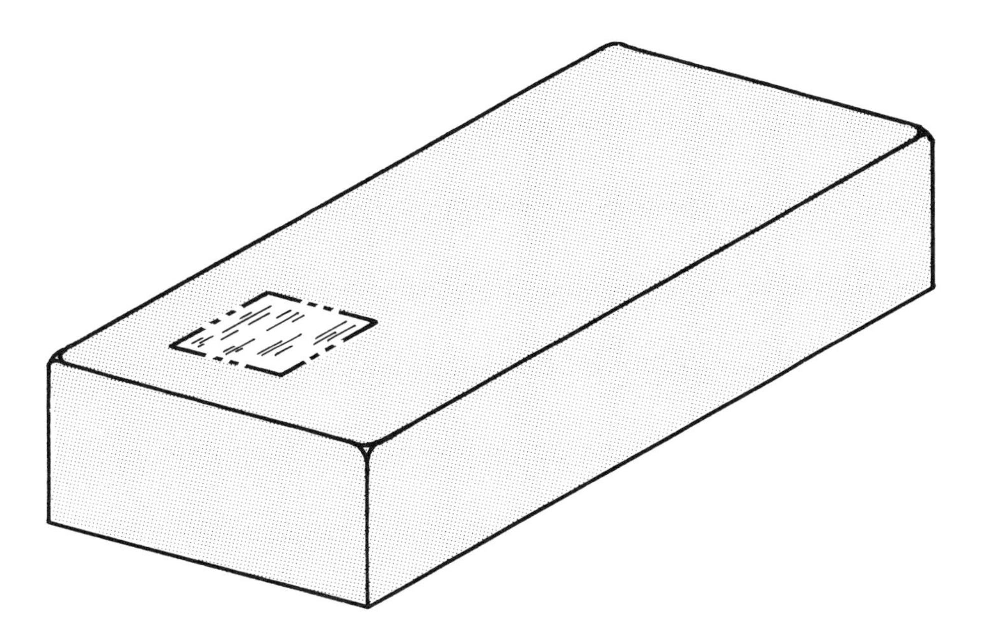

Example 5-9. A pictorial drawing of specific area flatness.

When specific area flatness is used, the specific area is outlined with phantom lines and then crosshatched or section lined within the area. The specific area is then located, preferably from datums, with basic or ± dimensions. The feature control frame is then connected to the area with a leader line as shown in Example 5-10.

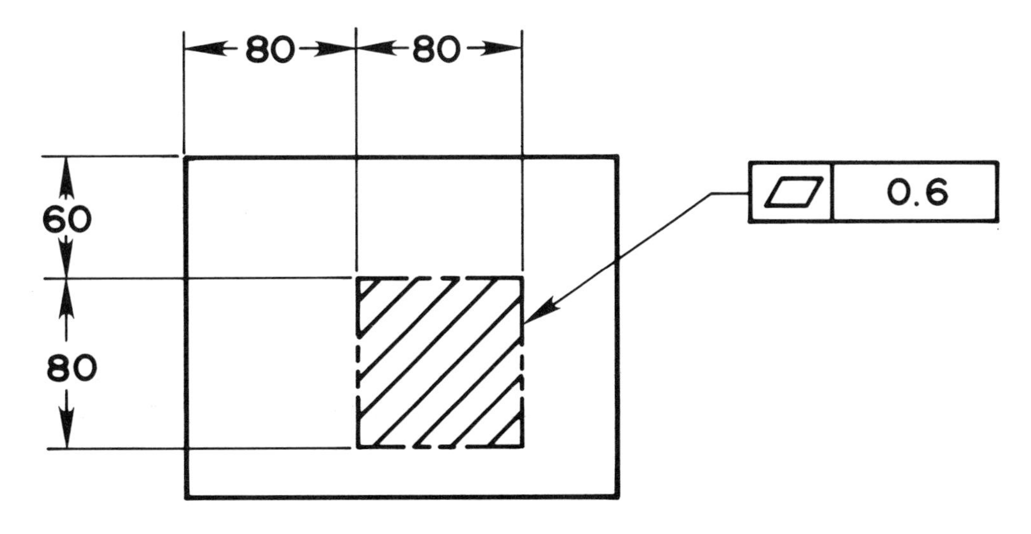

Example 5-10. Specific area flatness.

UNIT FLATNESS may be specified when it is desirable to control the flatness of a given surface area as a means of controlling an abrupt surface variation within a small area of the feature. The unit flatness specification may be used alone or in combination with a total tolerance. Most applications use unit flatness in combination with a total tolerance over the entire surface so that the unit callout is not allowed to get out of control. When this is done, the height of the feature control frame is doubled with the total tolerance placed in the top half and the unit tolerance plus the size of the unit area placed in the bottom half. The unit tolerance must be smaller than the total tolerance, as shown in Example 5-11.

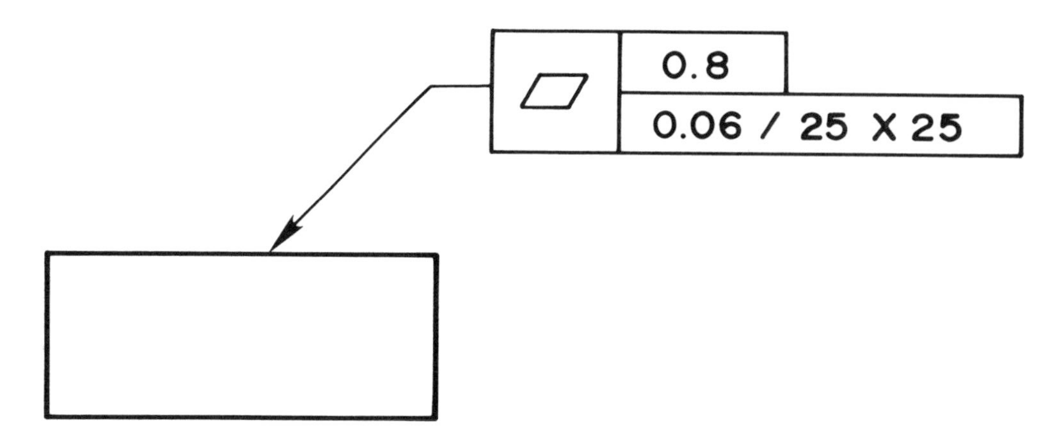

Example 5-11. Total flatness with unit flatness and area size given.

CIRCULARITY TOLERANCE

CIRCULARITY is characterized by any given cross section taken perpendicular to the axis of a cylinder or cone, or through the common center of a sphere. Circularity is a form tolerance. The CIRCULARITY GEOMETRIC TOLERANCE is formed by a radius zone creating two concentric circles within which the actual surface must lie. The circularity feature control frame is detailed in Example 5-12. The tolerance applies to only one sectional element at a time, as shown exaggerated in Example 5-13.

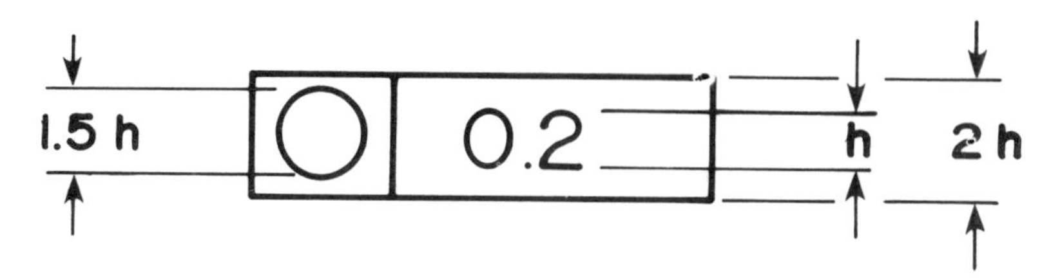

Example 5-12. Circularity feature control frame.

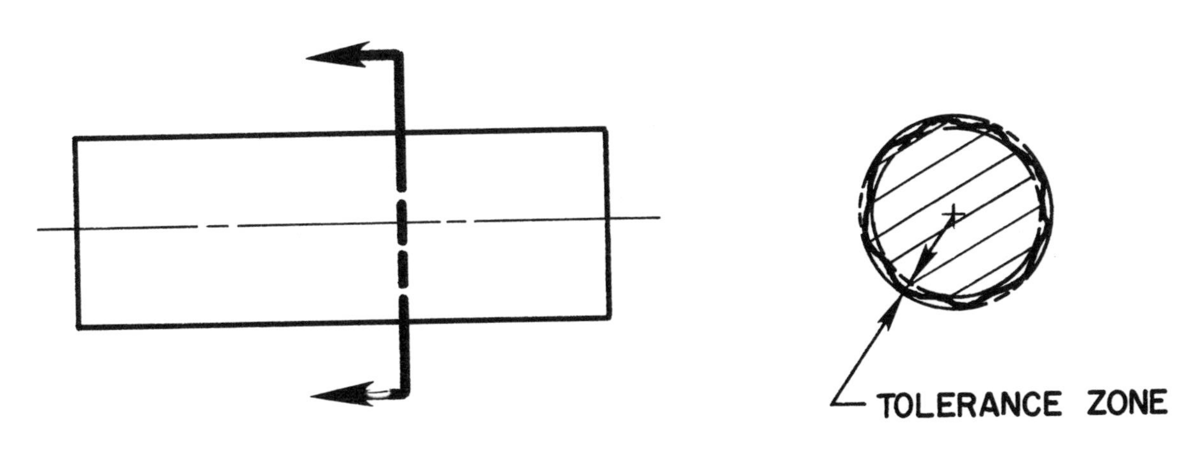

Example 5-13. Circularity geometric tolerance.

The circularity feature control frame is connected with a leader to the view where the feature appears as a circle or in the longitudinal view as shown in Example 5-14.

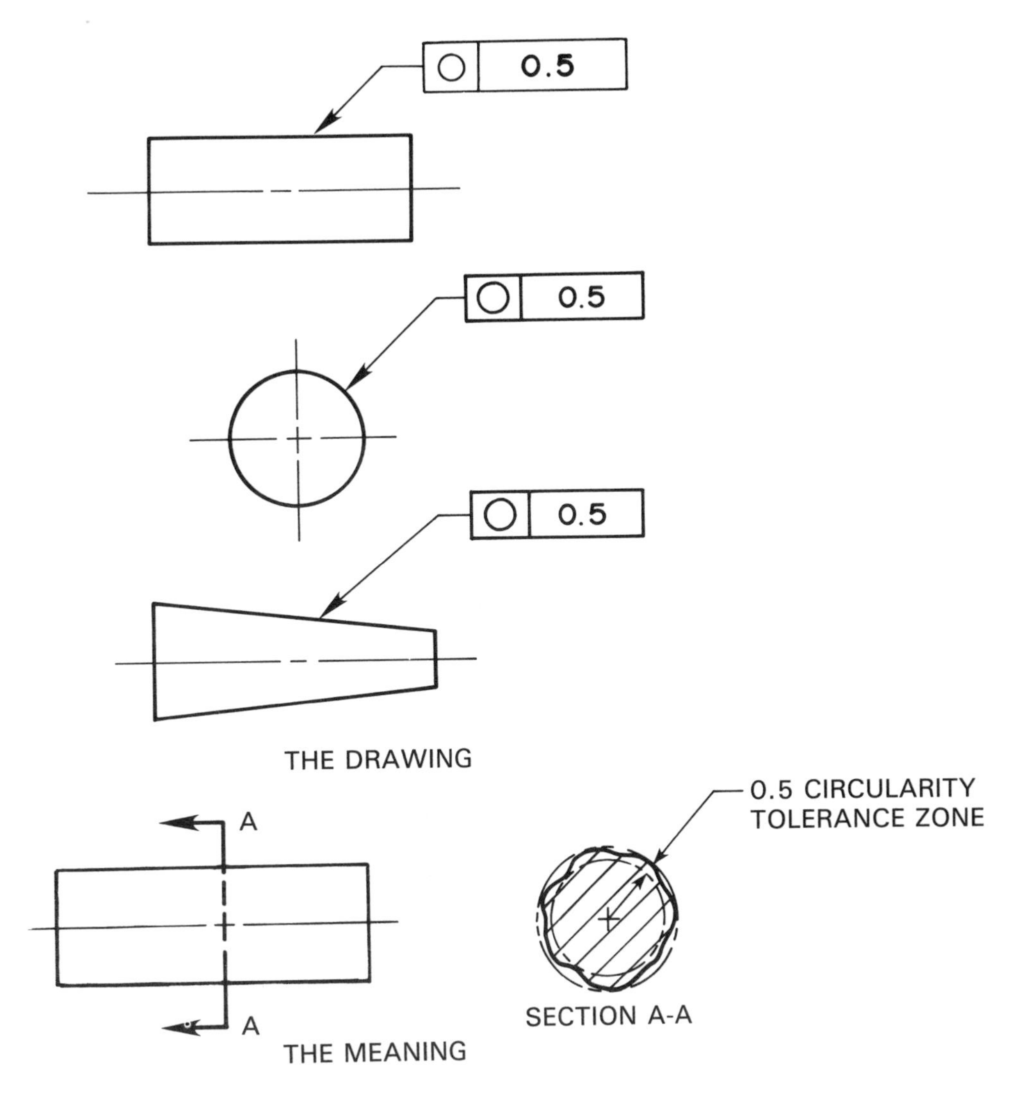

Example 5-14. Circularity tolerance specification.

CYLINDRICITY TOLERANCE

CYLINDRICITY is identified by a radius tolerance zone that establishes two perfectly concentric cylinders within which the actual surface shall lie. Cylindricity is a form tolerance. The cylindricity feature control frame is detailed in Example 5-15. The feature control frame showing the cylindricity tolerance specification is connected by a leader to either the circular or longitudinal view, as shown in Example 5-16.

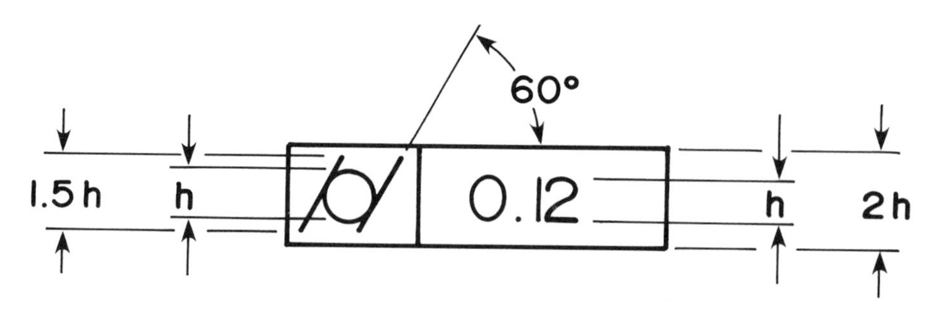

Example 5-15. Cylindricity feature control frame.

The difference between circularity and cylindricity is that circularity (circle) is a single cross sectional tolerance representing a zone between two concentric circles, while the cylindricity (cylinder) is tolerance that covers both circular and longitudinal elements at the same time representing a zone between two concentric cylinders. Cylindricity could be characterized as a blanket tolerance that covers the entire feature.

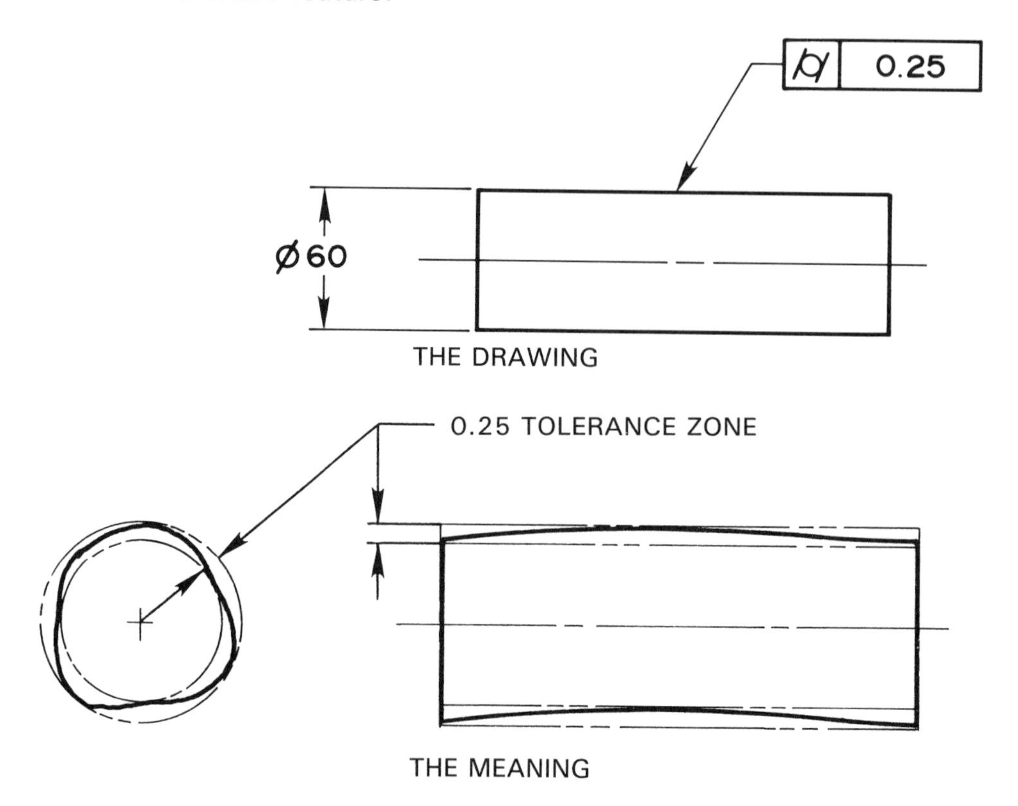

Example 5-16. Cylindricity tolerance specification.

PROFILE

PROFILE may be characterized by the outline of an object either represented by an external view or a cross section through the object. The true profile or actual desired shape of the object is the basis of the profile tolerance. The true profile should be defined by basic dimensions in most applications. The PROFILE TOLERANCE specifies a uniform boundary along the true profile within which the elements of the surface must lie. Profile may be used to control form, or combinations of size, form, and orientation. When used as a refinement of size, the profile tolerance must be contained within the size tolerance.

Profile tolerances may be bilateral, which means the tolerance zone is split equally on each side of the true profile. The profile tolerance zone may also be unilateral where the entire zone is on one side of the true profile. The profile tolerance may also be specified between two given points or all around the object. A profile tolerance is specified by connecting the feature control frame, using a leader, to the view or section that clearly depicts the intended profile. There are two profile characteristics: profile of a line and profile of a surface.

PROFILE OF A LINE

The PROFILE OF A LINE TOLERANCE is a two dimensional or cross sectional tolerance that extends along the length of the feature. The profile of a line symbol and associated feature control frame is shown in Example 5-17. Profile of a line is used where it is not necessary to control the profile of the entire feature.

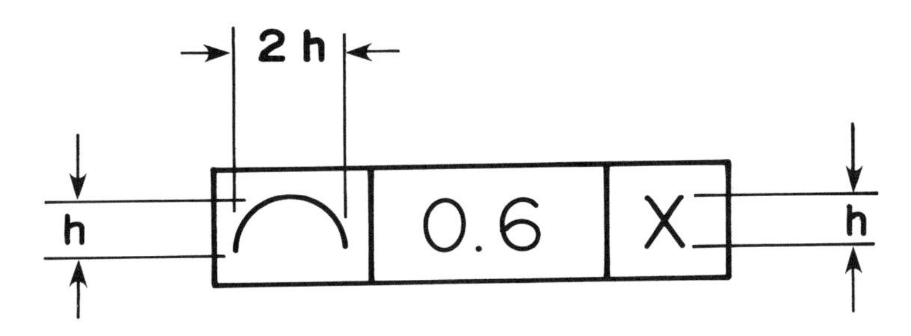

Example 5-17. Profile of a line in a feature control frame.

This tolerance is used in situations where parts or objects have changing cross sections throughout the length. An example is an aircraft wing. Datums may be used in some situations, but are not necessary when the only requirement is the profile shape taken at various cross sections. When the leader from the profile feature control frame extends to the related surface without any additional clarification, the profile tolerance is assumed to be bilateral. Bilateral tolerances are equally displaced on each side of the basic dimensions that establish true form.

The profile tolerance may be between two given points of the object. This specification may be presented as the note: BETWEEN X & Y under the feature control frame. Any combination of letters may be used, such as A & B, or C & D. The true profile may be established by a basic or tolerance dimension. See Example 5-18.

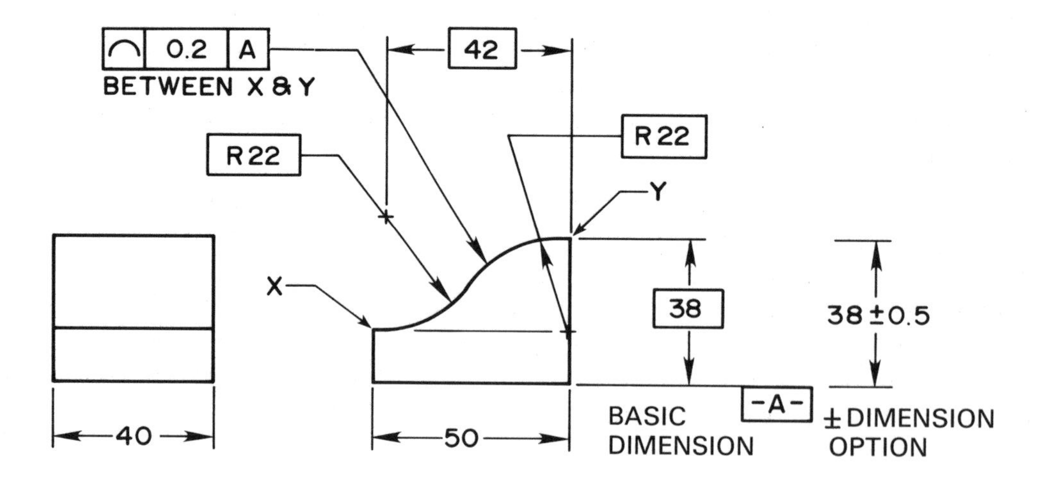

THE DRAWING

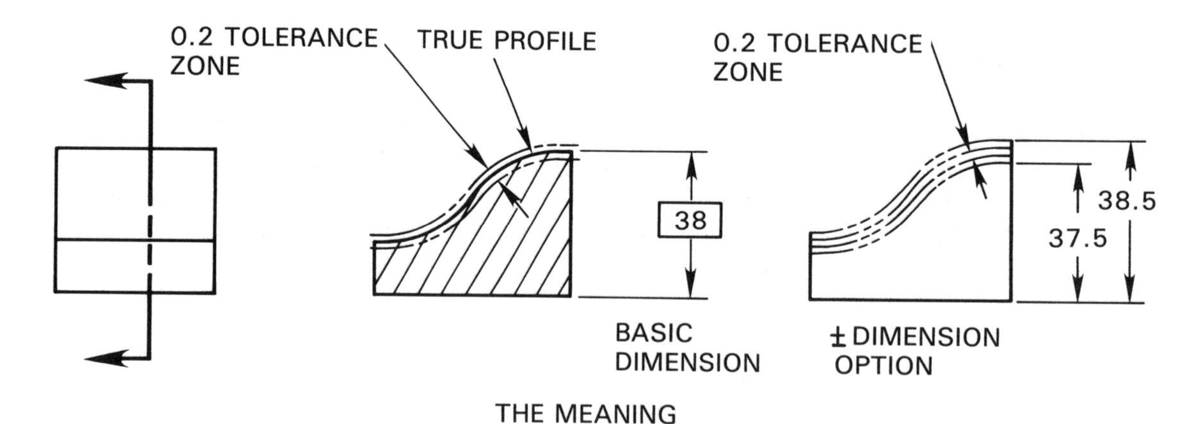

THE MEANING

Example 5-18. Profile of a line, between two given points.

Profile of a line may also specify a tolerance zone that goes around the entire object. When this is desired, the feature control frame is connected to the object with a leader as previously discussed and the all around symbol is placed on the leader as shown in Example 5-19. Any note specifying between two points is excluded.

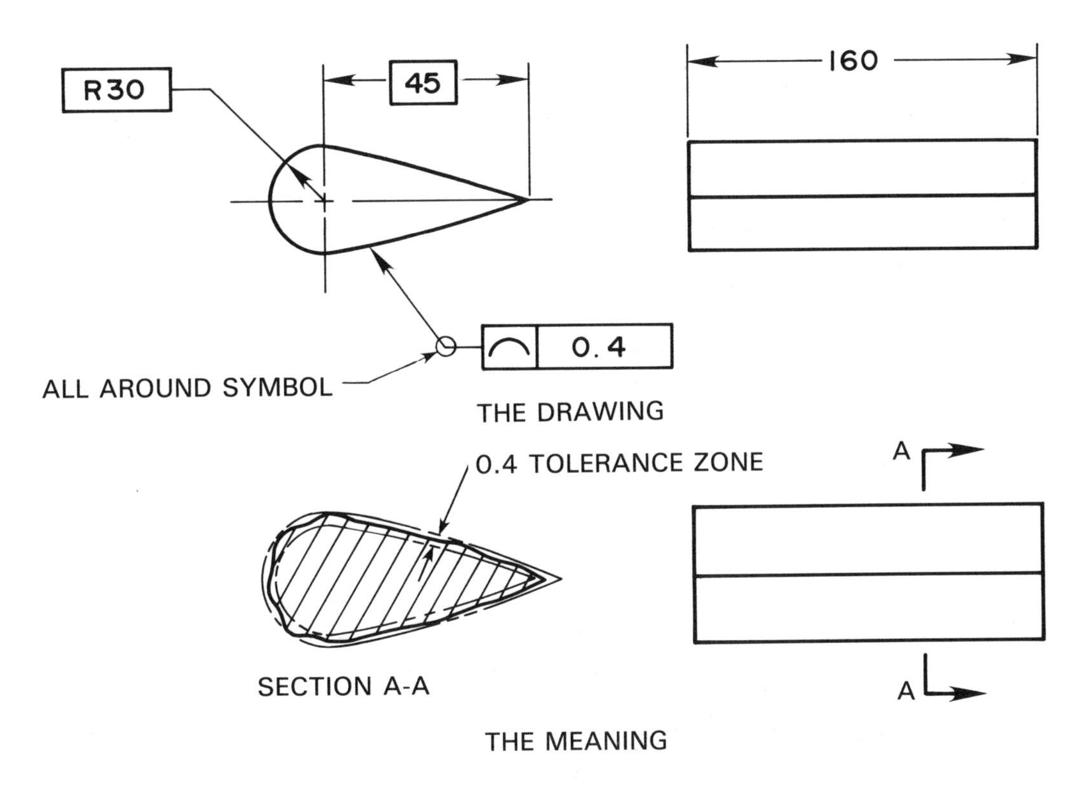

Example 5-19. Profile of a line all around.

PROFILE OF A SURFACE

PROFILE OF A SURFACE is used where it is desired to control the entire surface as a single entity. The profile of a surface geometric characteristic and associated feature control frame is detailed in Example 5-20. Profile of a surface is a blanket tolerance that is three dimensional extending along the total length and width or circumference of the object or feature(s). In most cases, the profile of a surface tolerance requires reference to datums for proper orientation of the profile.

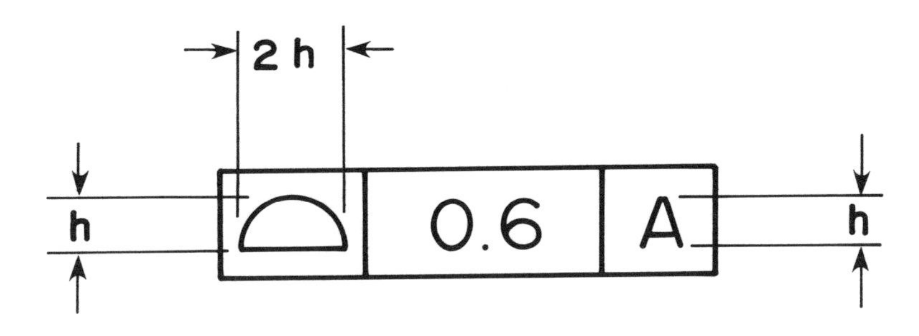

Example 5-20. Profile of a surface symbol in a feature control frame.

The profile of a surface tolerance zone may be bilateral or unilateral, just as for profile of a line. A BILATERAL PROFILE means that the tolerance zone is split equally on each side of the true profile, as shown in Example 5-21.

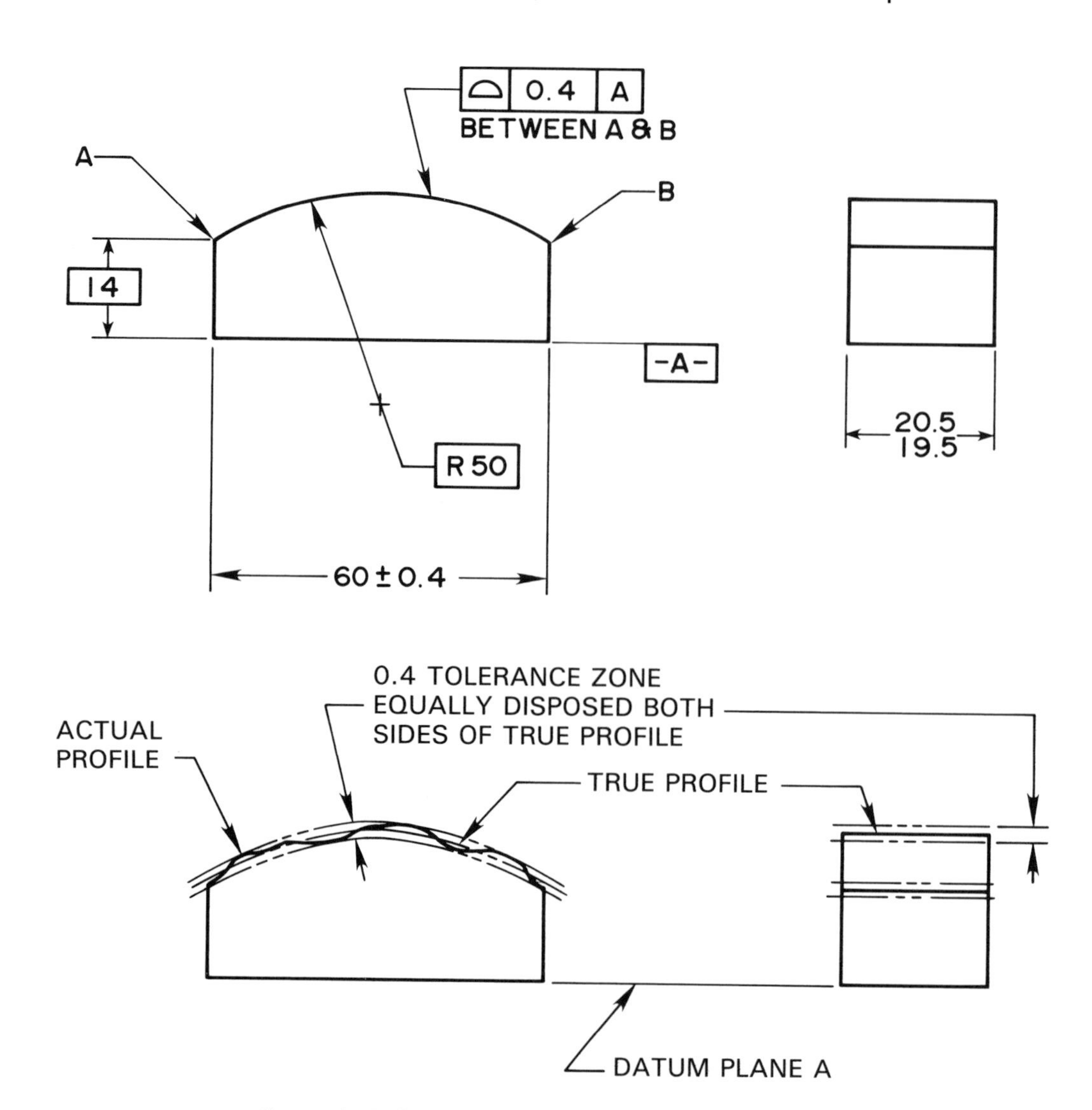

Example 5-21. Bilateral profile of a surface.

A bilateral profile tolerance is assumed unless unilateral specifications are provided. A UNILATERAL PROFILE is where the entire tolerance zone is on one side of the true profile. When a unilateral profile tolerance is required, a short phantom line is drawn parallel to the true profile on the side of the intended unilateral tolerance. A dimension line with arrow head is placed on the far side and a leader line connects the feature control frame on the other side, as shown in Example 5-22.

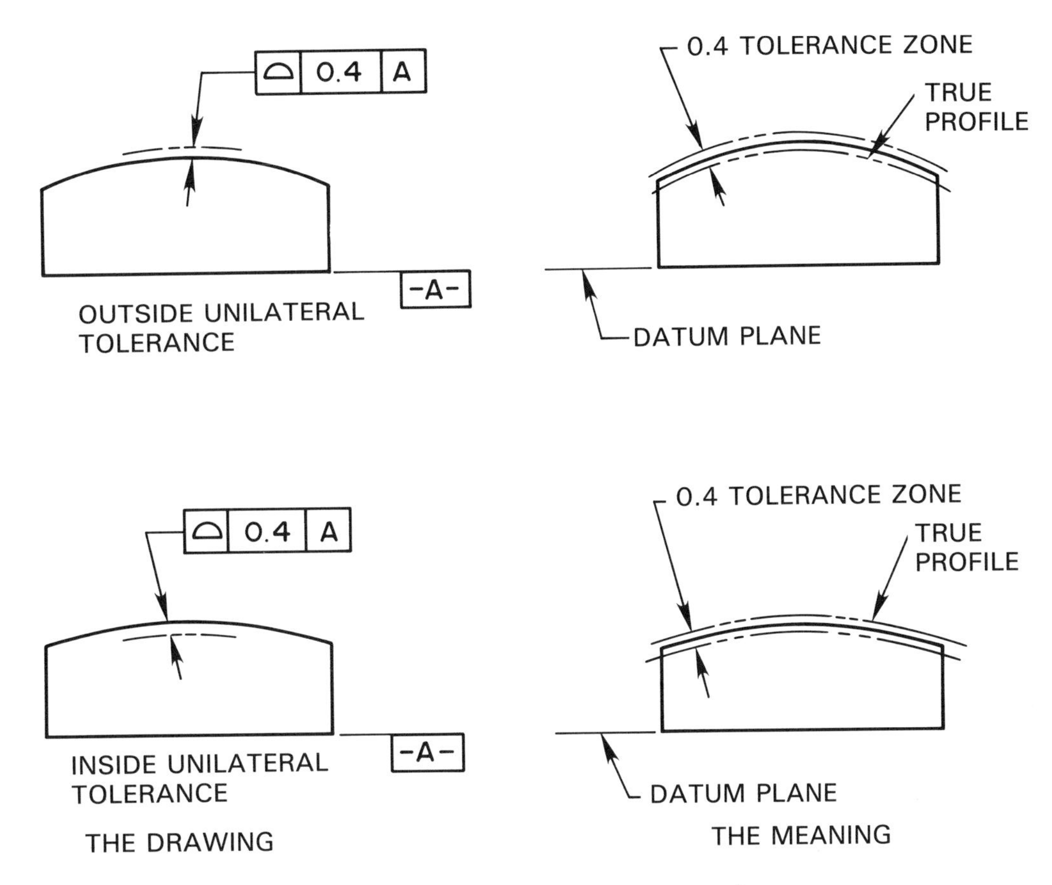

Example 5-22. Unilateral profile of a surface.

To save time when an extensive number of basic dimensions are used on a drawing, a general note may be used to specify basic dimensions. Use UNTOLERANCED DIMENSIONS ARE BASIC rather than using the customary rectangular block around the dimension to denote basic. See Example 5-23.

Surface profile may also be applied to completely blanket objects that have a constant uniform cross section by placing the all around symbol on the leader line. When this is done, surfaces all around the object outline must lie between two parallel boundaries equal in width to the given geometric tolerance. The tolerance zone should also be perpendicular to a datum plane. See Example 5-23.

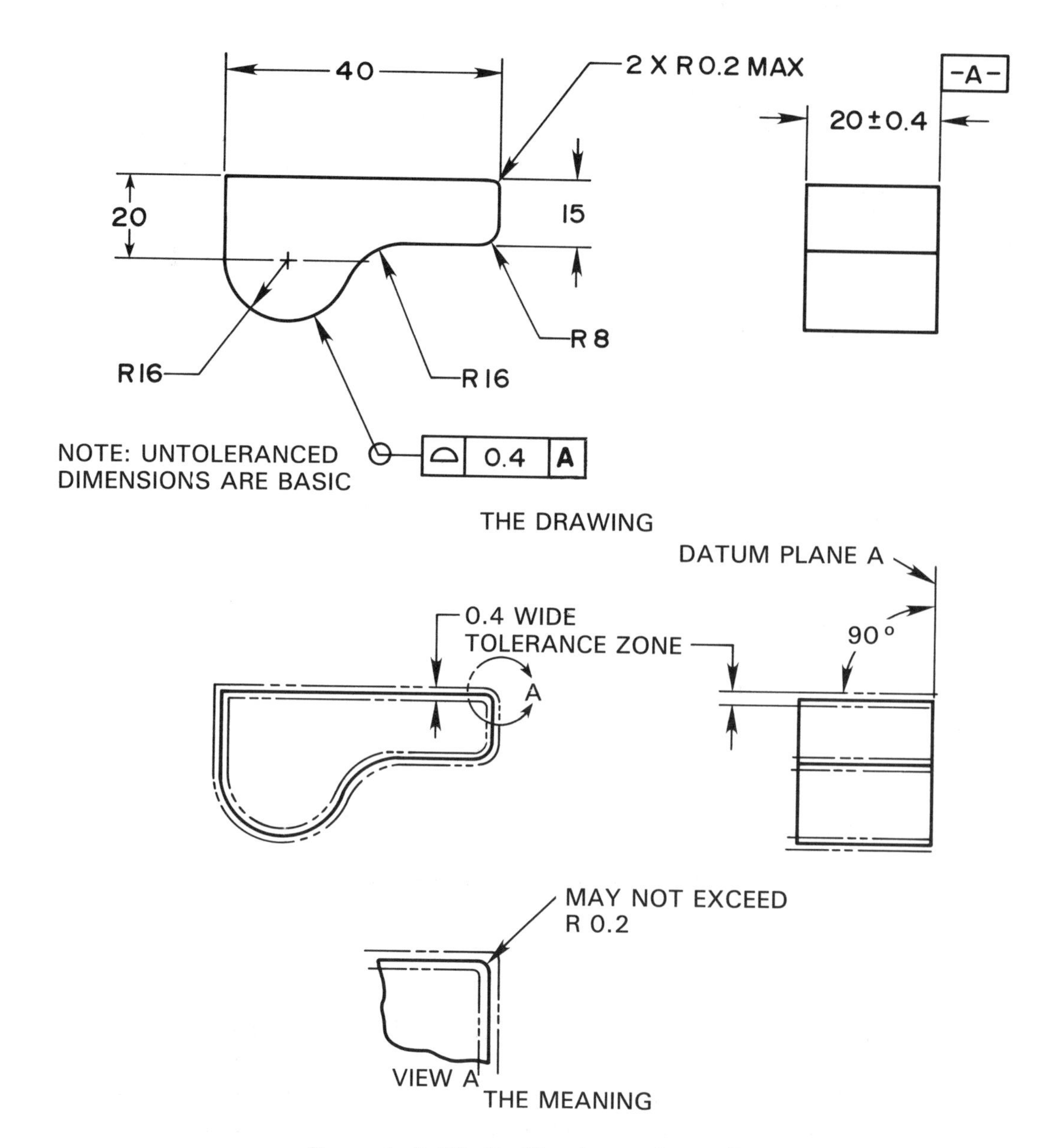

Example 5-23. Profile of a surface all around.

When a profile tolerance is at a sharp corner, the tolerance zone extends to the intersection of the boundary lines. In these situations, a rounded corner may occur, because the actual surface may be anywhere within the tolerance zone boundary. If this roundness must be controlled, then a maximum radius note shall be added to the drawing. Notice, R 0.2 MAX as shown in Example 5-23.

PROFILE OF COPLANAR SURFACES

COPLANAR SURFACES are two or more surfaces on a part that are on the same plane. A COPLANAR PROFILE TOLERANCE may be used when it is desirable to treat two or more separate surfaces, that lie on the same plane, as one surface. To control the profile of these surfaces as a single surface, place a phantom line between the surfaces in the view where the required surfaces appear as lines. Connect a leader from the feature control frame to the phantom line and add a note identifying the number of surfaces below the feature control frame. Refer to 2 SURFACES as shown in Example 5-24.

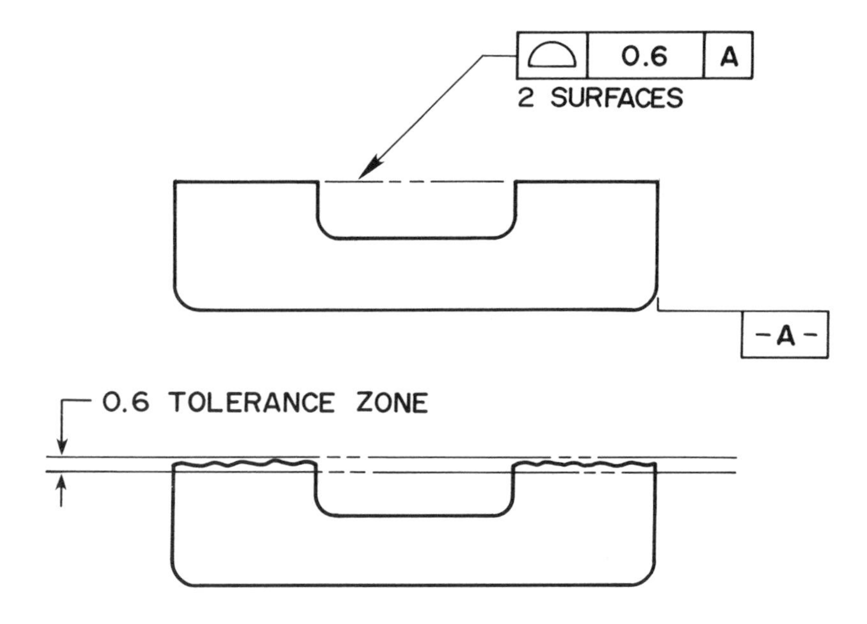

Example 5-24. Coplanar profile.

PROFILE OF PLANE SURFACES

Profile tolerancing may be used to control the form and orientation of plane surfaces. For example, profile of a surface may be used to control the angle of an inclined surface in relationship to a datum, as shown in Example 5-25. Notice that the required surface must lie between two parallel planes 0.1 apart equally split on each side of a true plane that has a basic angular orientation to a datum.

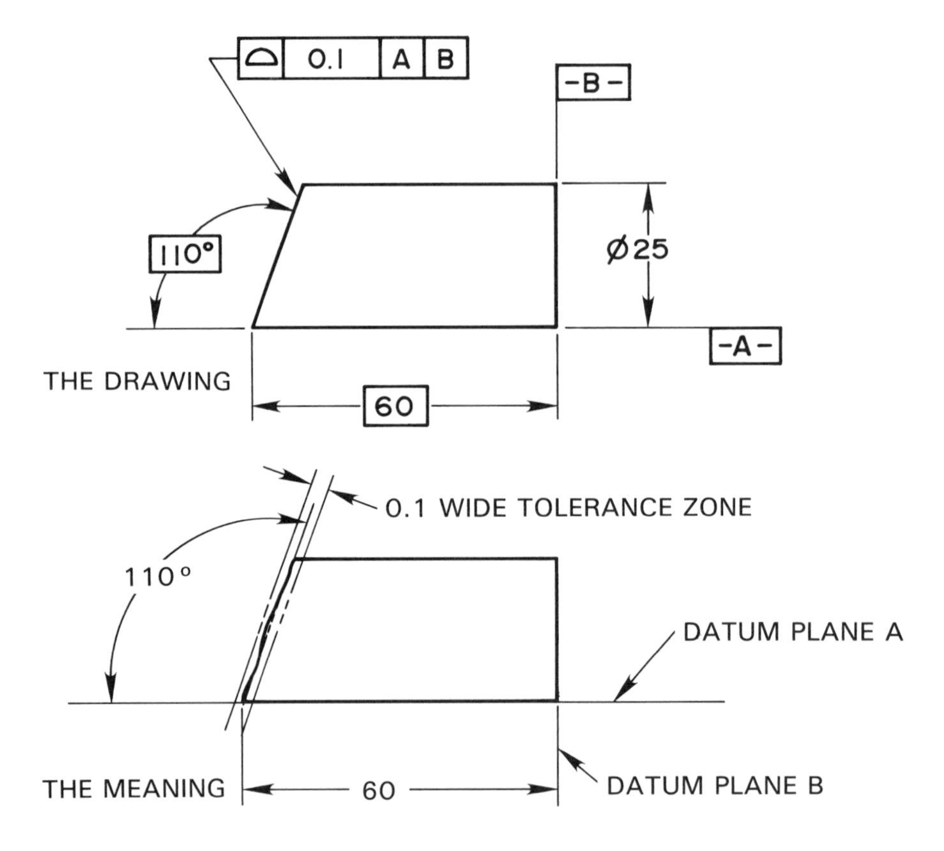

Example 5-25. Specifying profile of a plane surface.

PROFILE OF CONICAL FEATURES

A profile tolerance may be used to control the form, or form and orientation of a conical surface. The feature may be controlled independently as a refinement of size or oriented to a datum axis. In either case the profile tolerance must be within the size tolerance. Conical profile requires that the actual surface lie between two coaxial (coaxial means having the same axis) boundaries equal in width to the specified geometric tolerance, having a basic included angle, and within the size limits. See Example 5-26.

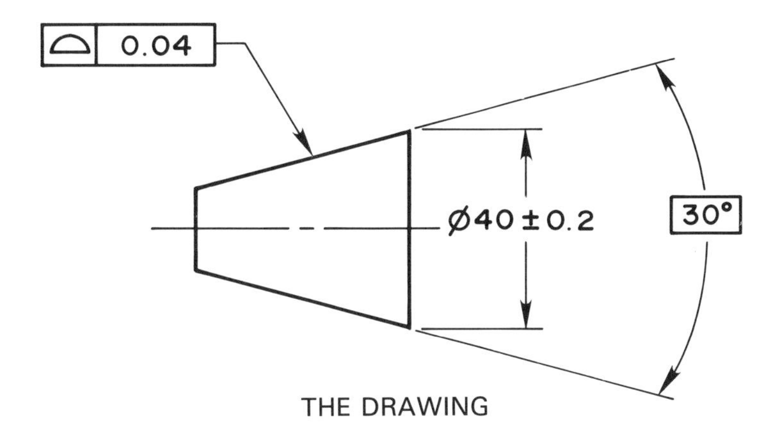

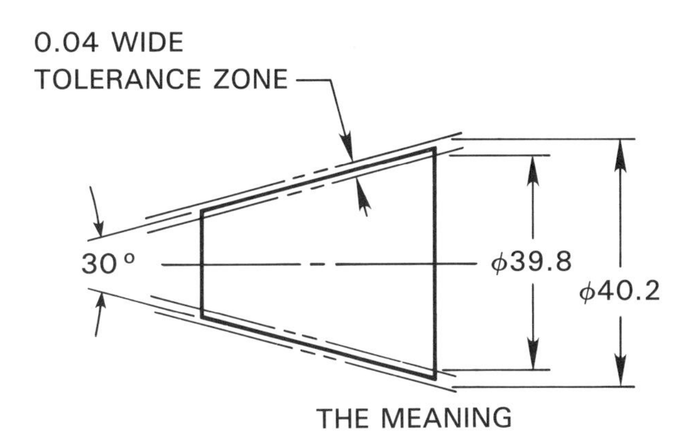

Example 5-26. Specifying profile of a conical feature.

When an excessive number of basic dimensions are on a drawing, a general note may be used to specify the basic dimensions rather than the customary rectangular block around the dimension. The general note, when used, reads NOTE: UNTOLERANCED DIMENSIONS ARE BASIC. Check with your industry or company standards before using this method.

TEST 5, TOLERANCES OF FORM AND PROFILE Name:____________________

1. The ____________________ tolerance specifies a zone within which the required surface element or axis must lie.
2. Explain the difference between the methods used to represent surface and axis straightness. ____
____
____
____
____
____
____
3. Axis straightness may be specified on an RFS or MMC basis. TRUE or FALSE?
4. Straightness may only be applied to cylindrical objects. TRUE or FALSE?
5. Straightness per unit of measure may be specified in an effort to help avoid abrupt surface variation within a relatively short length of a feature. TRUE or FALSE?
6. A ____________________ tolerance zone establishes the distance between two parallel planes within which the surface must lie.
7. Specific area flatness and unit flatness are the same. TRUE or FALSE?
8. ____________________ is characterized by any given cross section taken perpendicular to the axis of a cylinder or cone, or through the common center of a sphere.
9. Specify the appropriate views and the method of connecting a circularity feature control frame in a drawing. ____________________
____
____
____
10. What is the difference between the circularity tolerance and the cylindricity tolerance. ____
____
____
____
____
____

11. Which geometric tolerance, cylindricity or circularity, would require more precise control? ______________________

12. The ______________ tolerance specifies a uniform boundary along the true profile within which the elements of the surface must lie.

13. Complete this statement: A profile tolerance is specified by connecting the feature control frame using a leader to ______________________
____

14. Name the two types of profile geometric characteristics.
____
____

15. What situations, or types of features or parts, frequently require the use of a profile of a line tolerance. ______________________
____
____

16. How is a profile tolerance shown to be specified between two given points? ______________________
____
____

17. How is a profile tolerance specified all around an object or feature, rather than between two given points? ______________________
____
____

18. Explain the difference between profile of a line and profile of a surface.
____
____
____
____
____
____
____

19. Either the profile of a line or the profile of a surface may be all around, between two given points, unilateral, or bilateral. TRUE or FALSE?

20. Define a bilateral profile. ______________________
____

21. A bilateral profile is assumed unless otherwise specified. TRUE or FALSE?

22. Define a unilateral profile. ______________________________

____

23. Explain how a unilateral profile tolerance zone is specified on a drawing.

____

____

____

____

____

24. Give the general note that may be used to specify basic dimensions are used to dimension true profile rather than using the customary rectangular block around the dimension to denote basic. This may be used to save time when an extensive number of basic dimensions are on the drawing. ______________________________

25. Explain how to represent as a single surface the surface profile of four (4) coplanar surfaces. ______________________________

____

____

____

____

26. Profile tolerancing may be used to control the form and orientation of a plane surface. TRUE or FALSE?

27. Profile of a surface may be used to control the angle of an inclined surface in relationship to a datum. TRUE or FALSE?

28. A profile tolerance may be used to control the form, or form and orientation of a conical surface. TRUE or FALSE?

Chapter 6

TOLERANCES OF ORIENTATION AND RUNOUT

This chapter explains the concepts and techniques of dimensioning and tolerancing to control the orientation and runout of geometric shapes. ORIENTATION GEOMETRIC TOLERANCES control:

1. Parallelism.
2. Perpendicularity.
3. Angularity.

RUNOUT is a combination of controls that may include:

1. The control of circular elements of a surface,

   or

2. Control the cumulative variations of circularity, straightness, coaxiality, angularity, taper, and profile of a surface,

   or

3. Control variations of perpendicularity and flatness.

When size tolerances provided in conventional dimensioning do not provide sufficient control for the functional design and interchangeability of a product, then form and/or profile tolerances should be specified. Size limits control a degree of form and parallelism, and locational tolerances control a certain amount of orientation. Therefore, the need for further form and orientation control should be evaluated before specifying geometric tolerances of form and orientation.

ORIENTATION TOLERANCES

Orientation tolerances control the relationship of features to one another. ORIENTATION TOLERANCES include parallelism, perpendicularity, angularity, and in some cases profile. When controlling orientation tolerances, the feature is related to one or more datum features. Relation to more than one datum should be considered if required to stabilize the tolerance zone in more than one direction. Parallelism, perpendicularity, and angularity tolerances control flatness in addition to their intended orientation control.

Orientation tolerances are total, which means that all elements of the related surface or axis fall within the specified tolerance zone. When less demanding requirements controlling only individual line elements of a surface is the design goal, then a note, EACH ELEMENT, or EACH RADIAL ELEMENT, should be shown below the associated feature control frame. This application permits individual elements of a surface to be controlled in relation to a datum rather than the total surface.

Orientation tolerances imply RFS. Therefore, MMC or LMC must be specified if any application other than RFS is intended.

PARALLELISM TOLERANCE

A PARALLEL TOLERANCE is two parallel planes which are parallel to a datum plane within which the axis of the feature must lie. The parallelism geometric tolerance and associated feature control frame is detailed in Example 6-1.

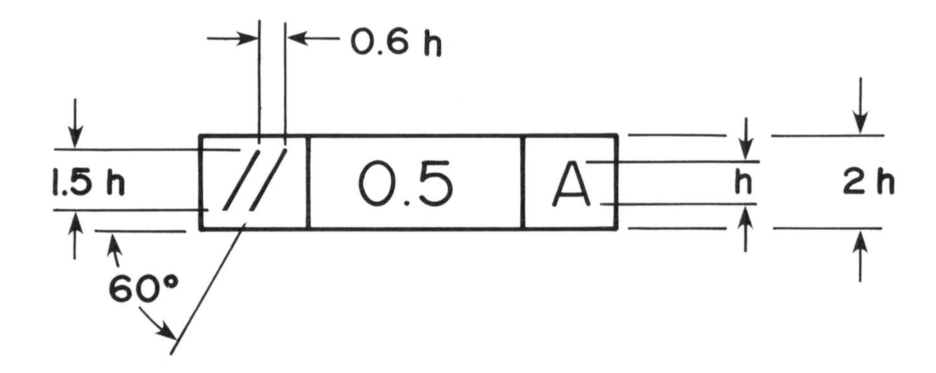

Example 6-1. Feature control frame with parallelism geometric tolerance.

When a surface is to be parallel to a datum, the feature control frame is either connected by a leader to the surface or connected to an extension line from the surface. The actual surface must be within the parallelism tolerance zone which is two parallel planes parallel to the datum. The parallelism tolerance zone must be within the specified size limits. See Example 6-2.

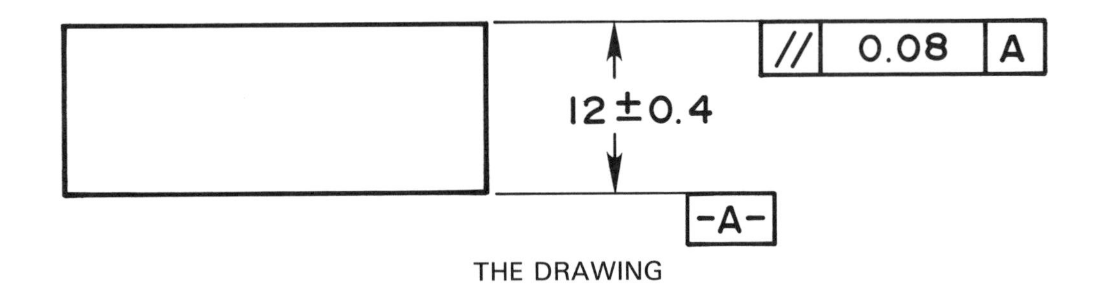

THE DRAWING

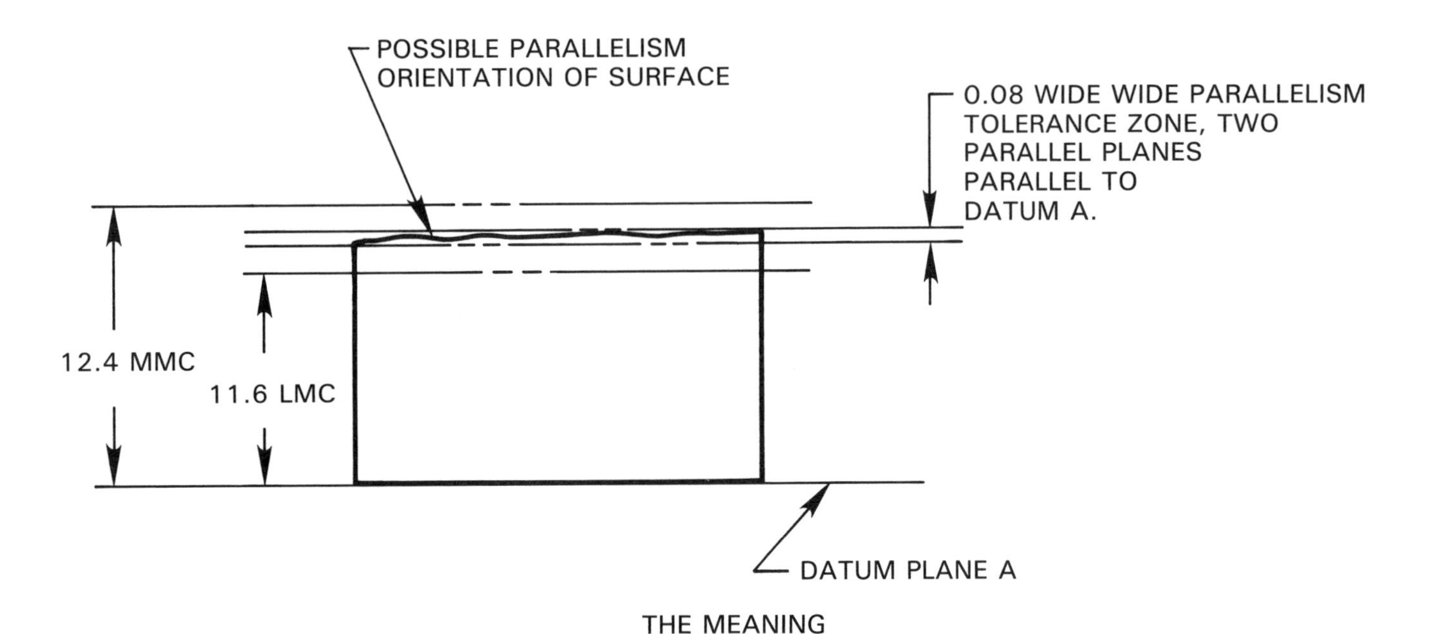

THE MEANING

Example 6-2. Surface parallelism.

A parallelism tolerance may be two parallel planes which are parallel to a datum plane within which the axis of a feature must lie. For example, the axis of a hole may be specified within a tolerance zone which is parallel to a given datum. This parallelism tolerance zone must also be within the specified locational tolerance. Thus, the feature control frame is placed with the diameter dimension as shown in Example 6-3. Remember, placing the feature control frame with a diameter dimension associates the related control with the feature axis.

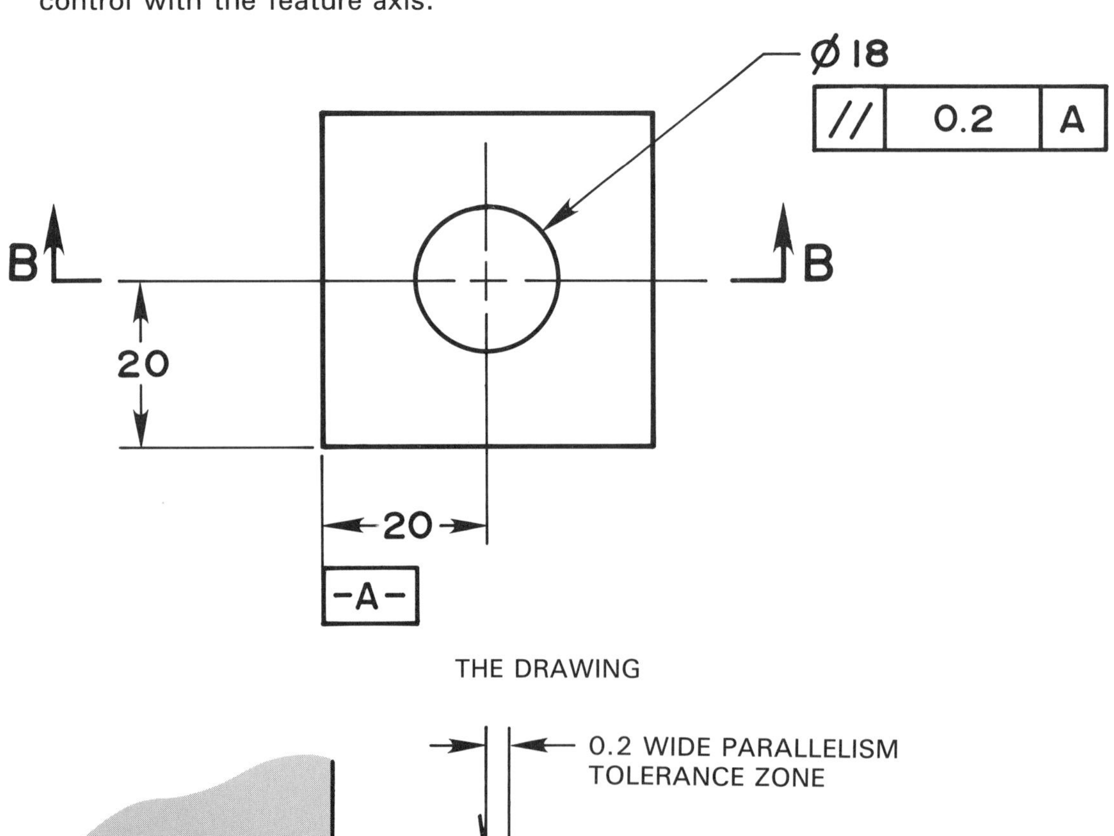

Example 6-3. Specifying axis parallelism.

Parallelism may also be applied to the axis of two or more features when a parallel relationship between the features is desired. Thus, the axis of a feature must lie within a cylindrical tolerance zone that is parallel to a given datum axis. This is a diameter tolerance zone as shown in Example 6-4.

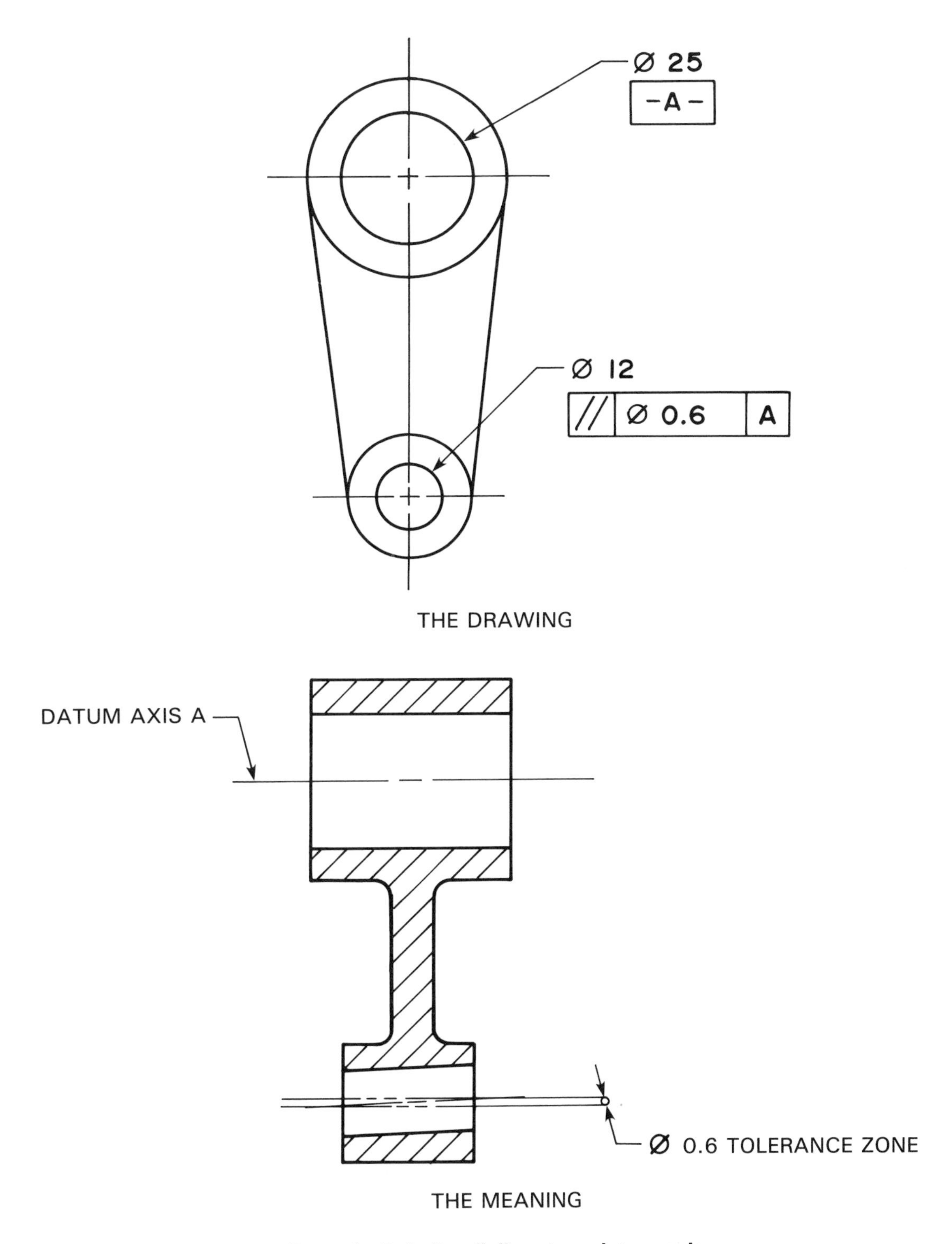

Example 6-4. Parallelism to a datum axis.

Orientation tolerances (parallelism, perpendicularity, angularity, and in some cases profile) are implied to be total where an axis or all elements of a surface must fall within the specified tolerance zone. Where it is desirable to control only individual line elements rather than the entire surface, the note EACH ELEMENT or EACH RADIAL ELEMENT is placed below the feature control frame as shown in Example 6-5.

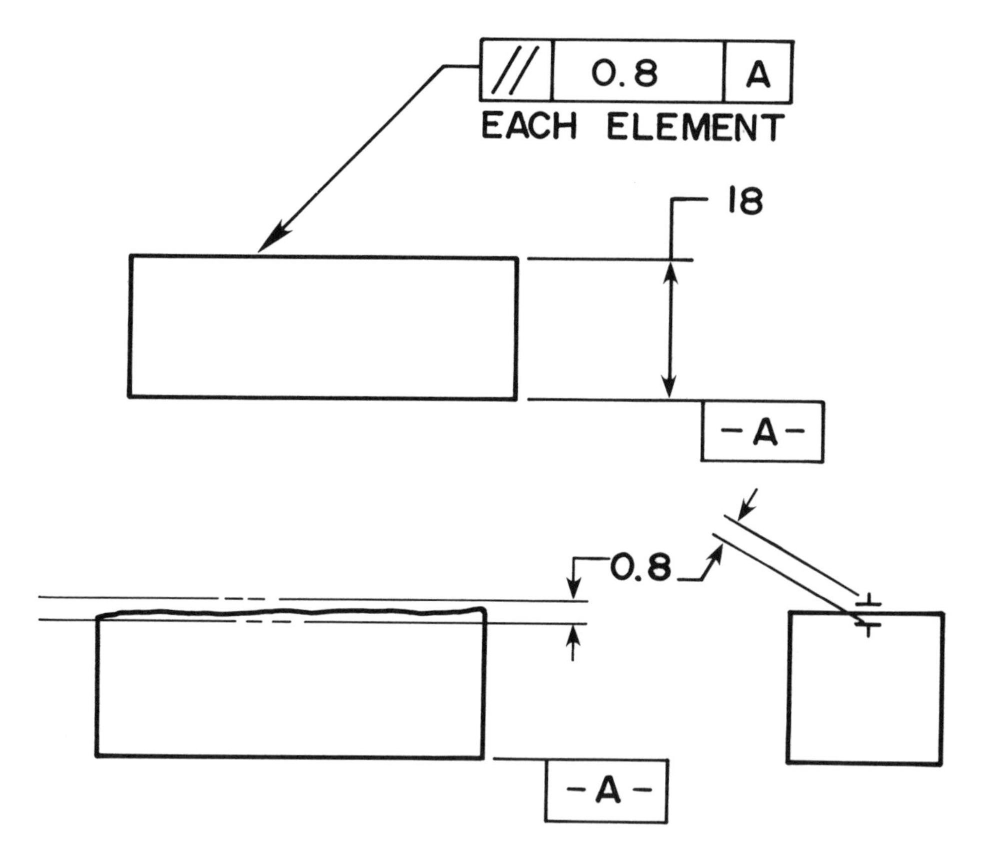

Example 6-5. Individual line elements parallel to given datum.

PERPENDICULARITY TOLERANCE

PERPENDICULARITY TOLERANCE is established by a specified tolerance zone made up of two parallel planes that are a basic 90° to a given datum plane or axis within which the actual surface may lie. The perpendicularity geometric characteristic symbol placed in a feature control frame is detailed in Example 6-6.

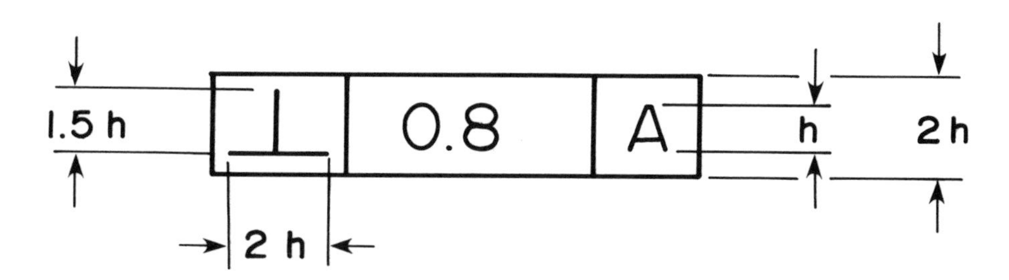

Example 6-6. Perpendicularity feature control frame.

The feature control frame may be connected to the surface with a leader or from an extension line as shown in Example 6-7.

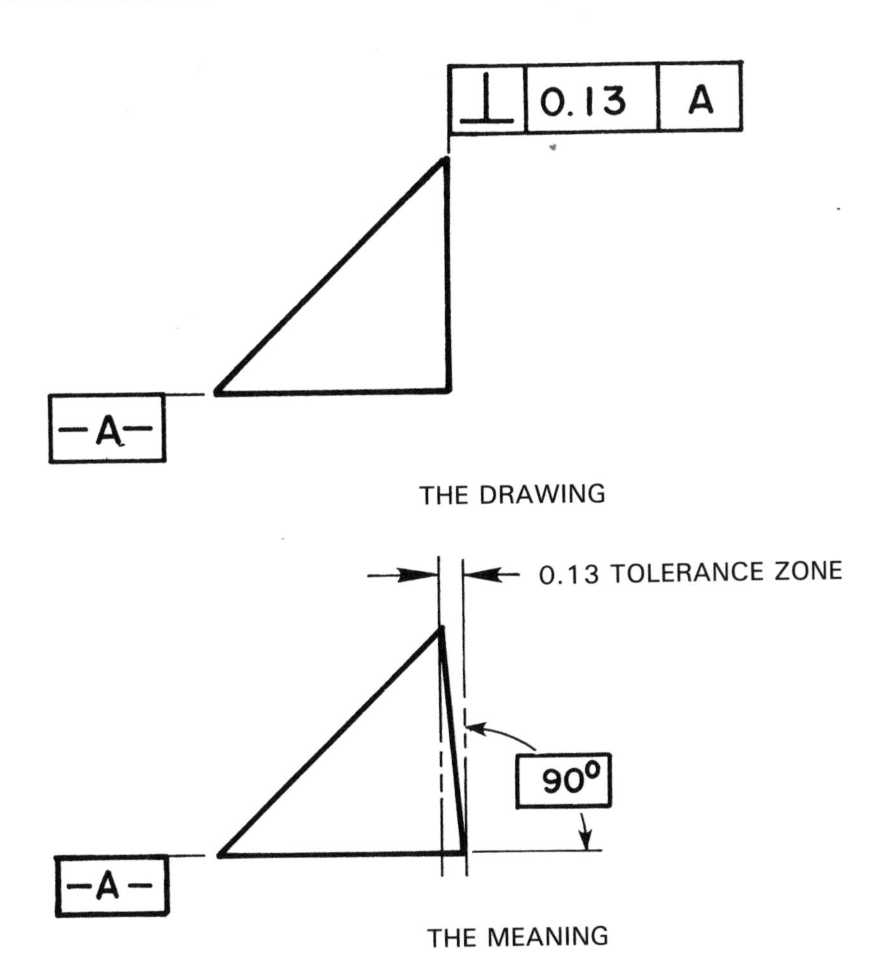

Example 6-7. Perpendicularity of a surface.

Perpendicularity may be a tolerance zone made up of two parallel planes perpendicular to a datum plane or axis within which the axis of a feature may lie. For example, the axis of a hole may be between two parallel planes which are perpendicular to a datum axis. In this application, both the datum axis and the feature control frame is placed adjacent to the diameter dimensions as shown in Example 6-8. RFS is implied unless MMC or LMC is placed in the feature control frame after the geometric tolerance.

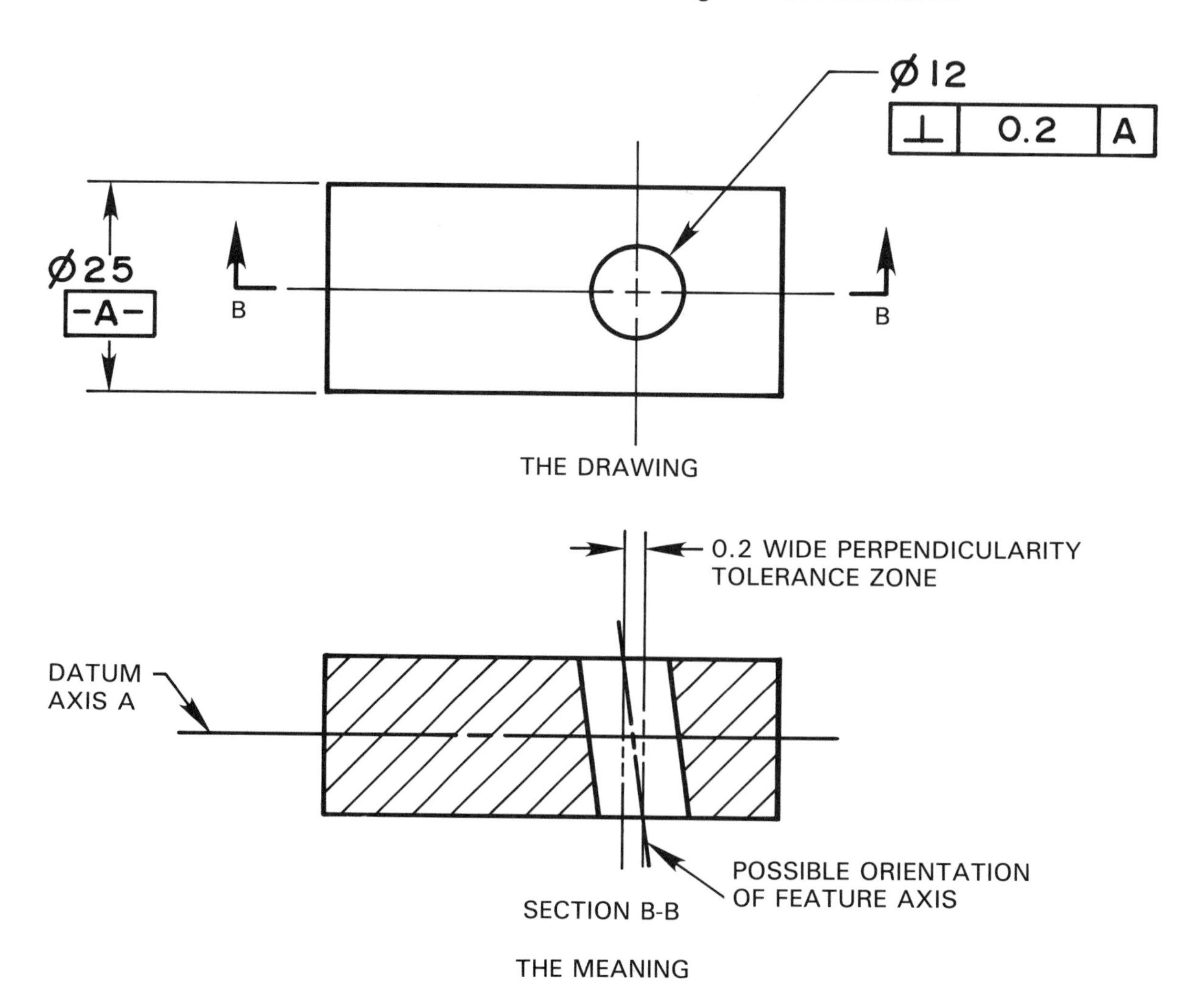

Example 6-8. Axis perpendicular to an axis.

Another application that may require a perpendicularity specification is a cylindrical feature such as a pin or stud. In this situation, the feature axis is within a cylindrical tolerance zone which is perpendicular to a datum plane. The feature control frame is attached to the diameter dimension and a diameter symbol preceeds the geometric tolerance to denote a cylindrical tolerance zone, as shown in Example 6-9.

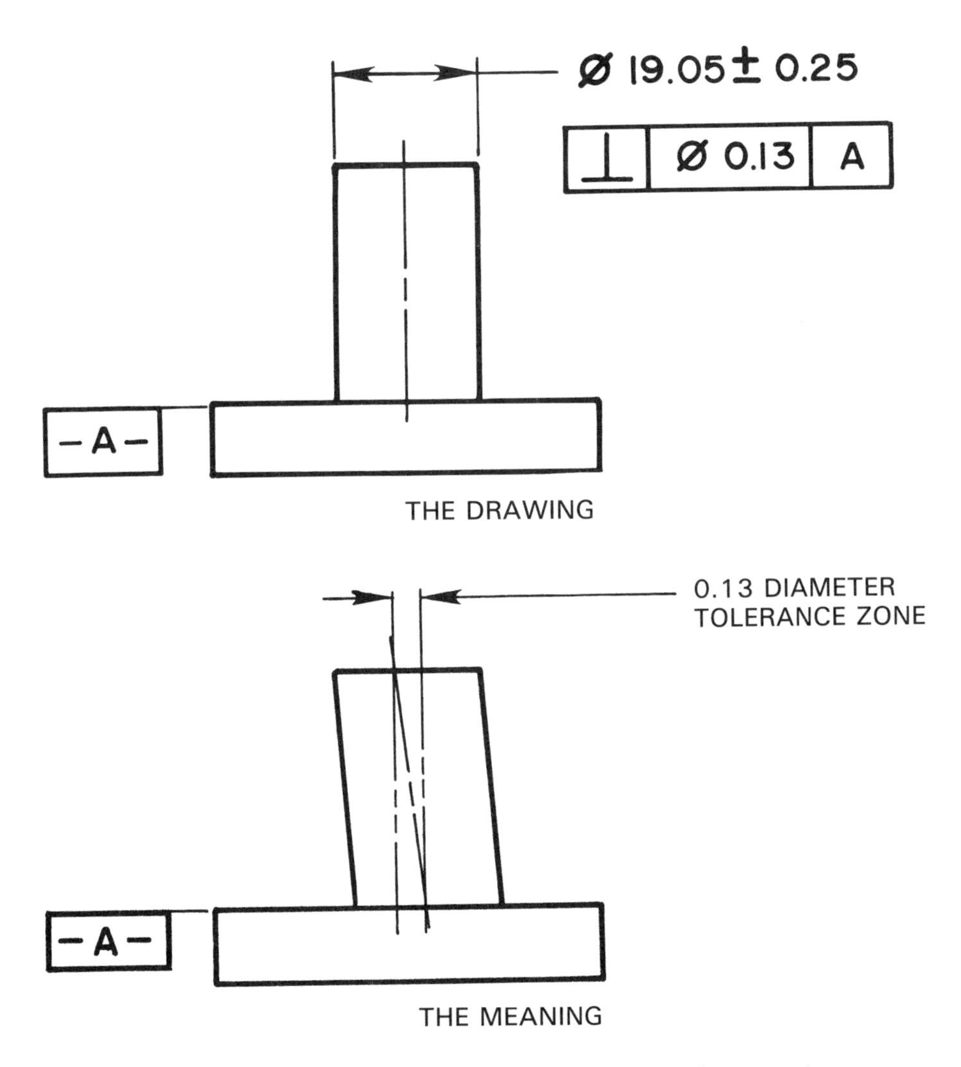

Example 6-9. Specifying perpendicularity for an axis.

A symmetrical feature such as a slot may be specified as perpendicular to a datum plane. In this application the feature center plane is held within two parallel planes which are perpendicular to a given datum plane. The center plane must also be within the specified locational tolerance. See Example 6-10.

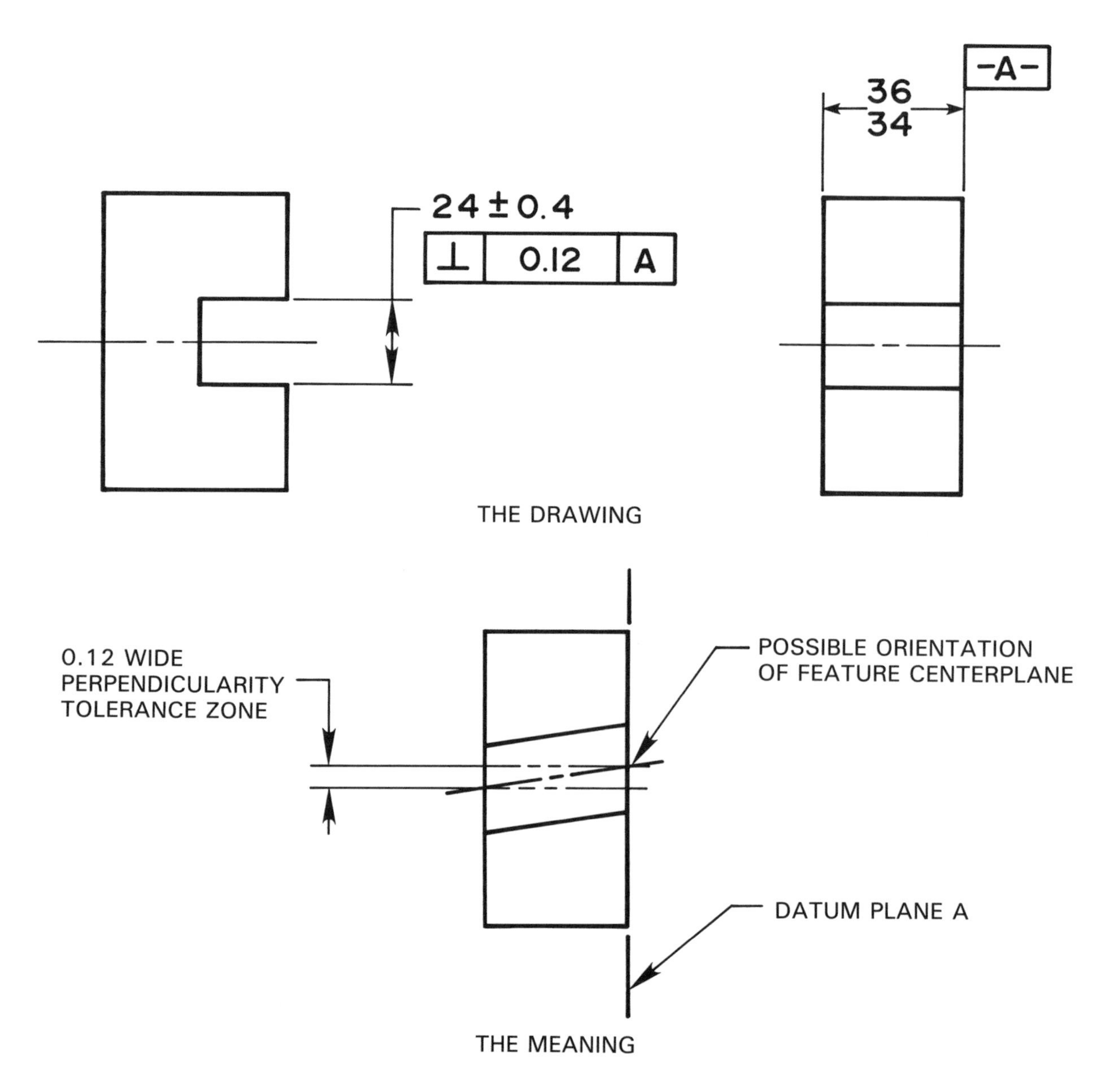

Example 6-10. Perpendicularity of a center plane.

Another possibility is that single line elements of a surface, rather than the entire surface, may be perpendicular to a given datum. When any single line element of the object shall be held perpendicular to a datum, the words EACH ELEMENT should be indicated below the feature control frame as shown in Example 6-11.

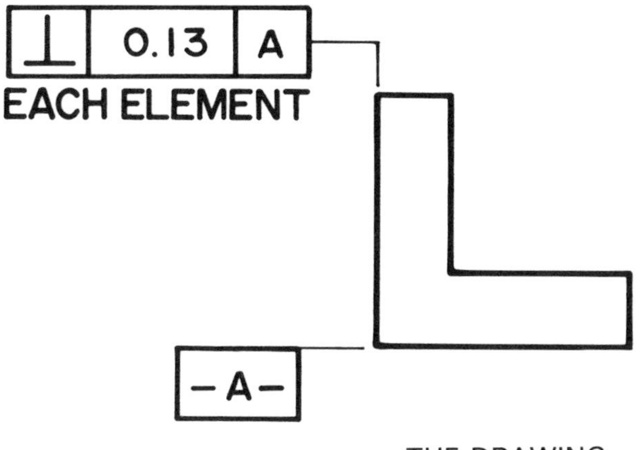

THE DRAWING

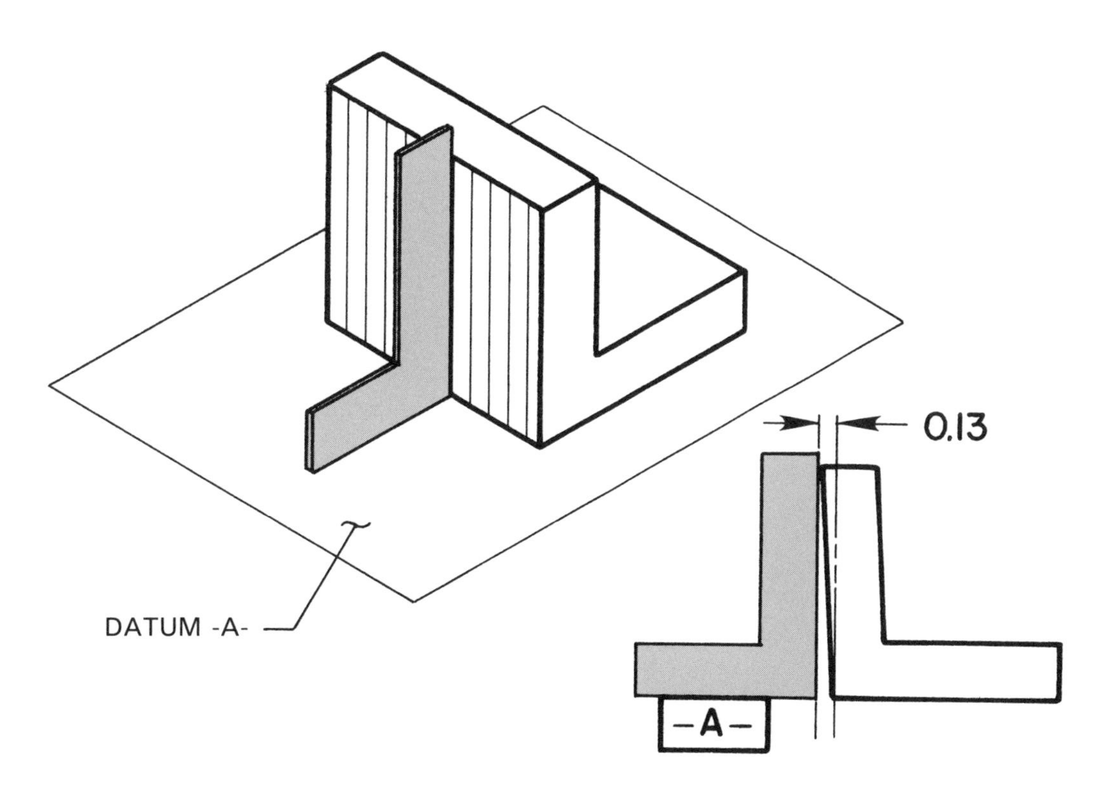

THE MEANING

Example 6-11. Perpendicularity of line elements.

ANGULARITY TOLERANCE

An ANGULARITY TOLERANCE zone is established by two parallel planes at any specified basic angle, other than 90°, to a datum plane or axis. The angularity geometric characteristic symbol placed in a feature control frame is detailed in Example 6-12.

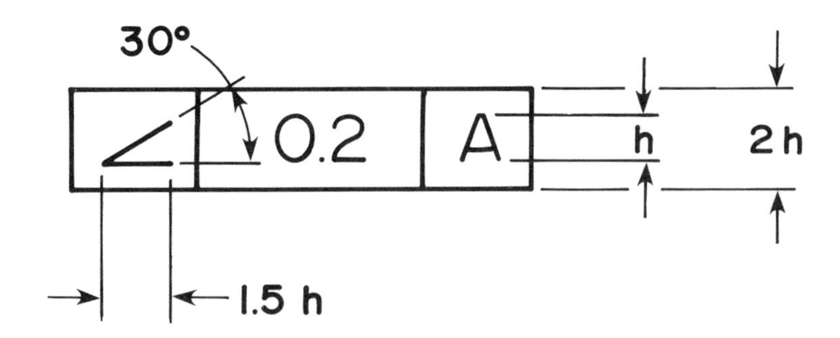

Example 6-12. Angularity feature control frame.

The feature control frame is normally connected to the surface by a leader. The specified angle must be basic and must be dimensioned from the datum plane as shown in Example 6-13. RFS is implied unless otherwise specified.

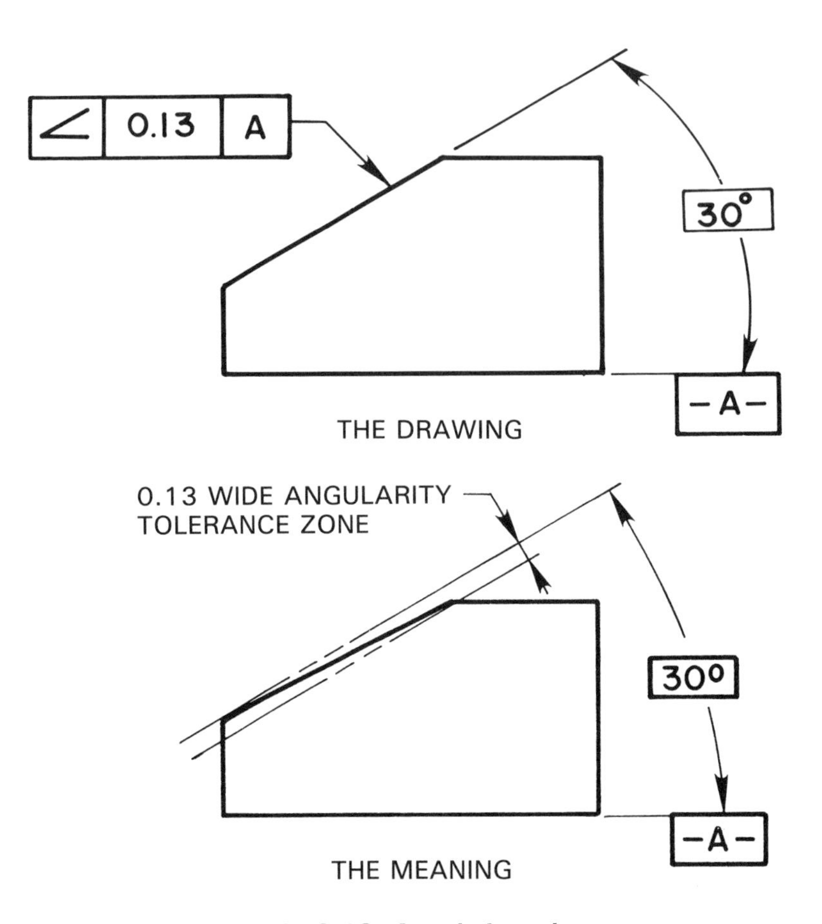

Example 6-13. Angularity tolerance.

The axis of a hole or other cylindrical feature can be dimensioned with an angularity tolerance if the feature is at an angle other than 90° to a datum plane or axis. This specification establishes two parallel planes spaced equally on each side of the specified basic angle from a datum plane or axis within which the axis of the considered feature must lie. This control applies only to the view on which it is specified. The feature control frame is shown adjacent to the feature diameter dimension to denote axis control, as shown in Example 6-14.

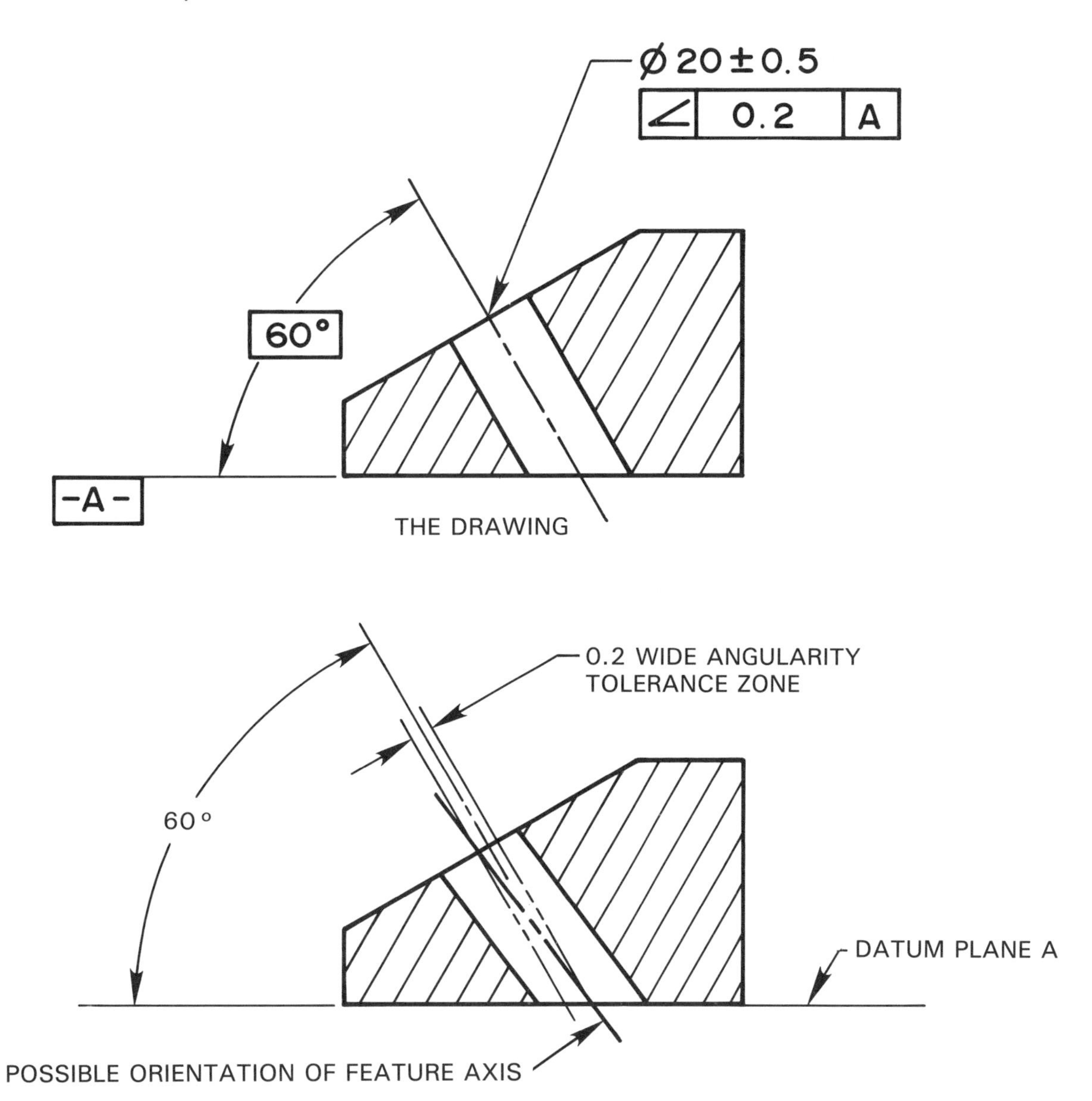

Example 6-14. Angularity of an axis.

RFS, MMC, AND ZERO TOLERANCE AT MMC

When a material condition symbol is not shown in the feature control frame, then RFS is implied. This means that the geometric tolerance is the same when the feature is manufactured at any produced size.

Placing the MMC material condition symbol after the geometric tolerance in the feature control frame denotes that the tolerance is held at the MMC produced size, and then the tolerance is allowed to increase equal to the amount of departure from MMC.

Another application is zero tolerance at MMC where in the geometric tolerance in the feature control frame is zero and the MMC material condition symbol is used. This means that at the MMC produced size, the feature must be perfect in orientation with respect to the specified datum. As the actual produced size departs from MMC, the geometric tolerance increases equal to the amount of departure.

Applications of RFS, MMC, and Zero tolerance at MMC are shown in Example 6-15.

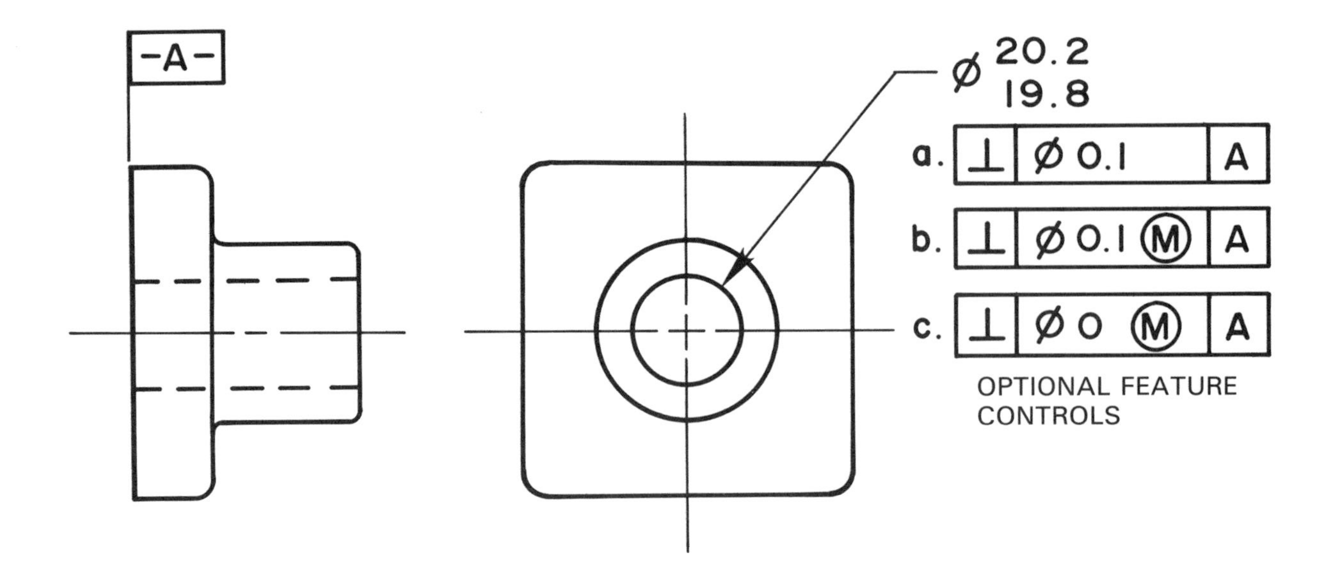

DIAMETER TOLERANCE ZONES ALLOWED

Possible Produced Sizes	a. RFS	b. MMC	c. Zero at MMC
19.8 MMC	0.1	0.1	0
19.9	0.1	0.2	0.1
20.0	0.1	0.3	0.2
20.1	0.1	0.4	0.3
20.2 LMC	0.1	0.5	0.4

Example 6-15. Applications of RFS, MMC, and Zero tolerance at MMC.

RUNOUT TOLERANCE

RUNOUT is a combination of tolerances used to control the relationship of one or more features of a part to a datum axis. Features that may be controlled by runout are either surfaces around or perpendicular to a datum axis. The datum axis should be selected as a diameter of sufficient length, as two diameters adequately separated on the same axis, or as a diameter and perpendicular surface. There are two types of runout, total runout and circular runout. The type of runout selected depends on design and manufacturing considerations. Circular runout is generally a less complex requirement than total runout. The feature control frame is connected by a leader line to the surface. Multiple leaders may be used to direct a feature control frame to two or more surfaces having a common runout tolerance. The runout geometric characteristic symbols are shown detailed in feature control frames in Example 6-16.

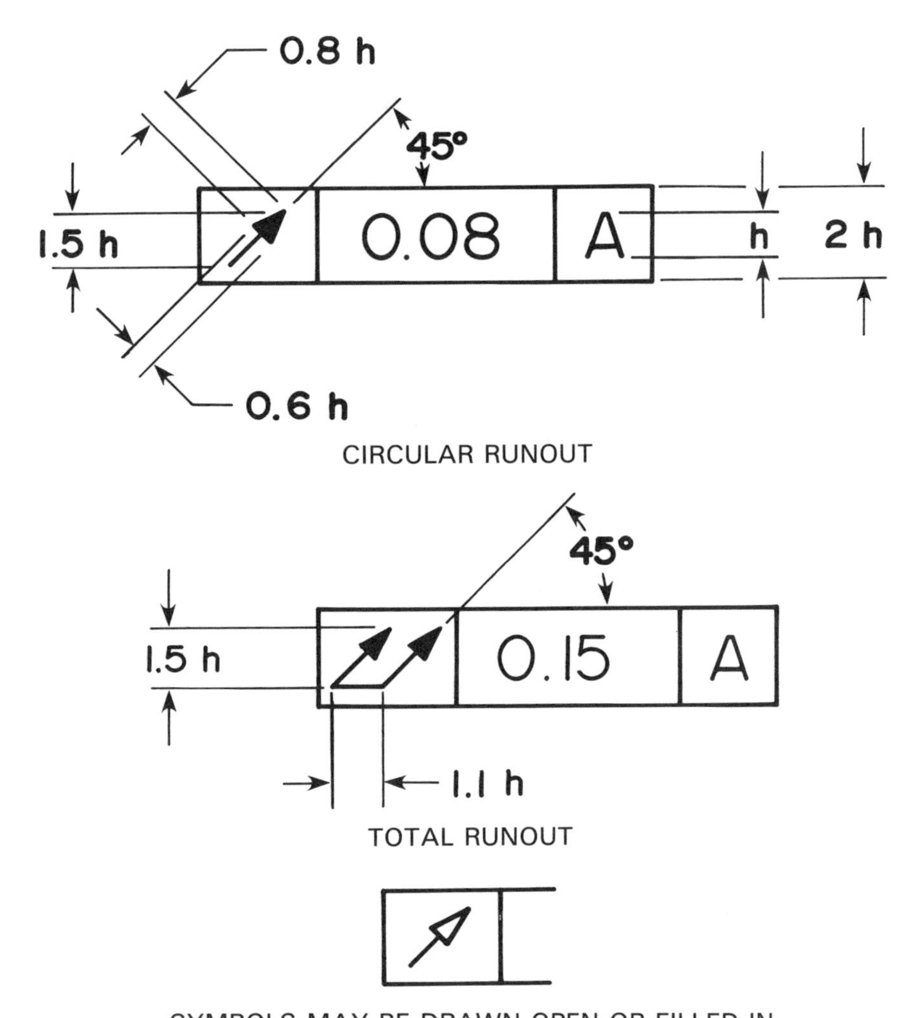

Example 6-16. Runout symbols.

CIRCULAR RUNOUT

CIRCULAR RUNOUT provides control of single circular elements of a surface. When applied to surfaces around a datum axis, circular runout controls circularity and coaxiality. COAXIALITY is a situation where two or more features share a common axis. When applied to surfaces at right angles to a datum axis, circular runout may be used to control wobbling motion. This tolerance is measured by the Full Indicator Movement (FIM) of a dial indicator placed at several circular measuring positions as the part is rotated 360°. FIM shows a total tolerance. An example of circular runout is shown in Example 6-17.

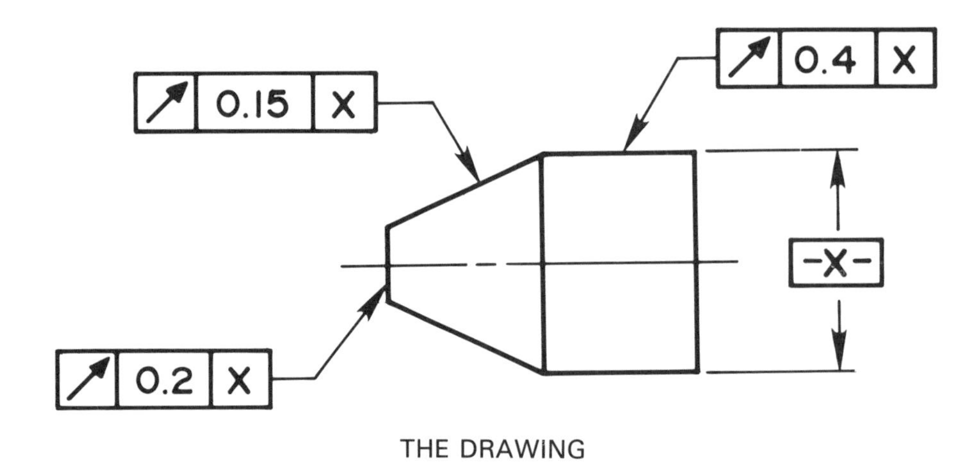

THE DRAWING

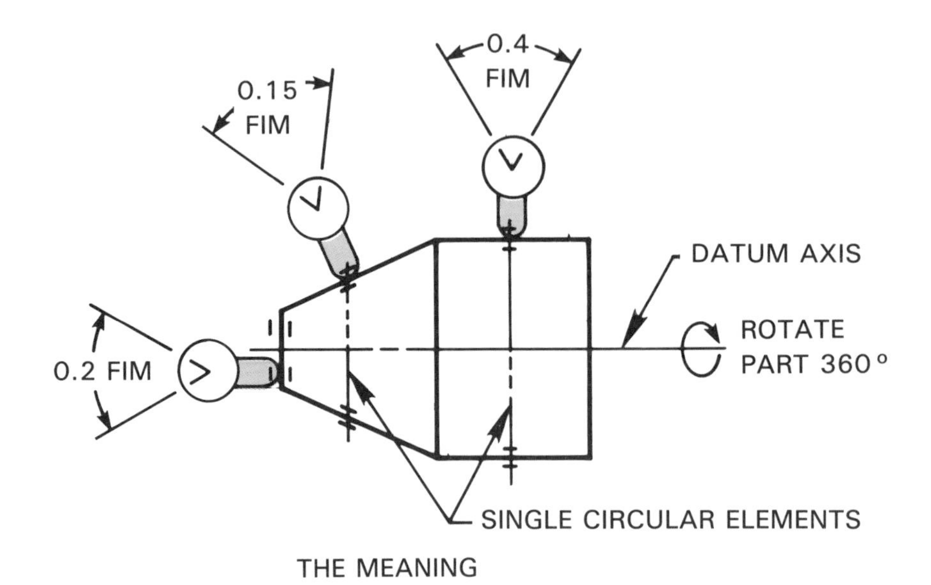

THE MEANING

Example 6-17. Circular runout.

TOTAL RUNOUT

TOTAL RUNOUT provides a combined control of a surface element. This is a tolerance that blankets the surface to be controlled. Total runout is used to control the combined variations of circularity, straightness, coaxiality, angularity, taper, and profile when applied to surfaces around a datum axis. Total runout may be used to control the combined variations of perpendicularity to control wobble and flatness and to control concavity or convexity when applied to surfaces perpendicular to a datum axis. The total runout tolerance zone encompasses the entire surface as the part is rotated 360°. The entire surface must lie within the specified tolerance zone. In order to determine this, the dial indicator is placed at selected locations along the surface as the part is rotated 360°. Total runout is shown in Example 6-18.

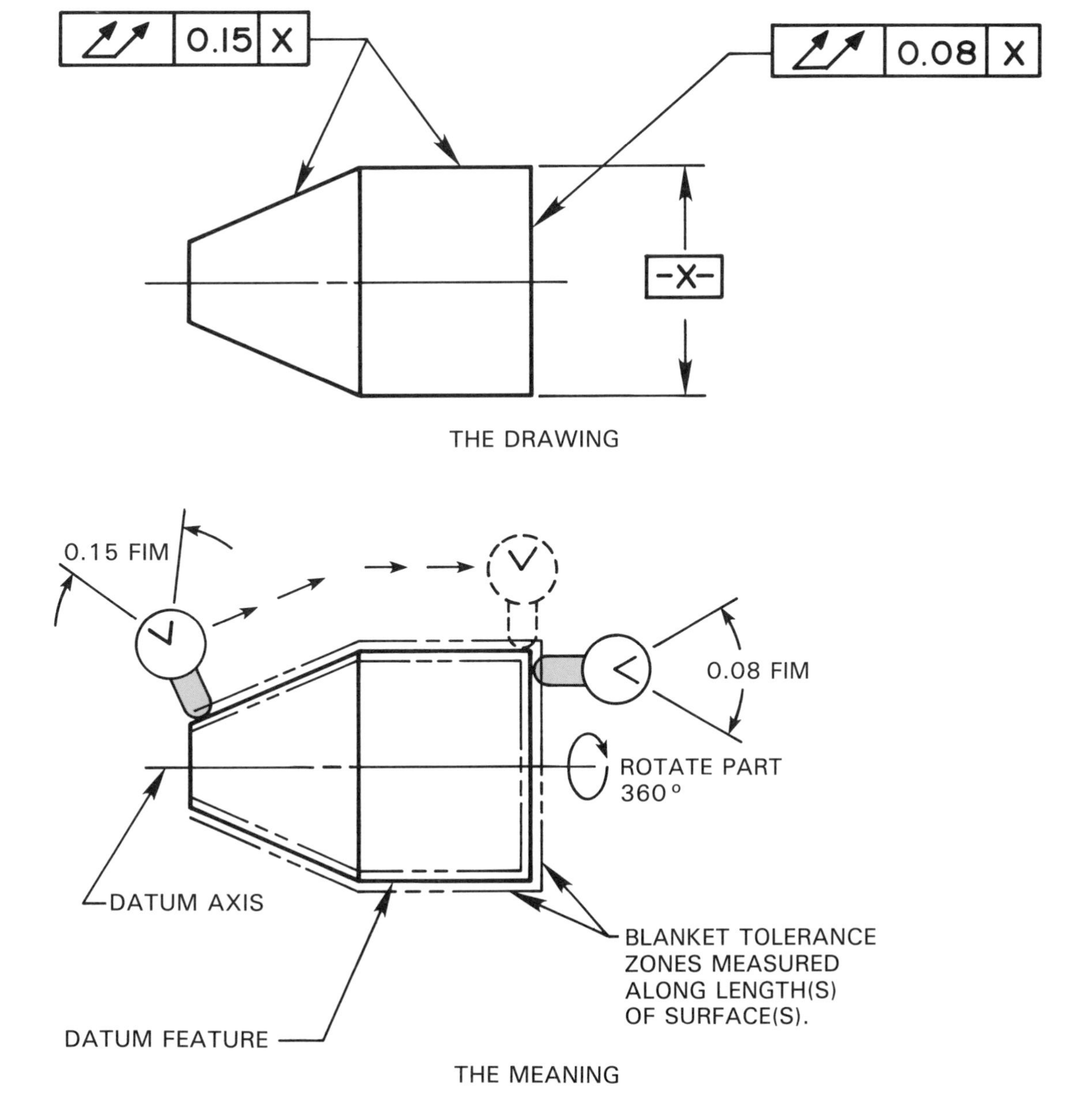

Example 6-18. Total runout.

A portion of a surface may have a runout tolerance specified if it is not desired to control the entire surface. This is done by placing a chain line located with basic dimensions in the linear view. The CHAIN LINE denotes the location of the profile tolerance around the object. The feature control frame is then connected by a leader line to the chain line. Runout tolerances may also be applied where two datum diameters collectively establish a single datum axis. This is done by placing the datum identifying letters separated by a dash in the feature control frame, for example G-H as shown in Example 6-19.

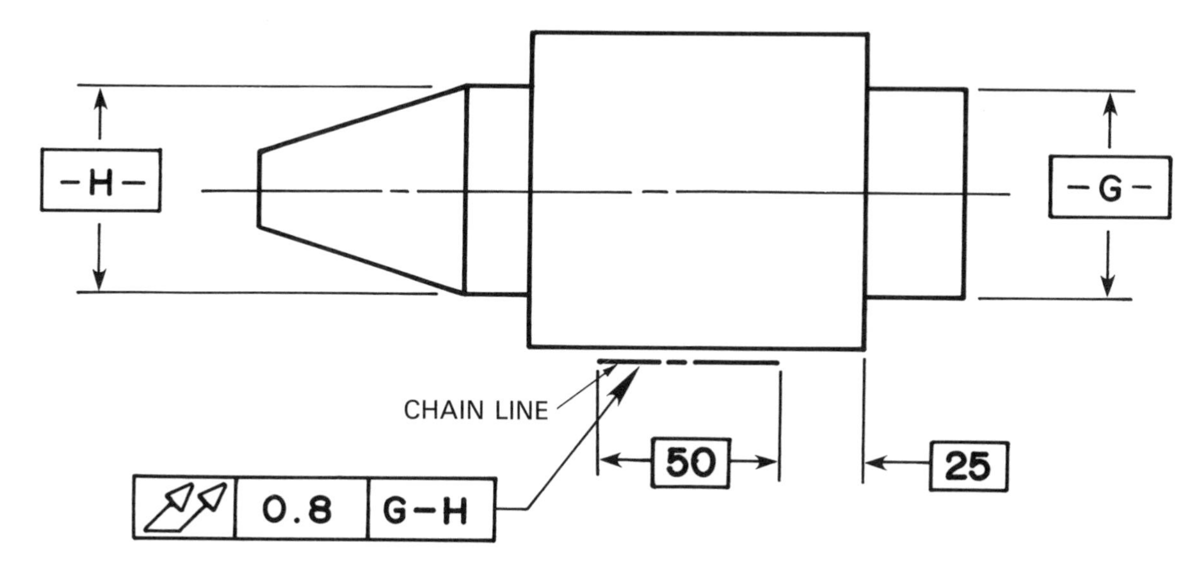

Example 6-19. Partial surface runout and specifying runout to two datum diameters.

COMBINATION OF GEOMETRIC TOLERANCES

Profile tolerancing may be combined with other types of geometric tolerances. For example, a surface may have a profile tolerance controlled within a specified amount of parallelism relative to a datum. When this is done the surface must be within the profile tolerance, and each line element of the surface must be parallel to the given datum by the specified parallelism tolerance, as shown in Example 6-20.

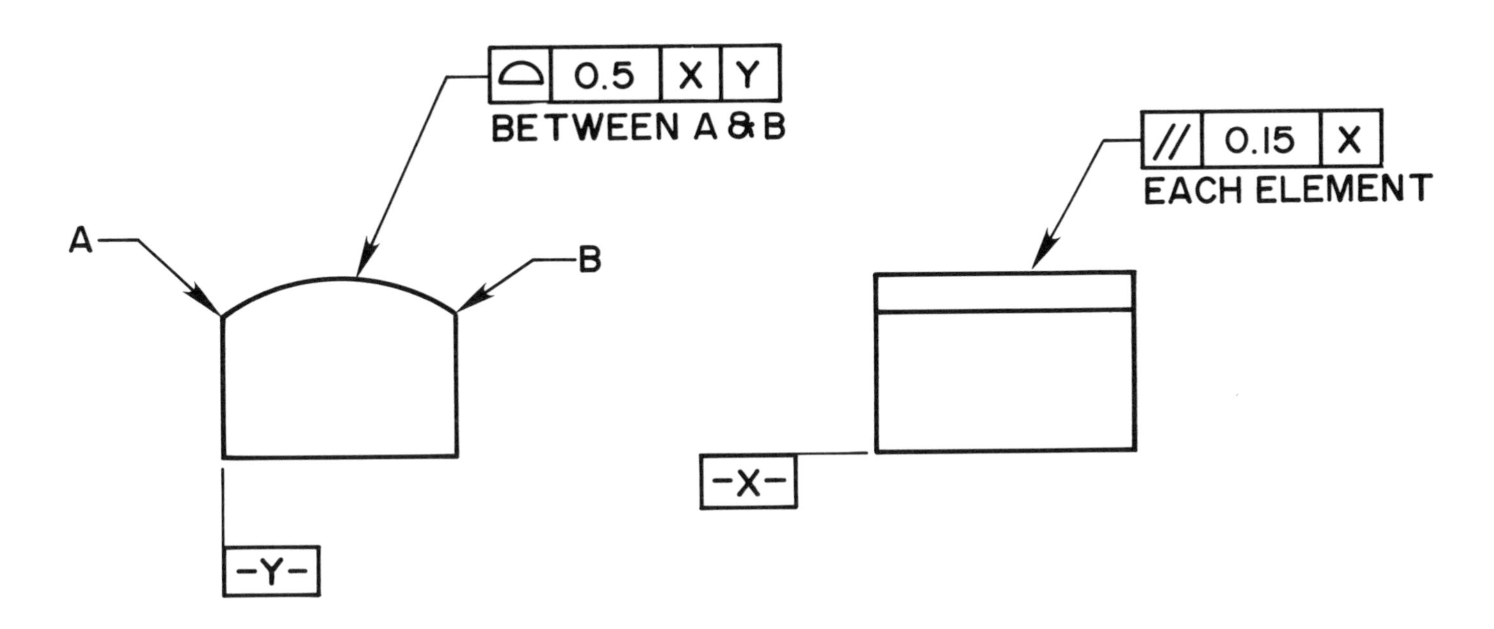

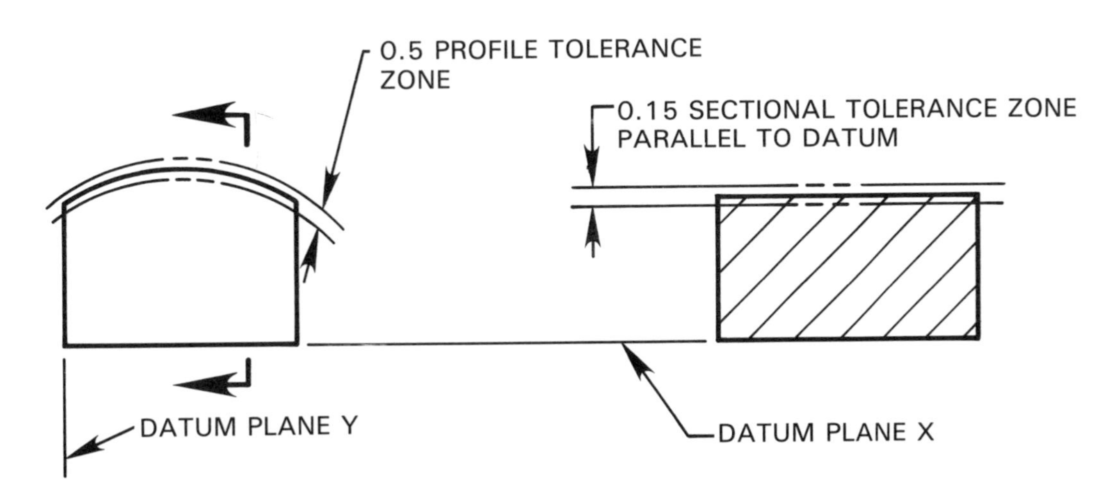

Example 6-20. Combined profile and parallelism.

A surface may also be controlled by profile and refined by runout. When this is done, as shown in Example 6-21, any line element of the surface must be within the profile tolerance, and any circular element of the surface must be within the specified runout tolerance.

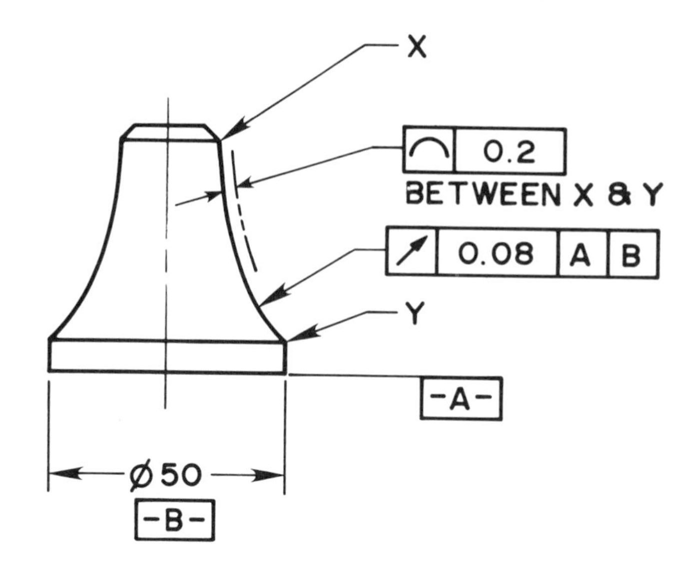

Example 6-21. Combined profile and runout tolerance.

In some situations, it may be necessary to control runout constrained by flatness, straightness, or cylindricity. A combination runout and cylindricity, as shown in Example 6-22, denotes that the datum surface must be controlled within the specified tolerances of runout and cylindricity. Notice that the different feature control frames are attached. This is different from the previous examples where the feature control frames were separate.

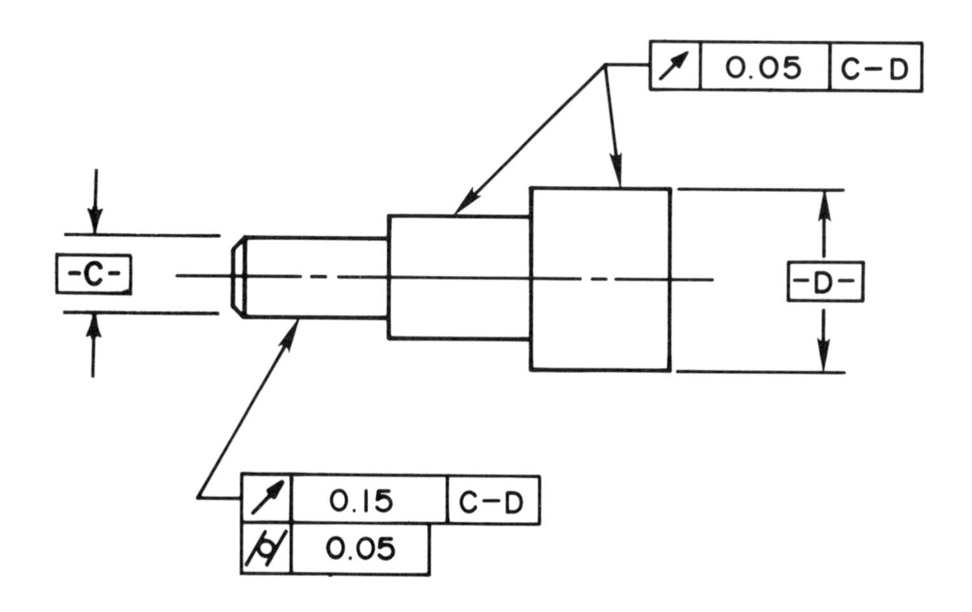

Example 6-22. Combination runout and cylindricity.

The combination of perpendicularity and parallelism may be achieved by combining the perpendicularity and parallelism controls as shown in Example 6-23. This allows versatility by providing uniform perpendicularity and parallelism to related datums.

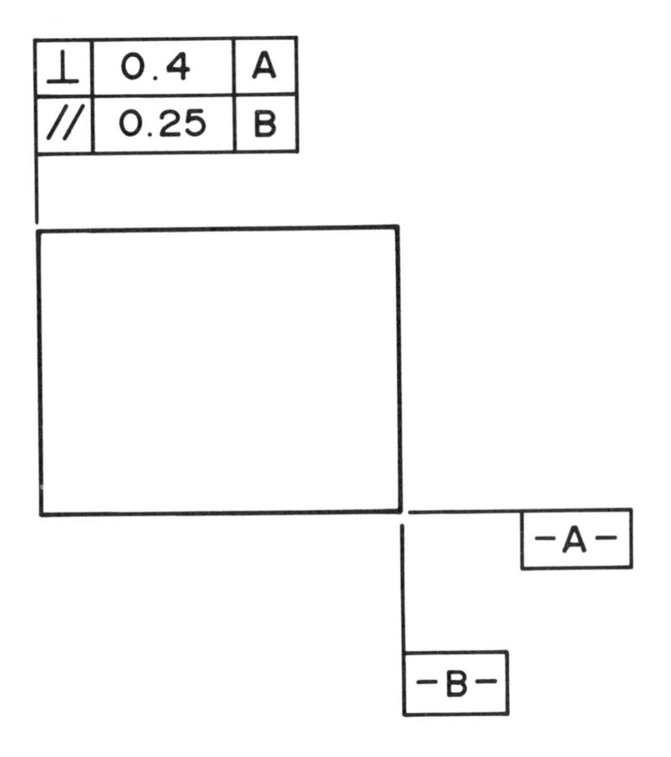

Example 6-23. Combined perpendicularity and parallelism.

The tolerance zones are often different, but when appropriate they may be the same as shown in Example 6-24.

⊥	0.25	A
//	0.25	B

Example 6-24. Combined feature control frame with equal tolerance zones.

TEST 6, TOLERANCES OF ORIENTATION AND RUNOUT

Name:____________________

1. Orientation geometric tolerances control ____________________,
   ____________________, and ____________________.
2. Runout is a combination of controls that may include:
   a. control of circular elements of a surface.
   b. control the cumulative variations of circularity, straightness, coaxiality, angularity, taper, and profile of a surface.
   c. control variations of perpendicularity and flatness.
   d. all of the above.
3. ____________________ tolerances control the relationship of features to one another.
4. Orientation tolerances must be related to one or more datum features. TRUE or FALSE?
5. Orientation tolerances control flatness. TRUE or FALSE?
6. Orientation tolerances imply MMC, RFS, or LMC? ____________________
7. A ____________________ tolerance zone is the distance between two parallel planes parallel to a datum.
8. What does the placement of a parallelism feature control frame below a diameter dimension denote? ____________________
   ____
   ____
   ____
9. Parallelism may be applied to the axis of two or more features when a parallel relationship between the features is desired. TRUE or FALSE?
10. Orientation tolerances are implied to be total. Therefore, how must a drawing be modified where it is desirable to control only individual line elements rather than the entire surface?
    ____
    ____
    ____
11. ____________________ is established by a specified tolerance zone made up of two parallel planes that are a basic 90° to a given datum plane or axis within which the actual surface must lie.
12. A symmetrical feature such as a slot may be specified as perpendicular to a datum plane. In this application the feature ____________________ is held within two parallel planes which are ____________________ to a given datum.

13. An ____________________ tolerance zone is established by two parallel planes at any specified basic angle, other than 90°, to a datum plane or axis. The specified angle must be ____________ and must be dimensioned from the ____________ plane.

14. Given the following drawing, a reference chart showing a range of possible produced sizes, and three (3) optional feature control frames that may be applied to the diameter dimension. Provide the geometric tolerance at each possible produced size for each feature control frame application. Suggestion: review Chapter 4, Material Condition Symbols.

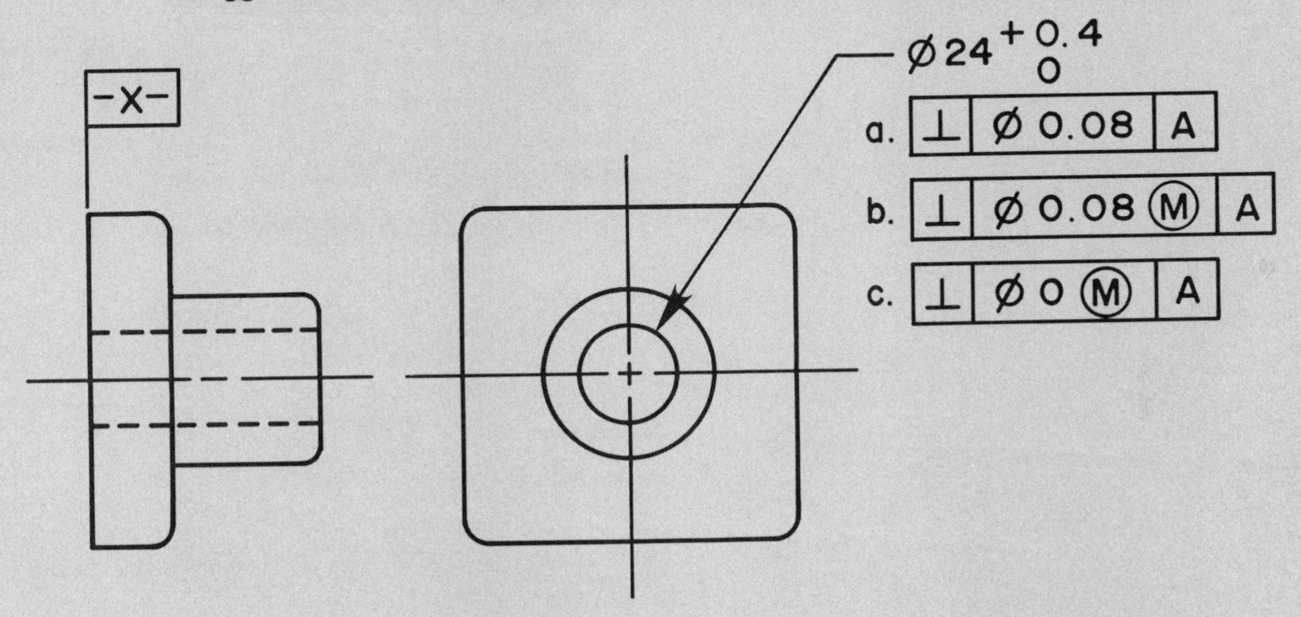

DIAMETER TOLERANCE ZONES ALLOWED

Possible Produced Sizes	a. RFS	b. MMC	c. Zero at MMC
24.0	________	________	________
24.1	________	________	________
24.2	________	________	________
24.3	________	________	________
24.4	________	________	________

15. ______________ is a combination of tolerances used to control the relationship of one or more features of a part to a datum axis.

16. Features that may be controlled by runout are either surfaces around or perpendicular to a datum axis. TRUE or FALSE?
17. Name the two types of runout. ________________ and ________________.
18. ________________________ provides control of single circular elements of a surface.
19. When applied to surfaces around a datum axis circular runout controls ________________ and ________________.
20. ________________ is a situation where two or more features share a common axis.
21. ________________________ provides a combined control of surface elements.
22. ________________________ is used to control the combined variations of circularity, straightness, coaxiality, angularity, taper, and profile when applied to surfaces around a ____________ axis.
23. ________________________ may be used to control the combined variations of perpendicularity to control wobble and flatness to control concavity or convexity when applied to surfaces perpendicular to a datum axis.
24. Explain the fundamental difference between how circular and total runout are established. ____
____
____
____
____
____
25. What does the chain line denote when specifying runout to a portion of a surface? ____
____

26. Explain the combination of geometric tolerances that exist in the following drawing.

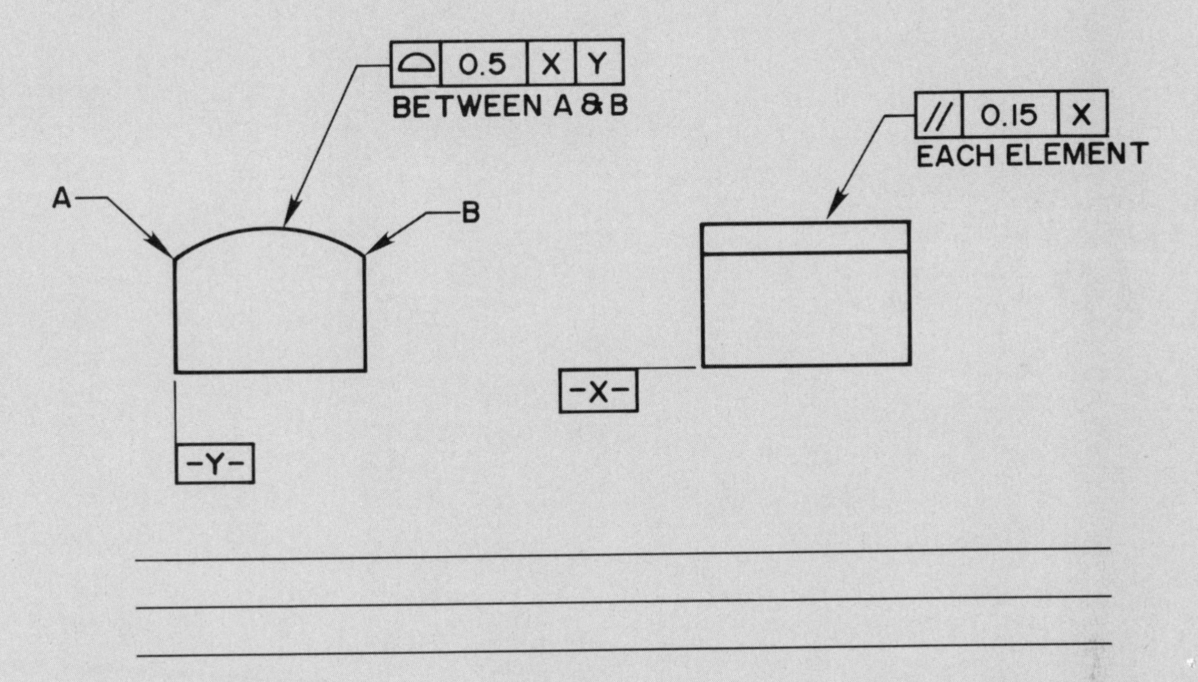

27. Explain the combination of geometric tolerances that exist in the following drawing.

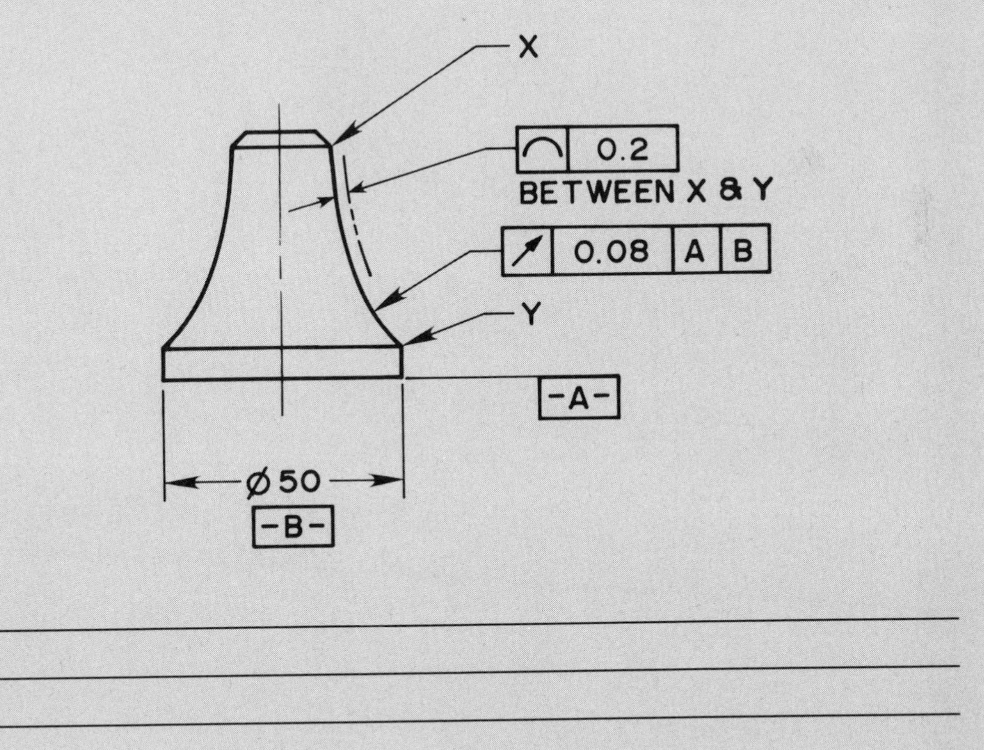

28. Given the drawing below, explain the meaning represented by the following specifications:
    a. Datum feature symbol -C-.
    b. Datum feature symbol -D-.
    c. Datum reference C-D in the feature control frame with the runout geometric tolerance.
    d. Combination runout and cylindricity.

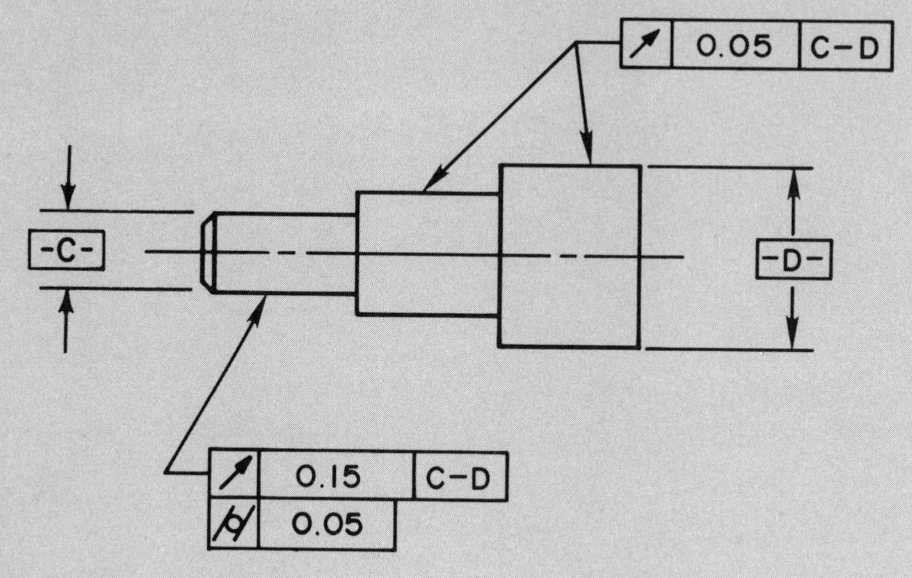

a. ______________________________

b. ______________________________

c. ______________________________

d. ______________________________

____

____

PRINT READING EXERCISES FOR CHAPTER 6

Name:____________________

The following print reading exercise is designed for use in programs for machining, welding, tool and die, dimensional inspection, and other manufacturing curriculums where the objective is the reading and interpretation of prints rather than the development of drafting skills. An actual industrial print is used with related questions that require you to read and interpret specific dimensioning and geometric tolerancing representations. The answers and interpretations should be based on the previously learned content of this book. The prints used are based on ANSI standards; however, company standards may differ slightly. When reading these prints or any other industrial prints, a degree of flexibility may be required to determine how individual applications correlate with the ANSI standard.

PRINT READING EXERCISE

Refer to the Hyster Company print of the FLYWHEEL-DSL found on page 218.

1. Interpret the feature control frame located at A. ____________________

2. Interpret the feature control frame located at B. ____________________

3. Interpret the feature control frame located at C. ____________________

4. Interpret the feature control frame associated with the ϕ376.81-376.76 dimension. ____________________

5. Interpret the feature control frame associated with the ϕ133.37-133.35 dimension. ____________________

Refer to the Hyster Company print of the CASE-DIFF found on page 217.

6. Interpret the feature control frame adjacent to datum feature symbol -A-.
____

7. Interpret the elements of the feature control frame labeled A.

____

____

8. Interpret the orientation of the surface that is 41.05-40.95 from datum A.

____

____

9. Completely interpret the elements of the feature control frame associated with the ϕ56.05-55.95 dimension.________________

____

____

____

Chapter 7

LOCATION TOLERANCES

This chapter covers location tolerances which include:

1. Position.
2. Concentricity.
3. Symmetry.

LOCATION TOLERANCES are used for the purpose of locating features from datums or for establishing coaxiality or symmetry.

POSITIONAL TOLERANCING is used to define a zone within which the center, axis, or center plane of a feature of size is permitted to vary from true position. TRUE POSITION is the theoretically exact location of a feature. BASIC DIMENSIONS are used with datum or chain dimensioning systems to establish the true position from specified datum features and between interrelated features.

LOCATION TOLERANCING is specified by a position or concentricity symbol, a tolerance, and appropriate datum references placed in a feature control frame. When position tolerancing is used, the MMC, RFS, or LMC material condition symbols must be specified after the tolerance and applicable datum reference in the feature control frame.

In comparison to conventional methods, the use of position tolerancing concepts provides some of the greatest advantages to mass production. The coordinate dimensioning system limits the actual location of features to a rectangular tolerance zone. Using position tolerancing, the location tolerance zone changes to a cylindrical shape thus increasing the possible location of the feature by about 57%. This improves the interchangeability of parts while improving manufacturing flexibility and reducing scrappage of parts. The use of MMC applied to the position tolerance allows the tolerance zone to increase in diameter as the feature size departs from MMC. This application also allows greater flexibility in the acceptance of mating parts.

POSITION TOLERANCE

The position geometric characteristic symbol is placed in the feature control frame as shown in Example 7-1. The next block of the feature control frame contains the diameter symbol, if a cylindrical tolerance zone is applied, followed by the specified tolerance and a material condition symbol. Additional compartments are used for datum reference.

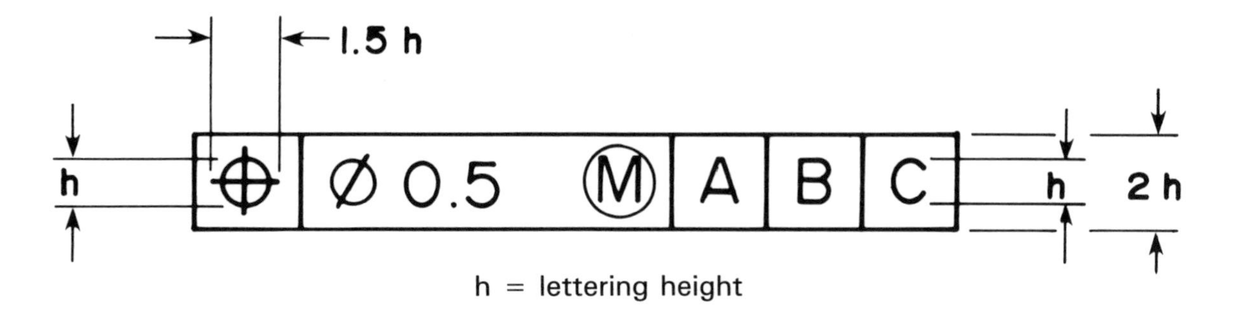

Example 7-1. Position tolerance in a feature control frame.

Some specific characteristics of position tolerances include:

- True position is the theoretically exact location of the centerline of a feature, as shown in Example 7-2.
- A material condition symbol for MMC, RFS, or LMC must follow the specified position tolerance, datum reference, or both as needed.
- Position tolerances control the location of a cylindrical tolerance zone within which the centerline of a feature is located, as shown in Example 7-2. Where a feature other than a cylindrical shape is located, then the tolerance value represents the distance between two parallel straight lines or planes, or the distance between two uniform boundaries.
- Position tolerances are established with a diameter tolerance zone descriptor unless the tolerance zone is between two parallel straight lines or planes, or between two uniform boundaries. See Example 7-2.

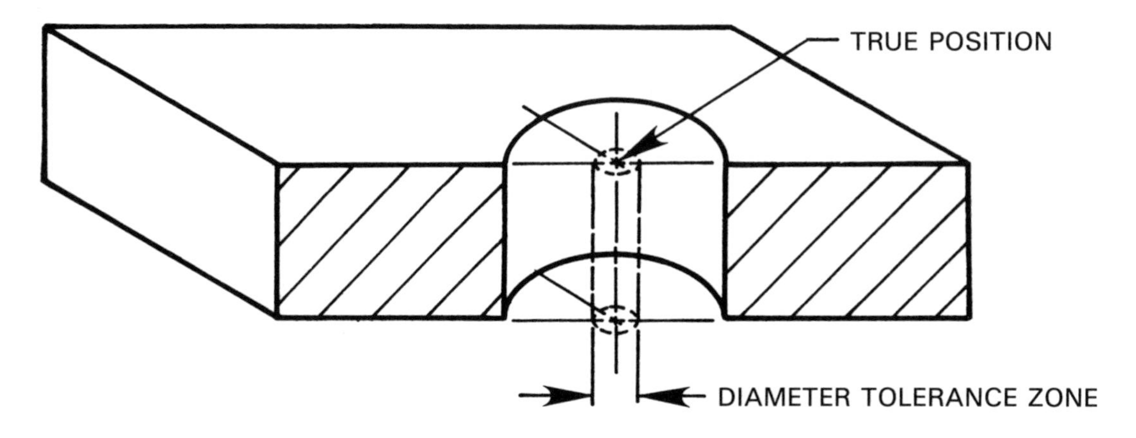

Example 7-2. Diameter position tolerance zone. Tolerance zone may also be between two parallel straight lines or planes, or between two uniform boundaries depending upon the application.

CONVENTIONAL TOLERANCING VS POSITION TOLERANCING

The term CONVENTIONAL TOLERANCING refers to the use of conventional coordinate dimensioning practices. A comparison between the use of conventional coordinate location dimensioning and position dimensioning and tolerancing will help you understand the function of geometric tolerancing for the location of features. The location of a hole using conventional dimensioning and tolerancing methods is shown in Example 7-3.

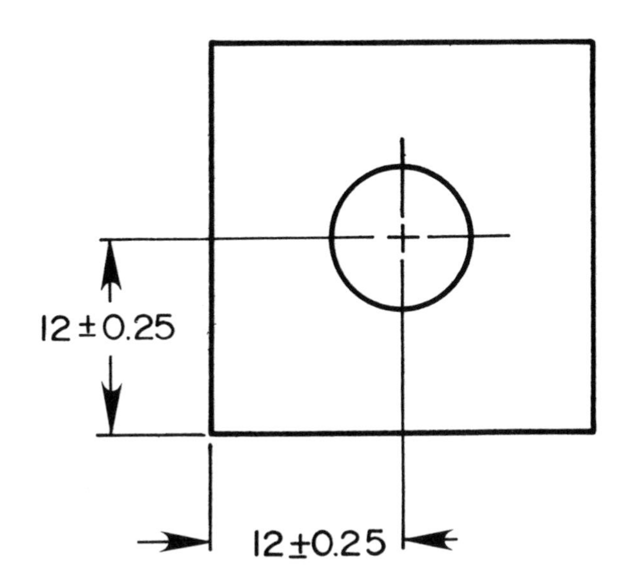

Example 7-3. Location of a hole using conventional coordinate dimensioning system.

The 12 ± 0.25 location dimensions in Example 7-3 establish a total tolerance zone of 0.5. This tolerance zone is square as you can see in Example 7-4.

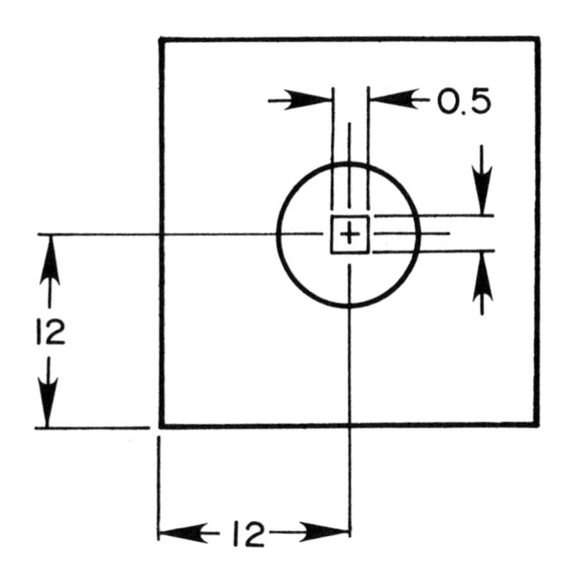

Example 7-4. Conventional tolerance zone.

Look at the 0.5 square tolerance zone a little closer to analyze what it really means. The tolerance zone in Example 7-5 means that the actual center of the hole can fall anywhere within the square area and be an acceptable part.

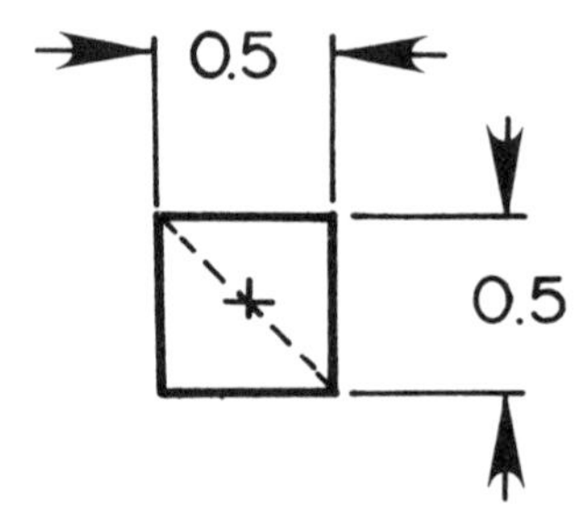

Example 7-5. A close look at the conventional location tolerance zone.

The application of position tolerancing to the same part will allow us to increase the acceptable tolerance zone. Consider the following points:

- The diagonal of the square tolerance zone, represented by the dashed line in Example 7-5, is the greatest distance that allows variation in the location of the center.
- The length of this diagonal is equal to a constant of 1.4 times the tolerance of the location dimensions.
- Using the tolerance zone from Example 7-5 will give a diagonal length of 1.4 × 0.5 = 0.7. See Example 7-6.

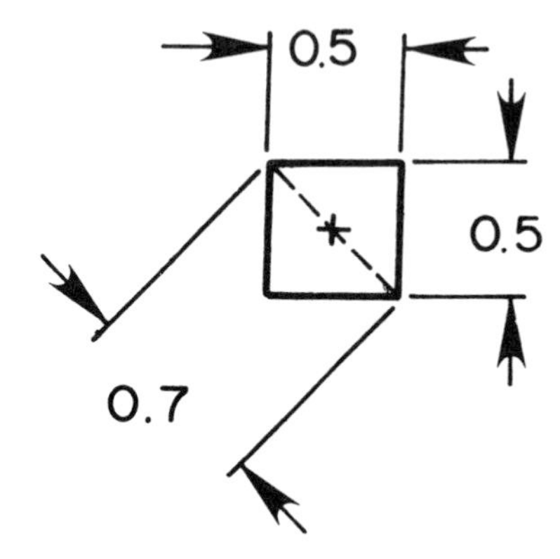

Example 7-6. The length of the diagonal of a square.

- In a position tolerance for the location of a hole, this 0.7 diagonal will become a diameter tolerance zone that is cylindrical in shape through the thickness of the part. This is how a conventional location tolerance may be converted directly to a position tolerance. It has been proven that the diagonal tolerance zone is acceptable in any direction, thus creating a circular tolerance zone. The result of this action is an increase of 57% of permissable area for the location of the hole. The relationship between the square conventional tolerance and the round position tolerance zone is shown in Example 7-7. With the use of position tolerancing there is an increase of acceptability of mating parts and a reduction in manufacturing costs.

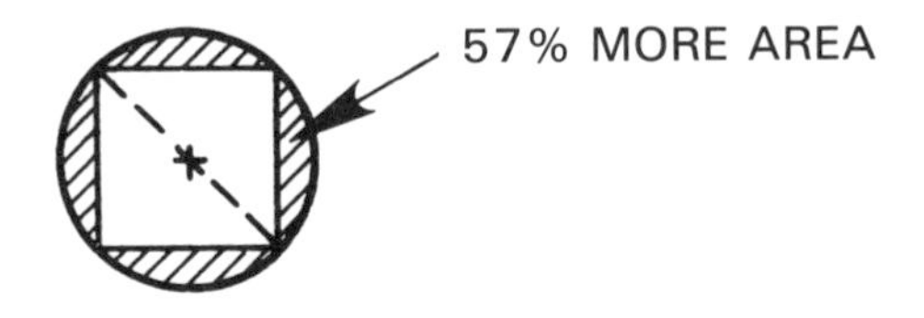

Example 7-7. The position tolerance circumscribed about the conventional tolerance.

When converting a drawing with conventional location dimensioning to a drawing with position tolerancing, use the following guide lines:

1. Add datums as appropriate. Notice in Example 7-8 the datums A, B, and C have been placed on the drawing. Perpendicularity of the true position centerline is controlled relative to datum A. Datums B and C control the location of the true position.
2. Change the location dimensions to basic dimensions as shown in Example 7-8. This locates the theoretically exact true position of the hole.
3. Add the feature control frame to the diameter dimension as shown in Example 7-8. Show the position symbol in the first section, followed by the diameter symbol, the calculated tolerance, and the MMC symbol unless otherwise specified. The last three blocks contain the datum references A, B, and C.

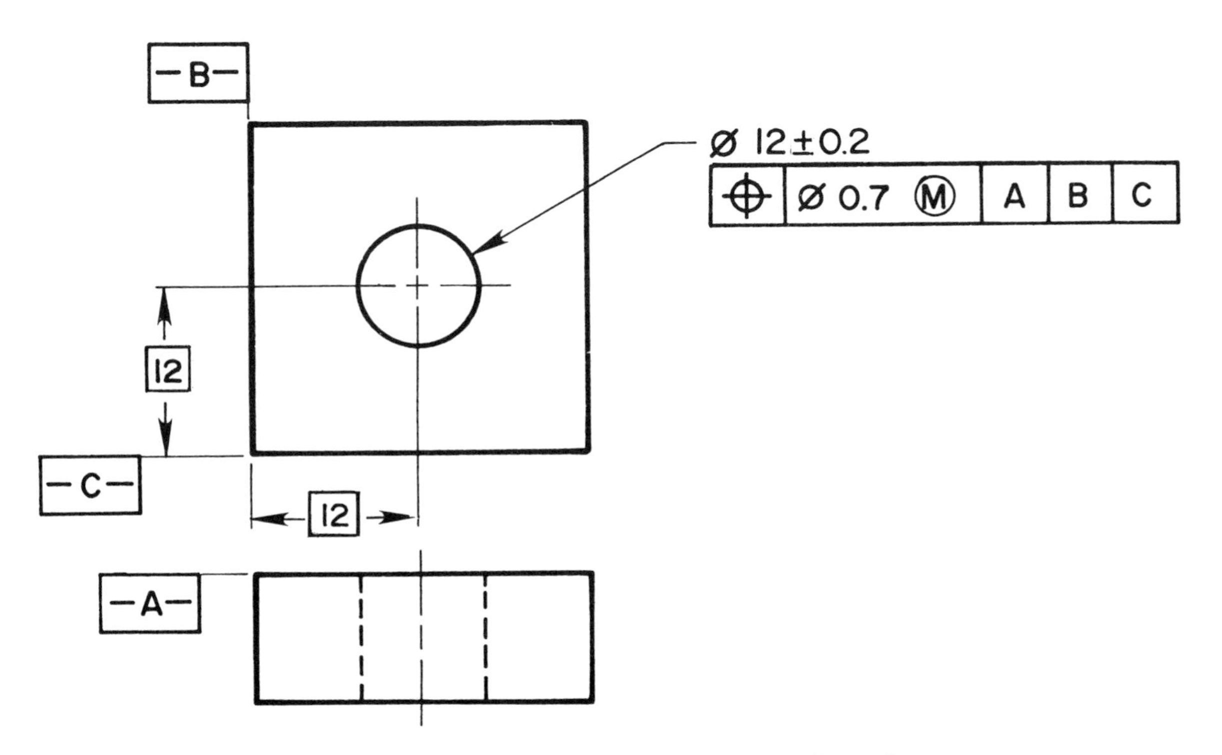

Example 7-8. A drawing with position tolerancing.

The previous discussion explained how to make a direct conversion of a drawing with conventional dimensioning to position tolerancing. In many situations the engineer will give the desired position tolerances. Only direct application to the drawing will be required and no conversion is necessary.

When locating holes using position tolerancing, the location dimensions must be basic. This may be accomplished by applying the basic dimension symbol to each of the basic dimensions, or specifying on the drawing or in a reference document the general note: UNTOLERANCED DIMENSIONS LOCATING TRUE POSITION ARE BASIC.

Position tolerances are often applied in regards to MMC. However, either MMC, RFS, or LMC must be indicated in the feature control frame to the right of the tolerance as applicable. Example 7-9 shows the cylindrical tolerance zone that is established by the position tolerance specified in Example 7-8. The true position centerline is perpendicular to datum A. The centerline of the hole can be anywhere within the diameter and length of the specified tolerance zone at the maximum material condition size of the hole.

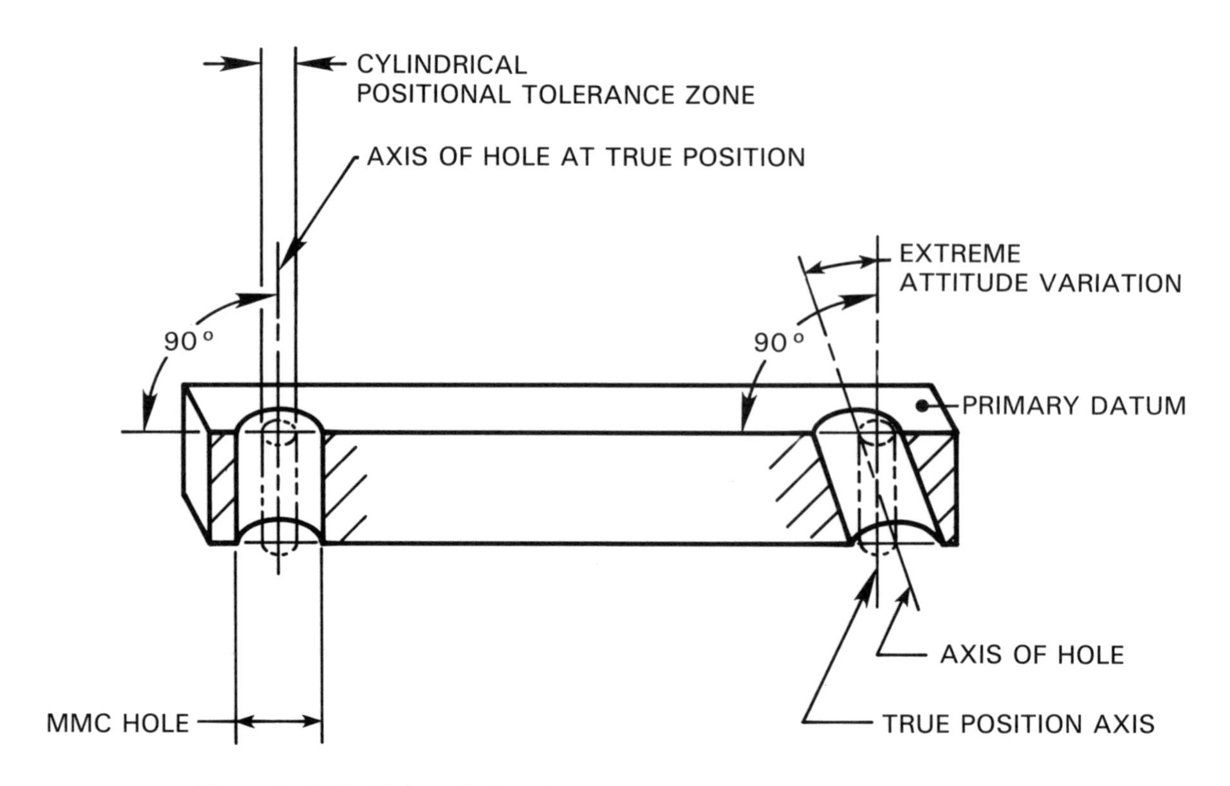

Example 7-9. Hole axis in relation to position tolerance zone.

POSITION TOLERANCE AT MMC

The maximum material condition (MMC) of a feature means that the actual size contains the maximum amount of material permitted by the toleranced size dimension for that feature. A hole or other internal feature is at MMC when the actual size is at the lower limit. A shaft or other external feature is at MMC when the actual size is at the upper limit.

A position tolerance should be applied at MMC which means that the specified position tolerance applies when the feature is manufactured at MMC. The axis of a hole must fall within a cylindrical tolerance zone whose axis is located at true position. The diameter of this cylindrical tolerance zone is equal to the specified position tolerance when the hole is manufactured at MMC. The position tolerance is then allowed to increase equal to the amount of change or departure from MMC. The maximum amount of positional tolerance is when the feature is produced at LMC as shown in the analysis provided in Example 7-10. When MMC is applied to a position tolerance the following formula may be used to calculate the position tolerance at any produced size:

Actual Size − MMC + Specified Position Tolerance = Actual Position Tolerance

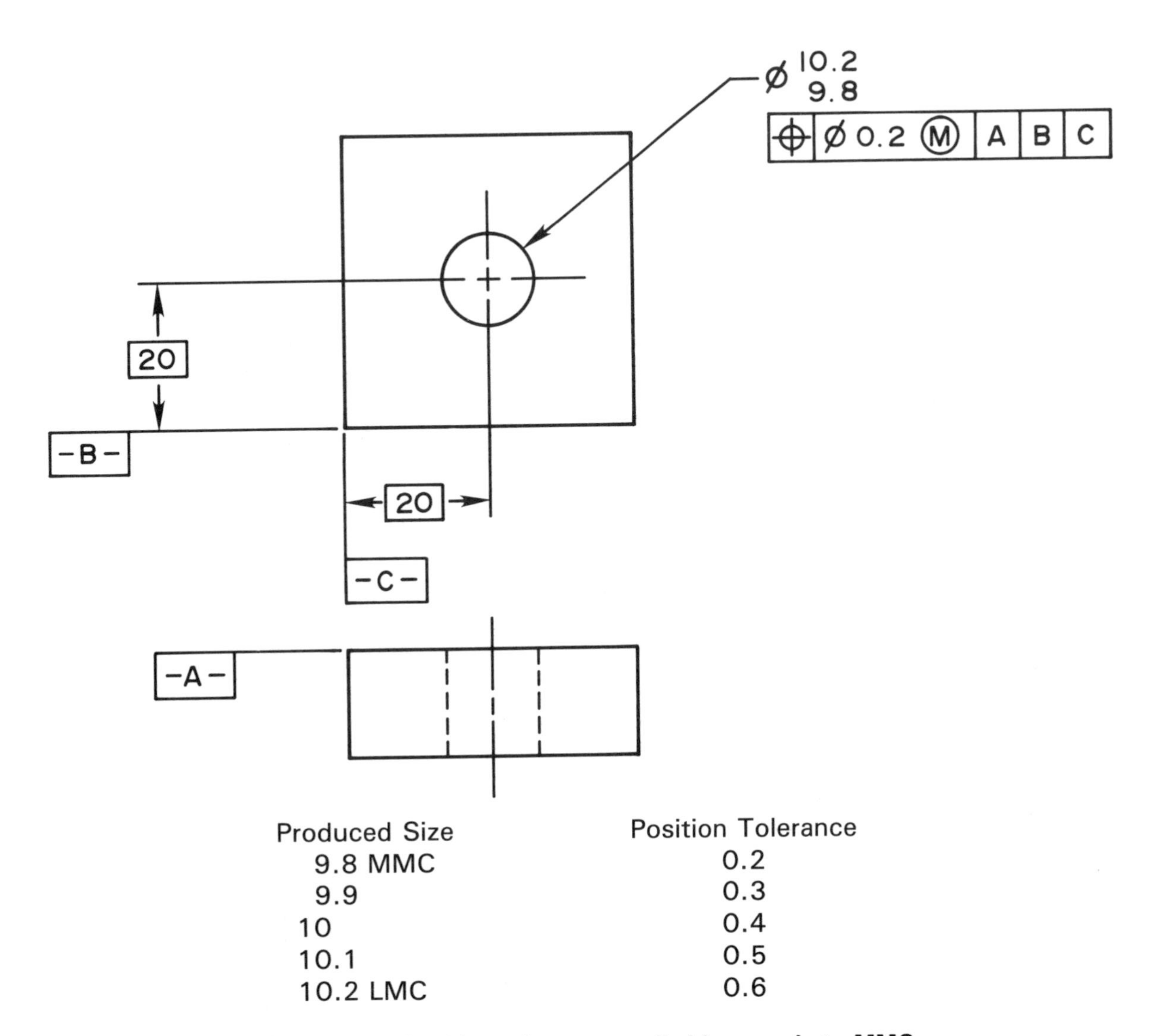

Produced Size	Position Tolerance
9.8 MMC	0.2
9.9	0.3
10	0.4
10.1	0.5
10.2 LMC	0.6

Example 7-10. Position tolerance applied in regards to MMC.

Position tolerance applied at MMC may also be explained in regards to the surface of the hole rather than the hole axis. All elements of the hole surface must be inside a theoretical boundary located at true position and the hole must be produced within the specified size limits as shown in Example 7-11.

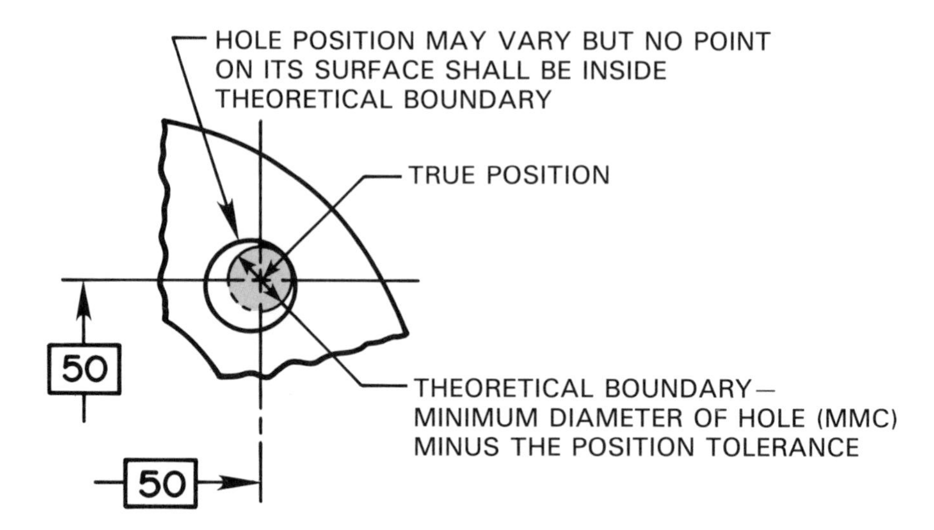

MMC HOLE—POSITIONAL TOLERANCE = THEORETICAL BOUNDARY

Example 7-11. Boundary for the surface of a hole at MMC.

ZERO POSITION TOLERANCING AT MMC

Zero geometric tolerancing was introduced in Chapter 6. This concept may also be applied to position tolerances. You have already seen that the application of position tolerance at MMC allows the tolerance zone to exceed the amount specified when the feature is produced at any actual size other than MMC. Zero position tolerance may be specified when it is important to insure that the tolerance be totally dependent on the actual size of the feature. When this is done, the position tolerance is zero when the feature is produced at MMC and must be located at true position. When the actual size of the feature departs from MMC, then the position tolerance is allowed to increase equal to the amount of departure. The total allowable variation in position tolerance is at LMC, unless a maximum tolerance is specified. Other than specifying zero position tolerance at MMC in the feature control frame, this is the same application explained in the previous discussion Position Tolerance at MMC. When zero position tolerance at MMC is specified, the engineer will normally apply the MMC of the hole at the absolute minimum required for insertion of a fastener when located at true position. See Example 7-12.

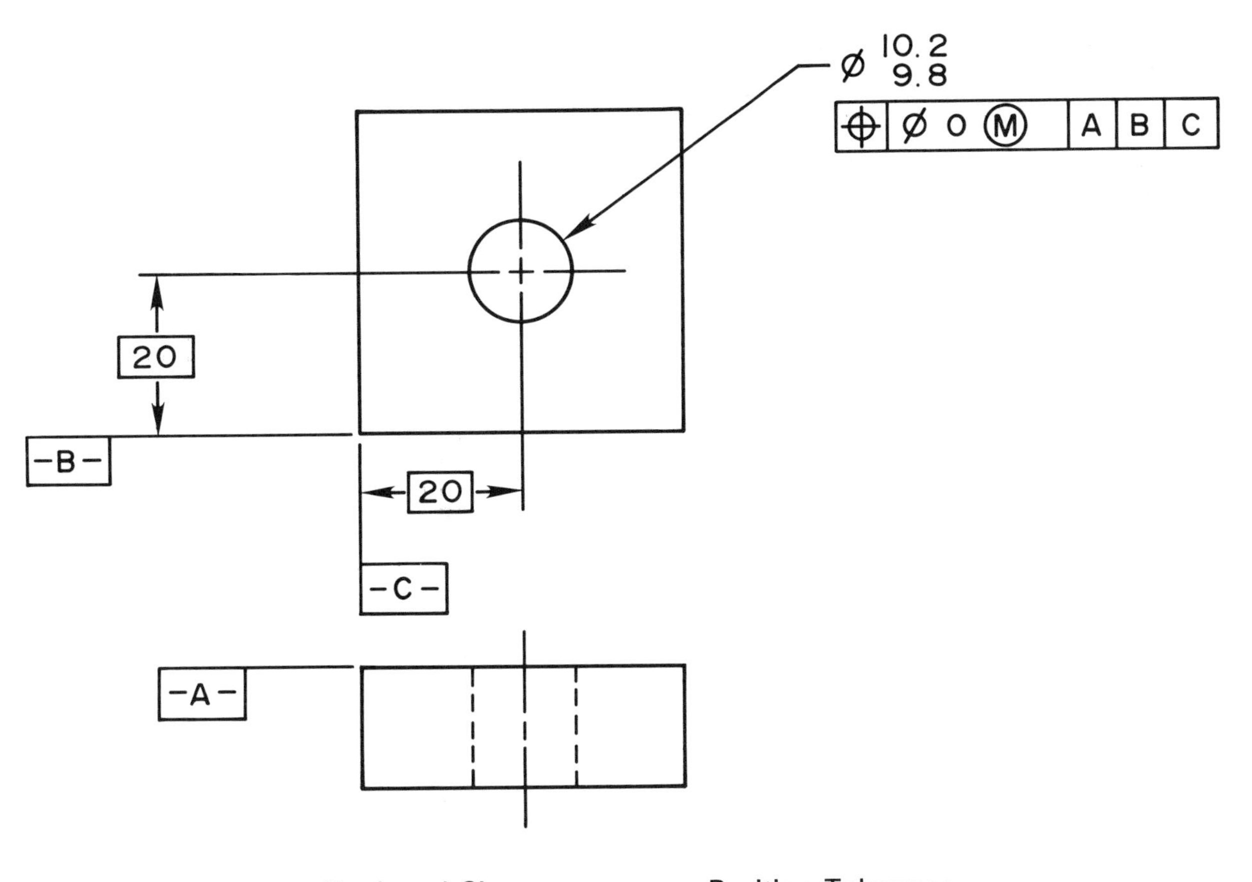

Produced Size	Position Tolerance
9.8 MMC	0
9.9	0.1
10	0.2
10.1	0.3
10.2 LMC	0.4

Example 7-12. Zero position tolerance at MMC.

POSITIONAL TOLERANCE AT RFS

The regardless of feature size (RFS) material condition symbol may be applied to the positional tolerance when it is desirable to maintain the given position tolerance at any produced size. The use of RFS may also be applied to the datum reference which must be maintained regardless of the actual feature sizes. The application of RFS requires closer controls of the features involved because the size of the tolerance zone may not get larger as when MMC is used. See Example 7-13.

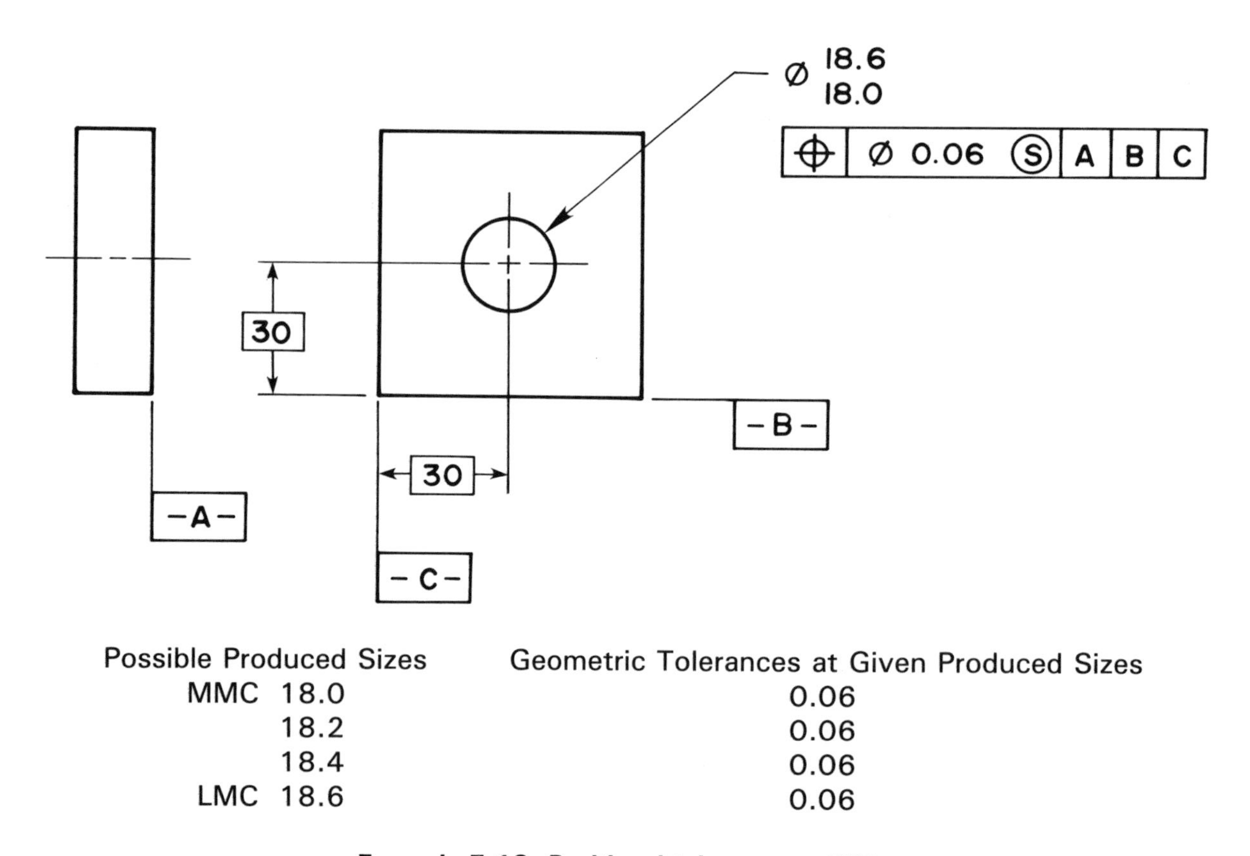

Possible Produced Sizes	Geometric Tolerances at Given Produced Sizes
MMC 18.0	0.06
18.2	0.06
18.4	0.06
LMC 18.6	0.06

Example 7-13. Positional tolerance at RFS.

POSITIONAL TOLERANCE AT LMC

Positional tolerance at least material condition (LMC) is used to control the relationship of the feature surface and the true position of the largest hole size. The function of LMC is generally used to control minimum edge distance. When the LMC material condition symbol is used in the feature control frame, the given position tolerance is held at the LMC produced size. Then as the produced size departs from LMC toward MMC, the position tolerance increases equal to the amount of change from LMC. The maximum amount of position tolerance is applied at the MMC produced size, as shown in Example 7-14. When using the LMC control, perfect form is required at the LMC produced size. RFS specifications are limited to position tolerances where the use of MMC does not give the desired control and RFS is too restrictive.

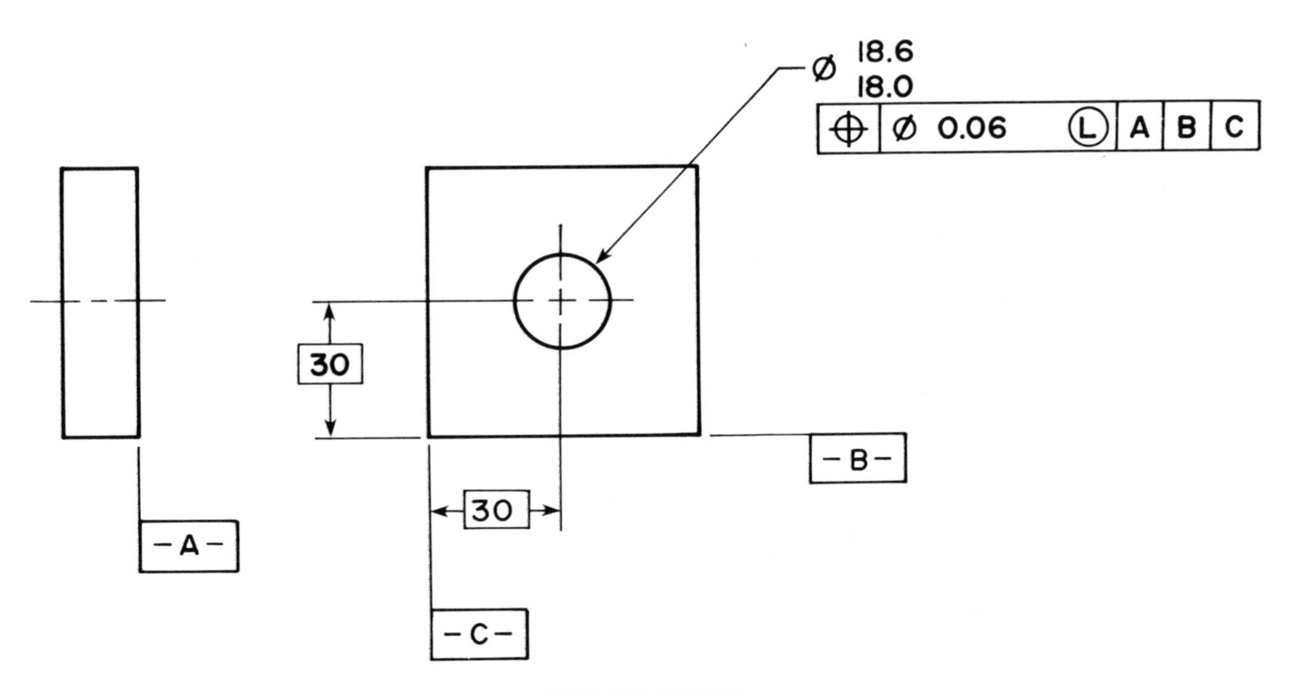

THE DRAWING

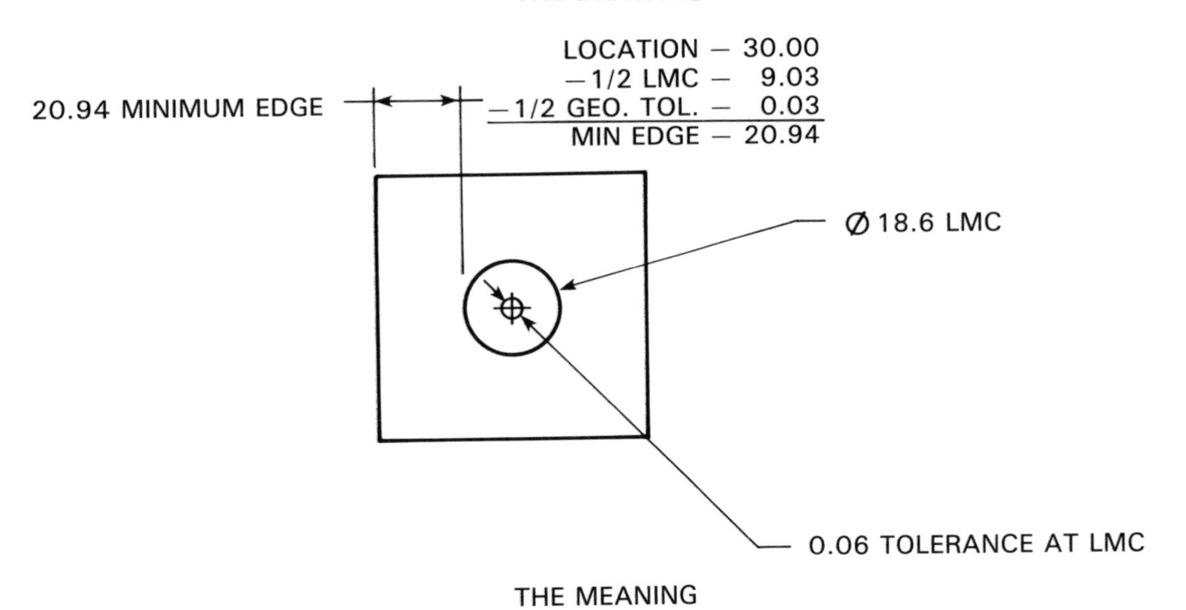

THE MEANING

Possible Produced Sizes	Geometric Tolerances at Given Produced Sizes
MMC 18.0	0.66
18.2	0.46
18.4	0.26
LMC 18.6	0.06

Example 7-14. Positional tolerance at LMC.

LOCATING MULTIPLE FEATURES

Multiple features of an object can be dimensioned using position tolerancing. When this is done, the location of the features must be dimensioned from datums and between features using datum or chain line dimensioning related to rectangular or polar coordinates as shown in Examples 7-15 and 7-16. When multiple features are located, the following guidelines apply:

- The pattern of features is located collectively in relationship to datum features that are not subject to size changes. The actual centers of all features in the pattern must lie on or be within the specified tolerance zone when measured from the given datums.
- Multiple patterns of features are considered a single composite pattern if the related feature control frames have the same datums in the same order of precedence with the same modifying symbols. When the relationship between multiple patterns of features are not the same, then the note SEP REQT (denoting like patterns are not considered multiple) is placed below the feature control frames that require separate inspection. This allows the patterns independent location relative to the specified datums.

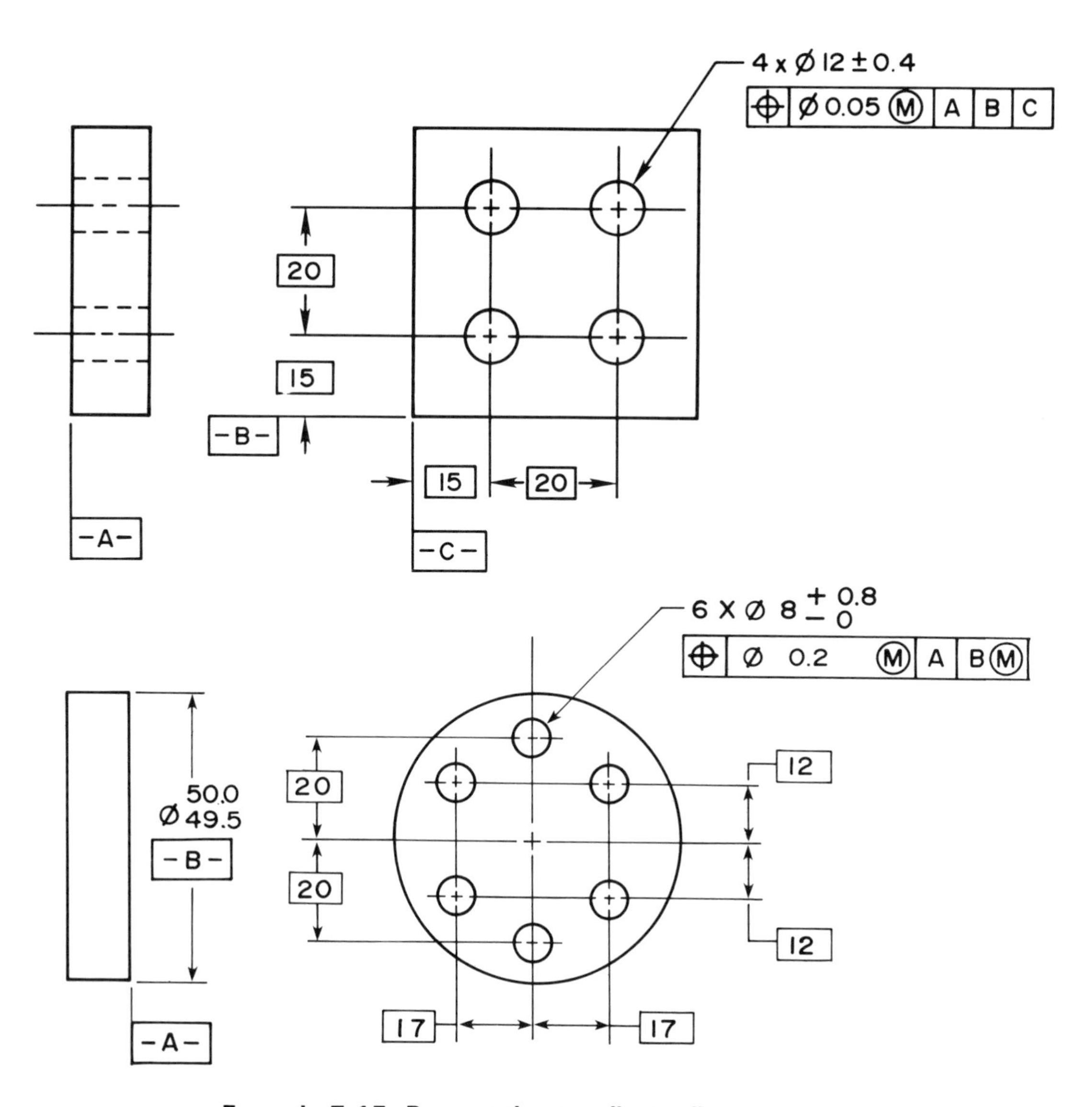

Example 7-15. Rectangular coordinate dimensioning.

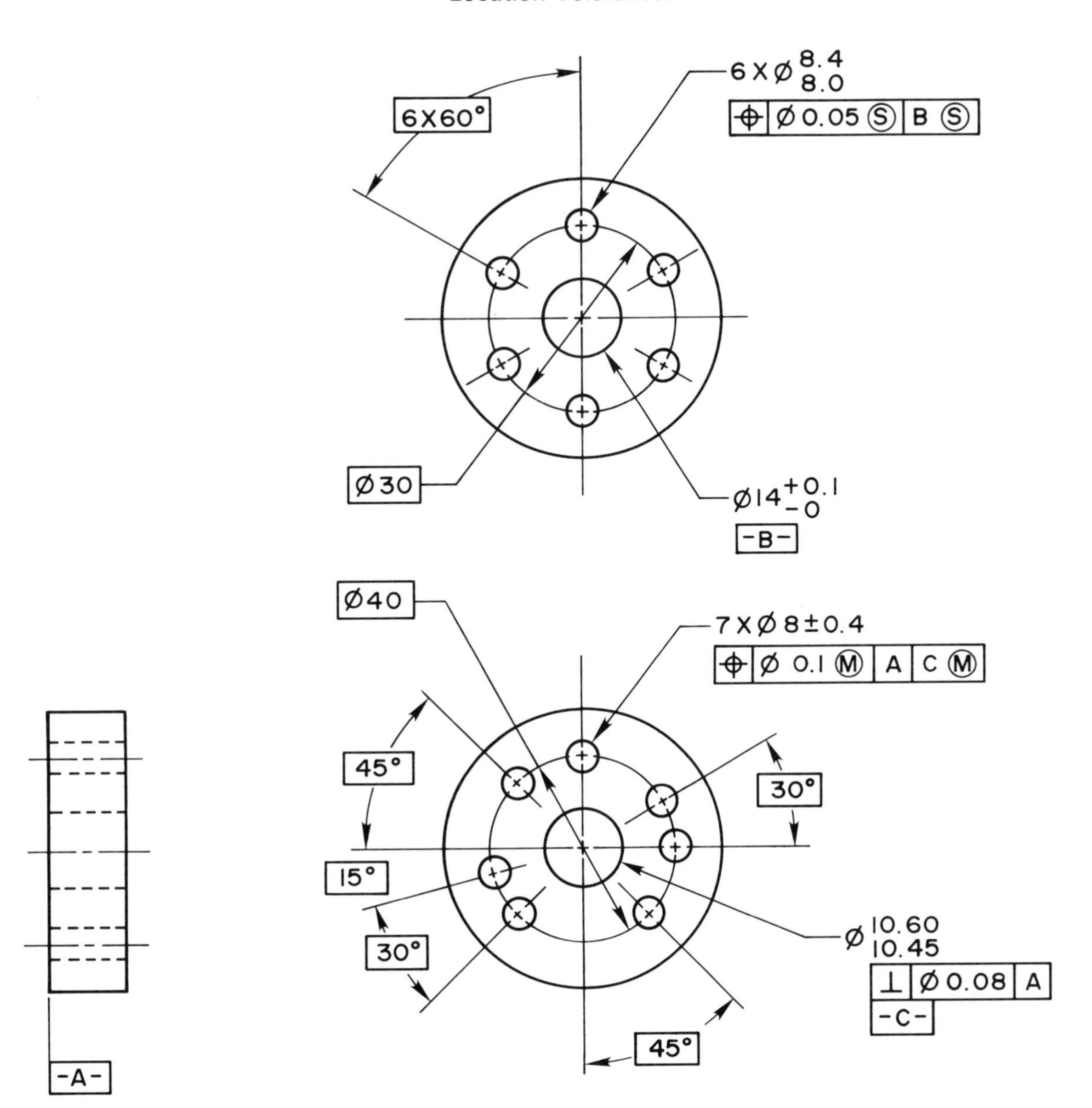

Example 7-16. Polar coordinate dimensioning.

LOCATING SLOTTED HOLES

Slotted holes may be located to their centers with basic dimensions from established datums. The feature control frame is added to both the length and width dimensions of the slotted features as shown in Example 7-17.

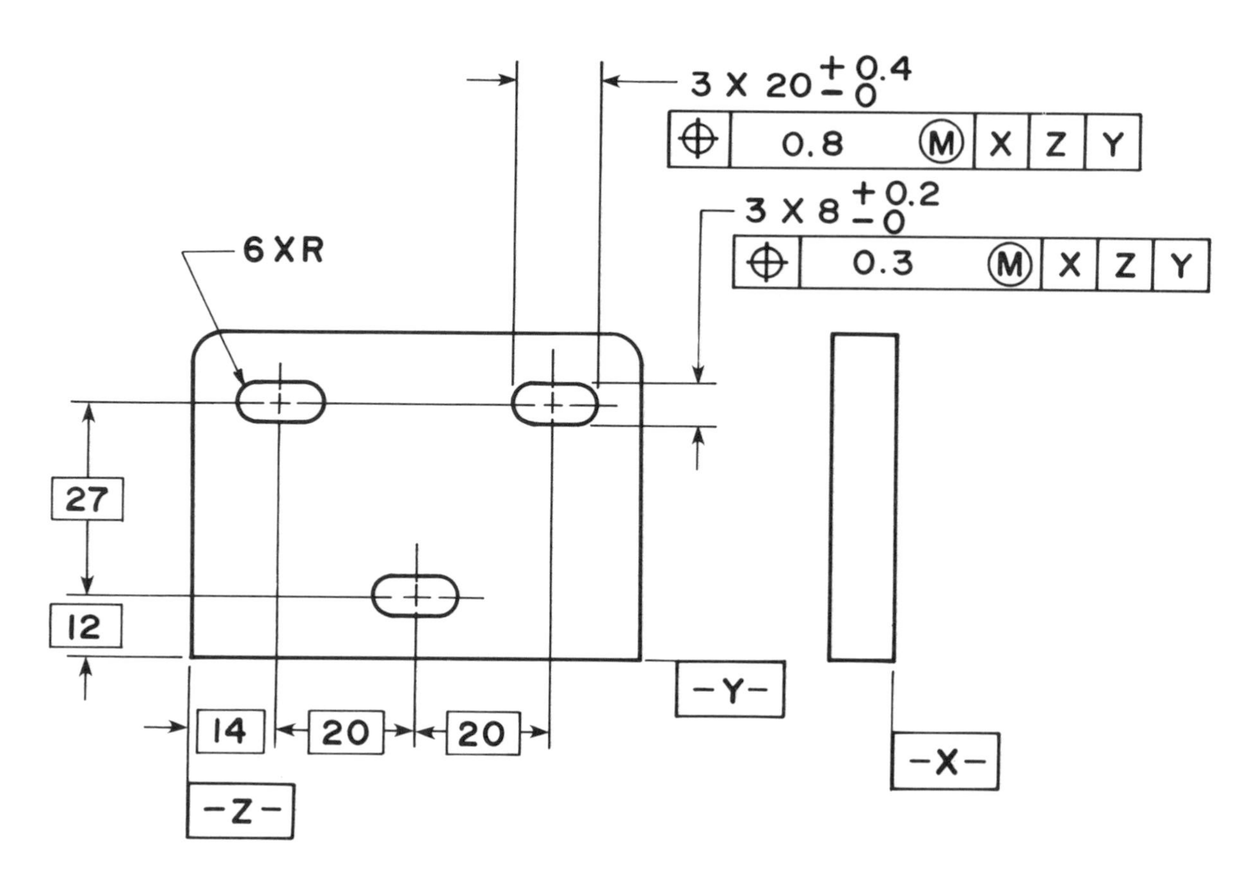

Example 7-17. Positional tolerance for slotted holes.

COAXIAL POSITION TOLERANCE

Coaxial features are two or more features that lie on a common axis. A COAXIAL POSITIONAL TOLERANCE may be used to control the alignment of two or more holes that share a common axis. This is used when a tolerance of location alone does not provide the necessary control of alignment of the holes and a separate requirement must be specified. When this is done, the positional tolerance feature control frame is doubled in height. The top half of the frame is used to specify the coaxial diameter tolerance zones at MMC located at true position relative to the specified datums within which the axes of the holes, as a group, must lie. The lower half of the feature control frame is used to designate the coaxial diameter tolerance zones at MMC within which the axes of the holes must lie relative to each other. See Example 7-18. Where coaxial holes are of different sizes and the same requirements apply to all holes, the feature control frame may be accompanied by the note: TWO (number of holes) COAXIAL HOLES.

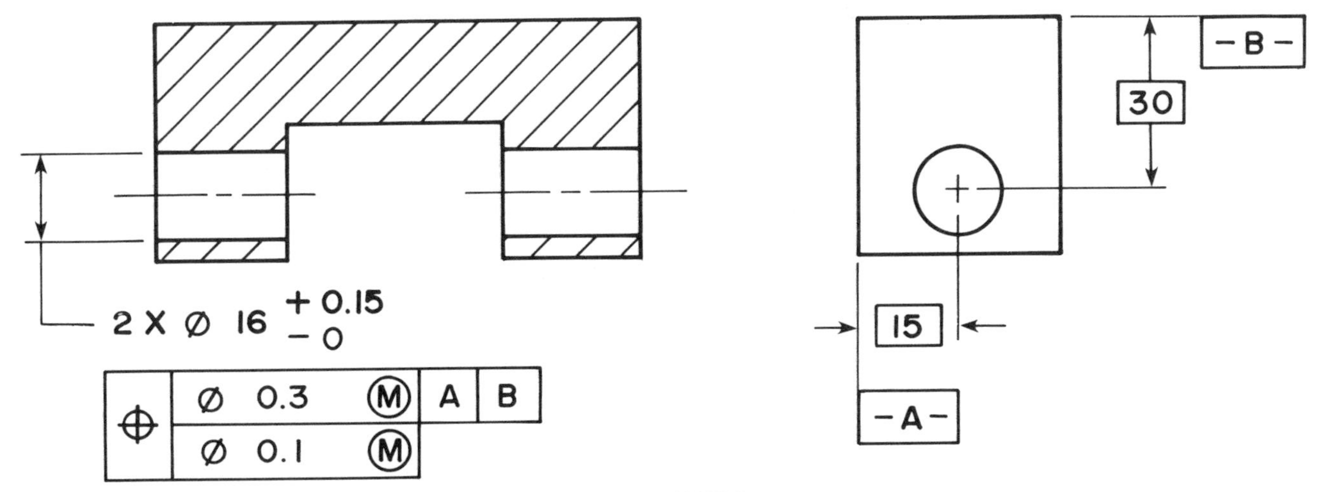

THE DRAWING
(Note: The diameter dimension and feature control frame may also point to the circular view of the hole using a leader line).

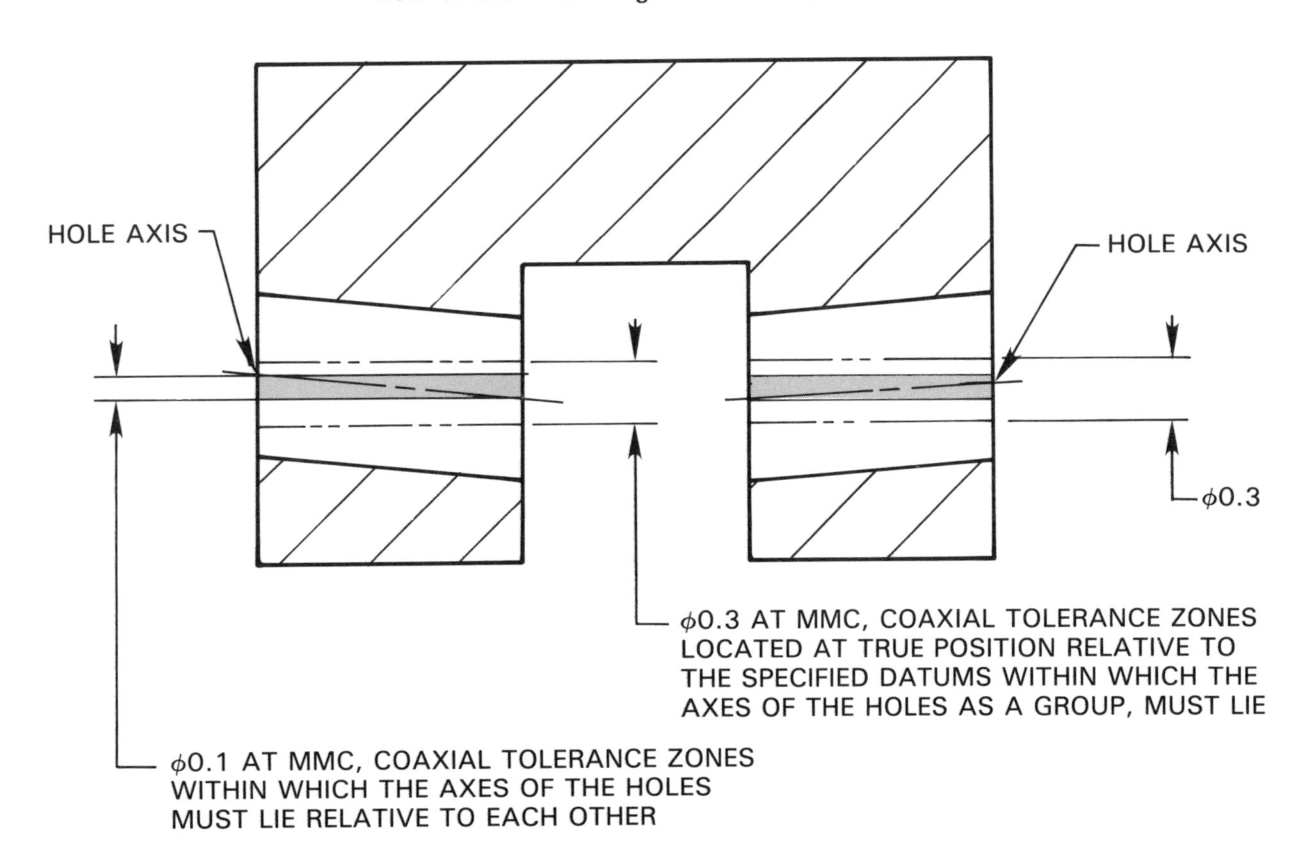

THE MEANING

Example 7-18. Positional tolerance for coaxial holes of the same size.

POSITION TOLERANCE OF COAXIAL FEATURES

Coaxial features are those features having a common axis, such as counterbores, countersinks, and counterdrills. When the position tolerance of the coaxial features is to be the same (for example the same for the counterbore and the associated hole) then the feature control frame is placed below the note specifying the hole and counterbore as shown in Example 7-19. When this is done, the tolerance zone diameters are identical for the hole and counterbore relative to the specified datums.

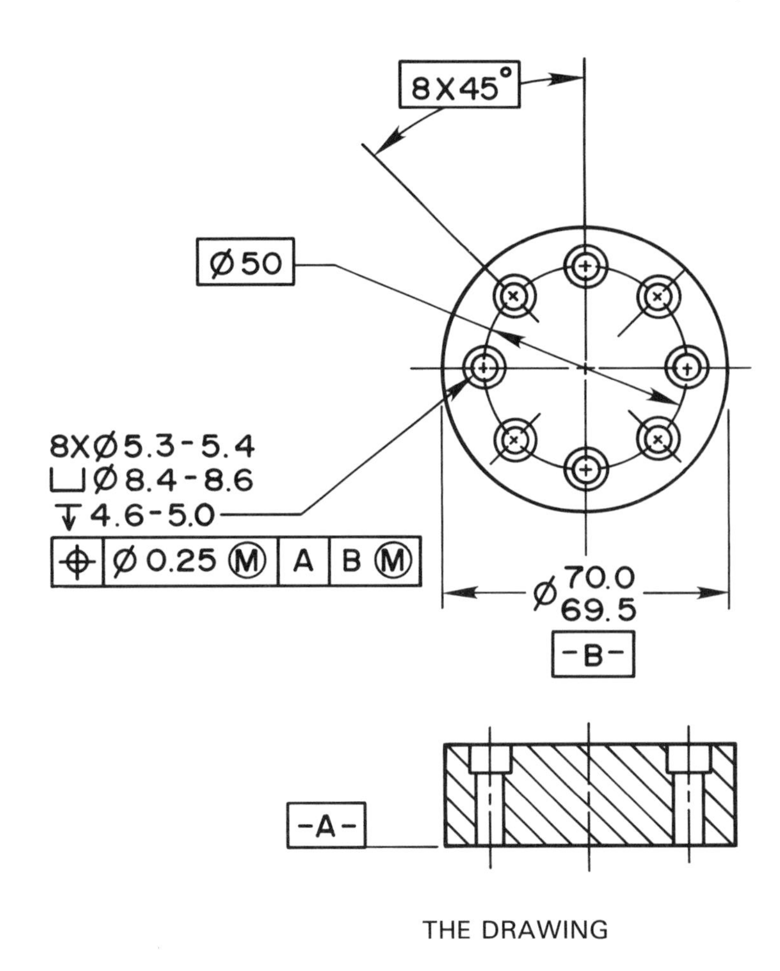

THE DRAWING

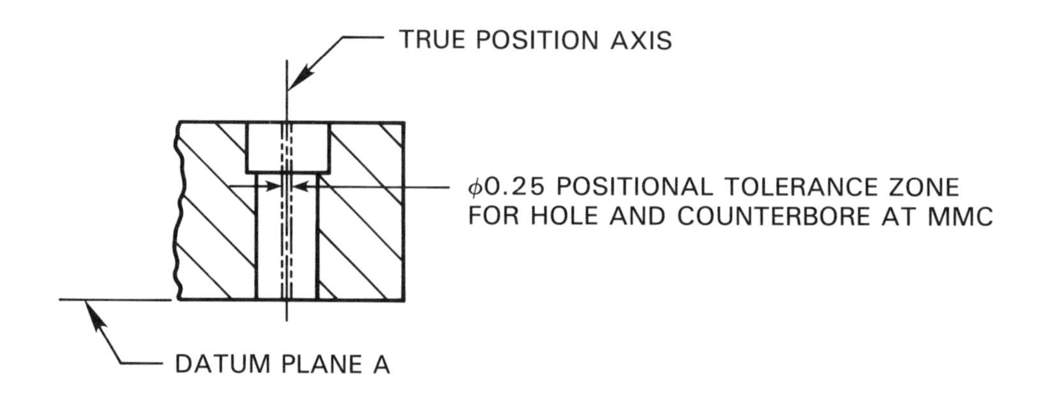

THE MEANING

Example 7-19. Same position tolerance for coaxial features.

Different position tolerances may be applied to coaxial features related to the same datum features. For example when the position tolerance is different for the counterbore and the related hole, then one feature control frame is placed under the note specifying the hole size and another feature control frame placed under the part of the note specifying the counterbore. See Example 7-20. This may be possible when the counterbore is a different tolerance than the hole. However, this is a very unlikely situation when the hole and counterbore are machined at the same set-up.

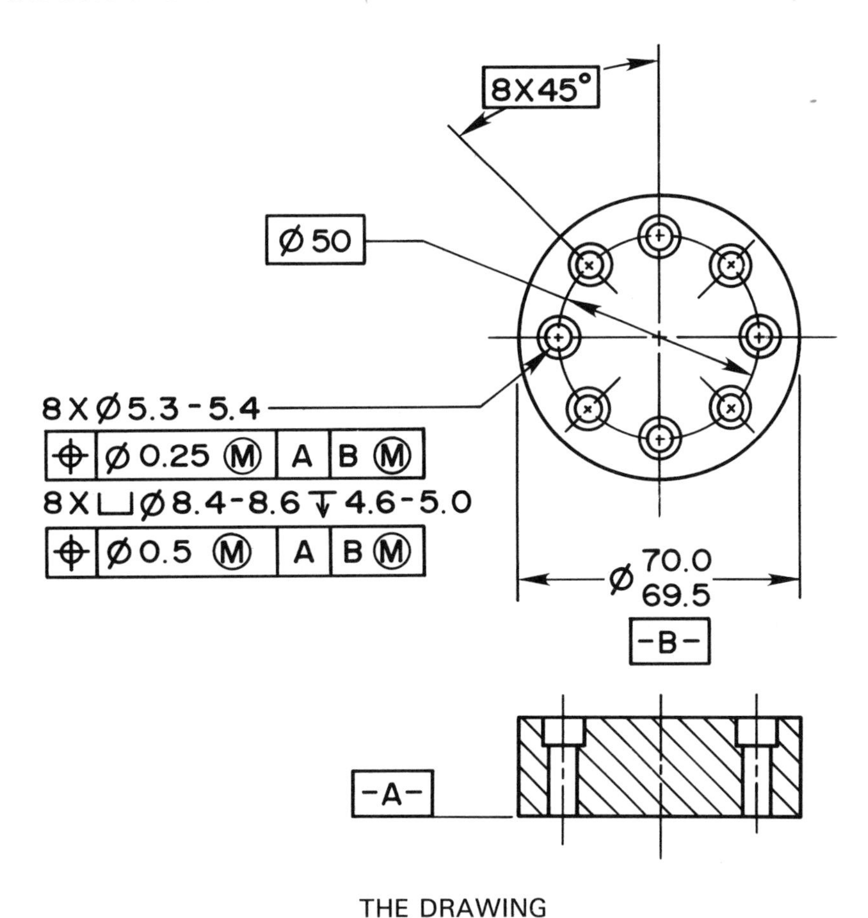

THE DRAWING

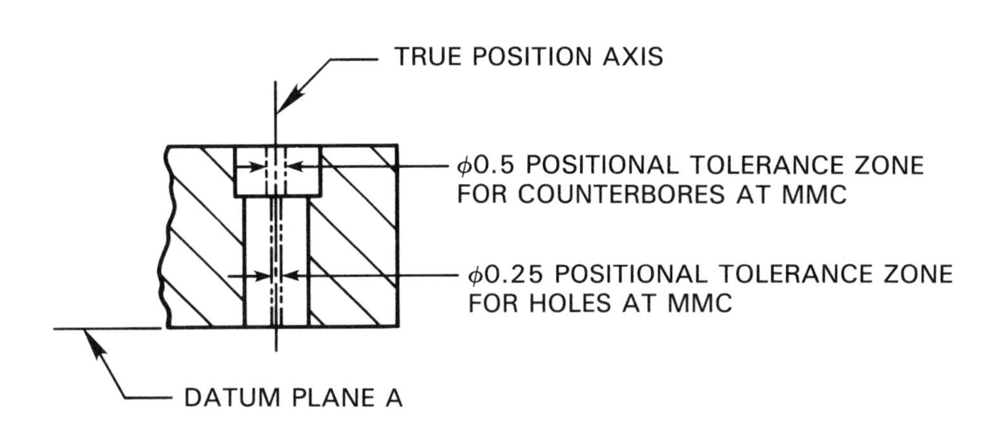

THE MEANING

Example 7-20. Different position tolerances for coaxial features.

Position tolerances may also be applied to coaxial features such as counterbore holes by controlling individual counterbore-to-hole relationships relative to different datum features. The same application is used as shown and detailed in Example 7-20 with the addition of a note placed under the datum feature symbol for the hole and under the feature control frame for the counterbore indicating the number of places each applies on an individual basis as shown in Example 7-21.

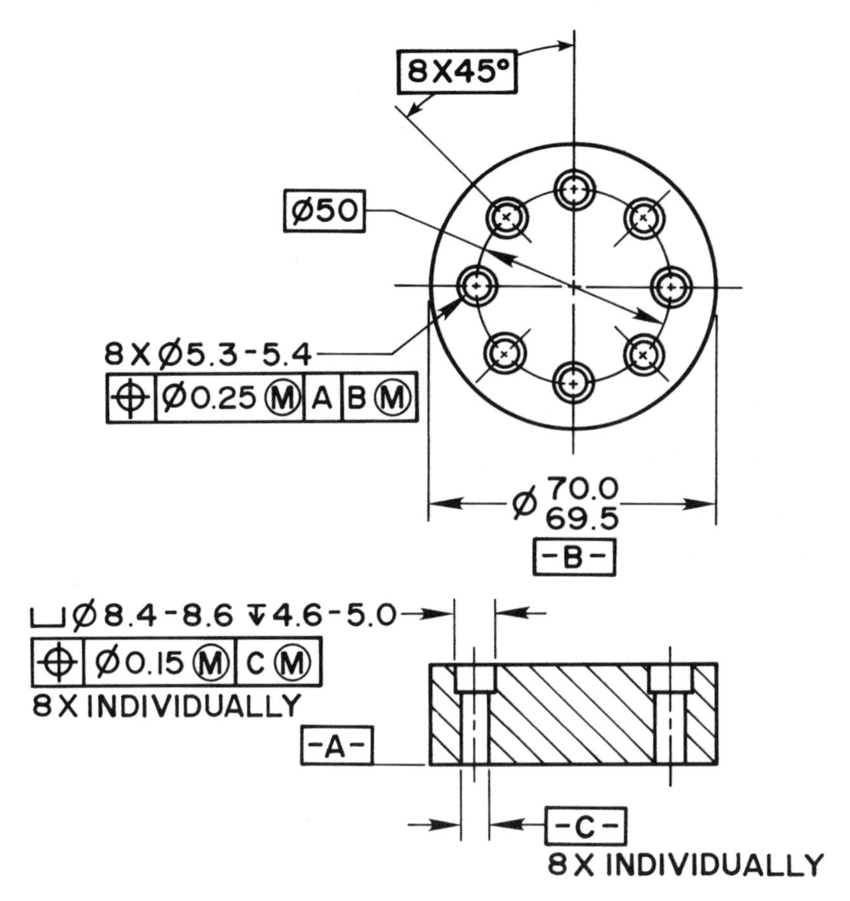

THE DRAWING

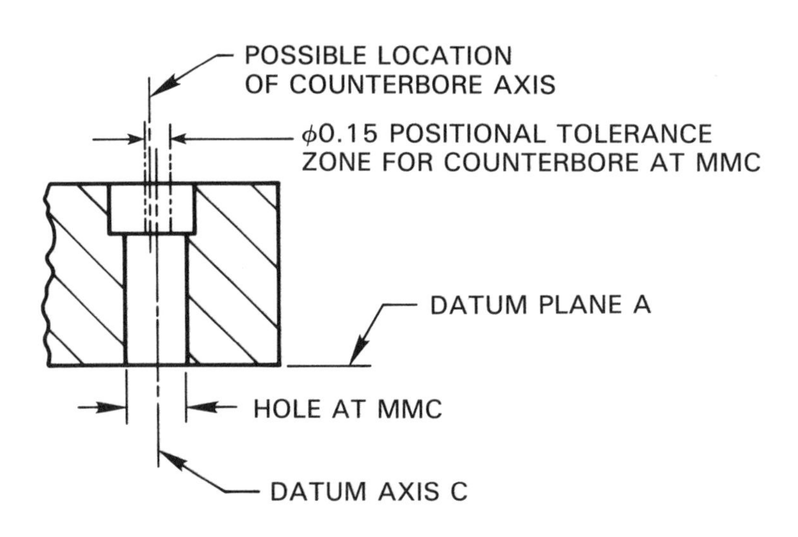

THE MEANING

Example 7-21. Position tolerance of coaxial features with different datum reference.

COMPOSITE POSITIONAL TOLERANCE

COMPOSITE POSITIONAL TOLERANCING is used when it is desirable to permit the location of a pattern of features to vary within a larger tolerance than the positional tolerance specified for each feature. For this application, the feature control frame is doubled in height and divided into two parts. The upper part is the PATTERN LOCATING CONTROL and specifies the larger positional tolerance for the pattern of features as a group. The lower entry is called the FEATURE RELATING CONTROL which specifies the smaller positional tolerance for the individual features within the pattern. The pattern locating control is located first with basic dimensions. The feature relating control is established at the actual position of the feature center. Only the primary datum is given in the feature relating control. The tolerance zone of an individual feature may extend partly beyond the group zone, but the feature axis may not fall outside the confines of both zones, Example 7-22.

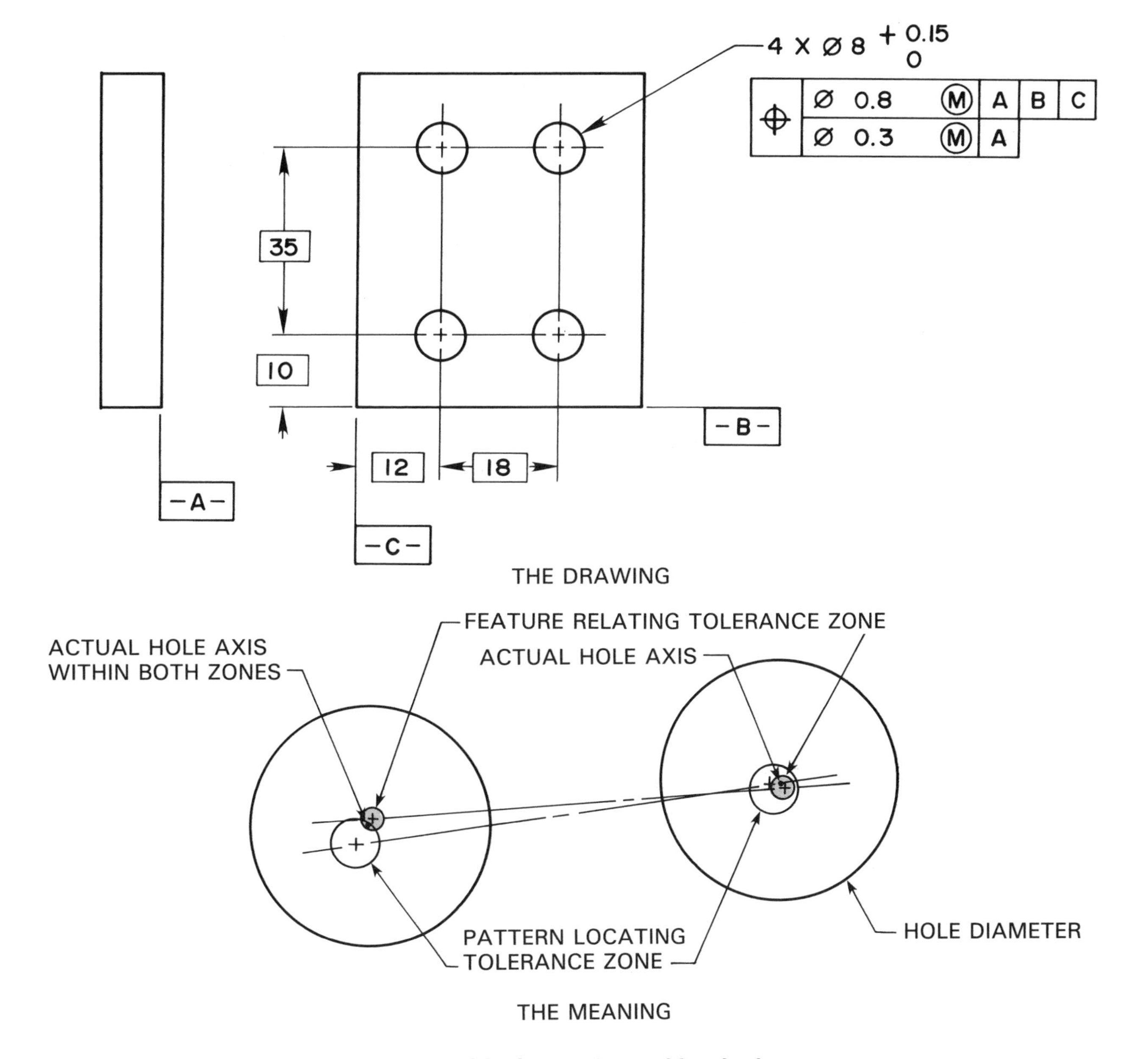

Example 7-22. Composite positional tolerance.

POSITION TOLERANCE OF NONPARALLEL HOLES

Position tolerances may be used in situations where the axes of the holes are not parallel to each other and where they may also be at an angle to the surface, as shown in Example 7-23.

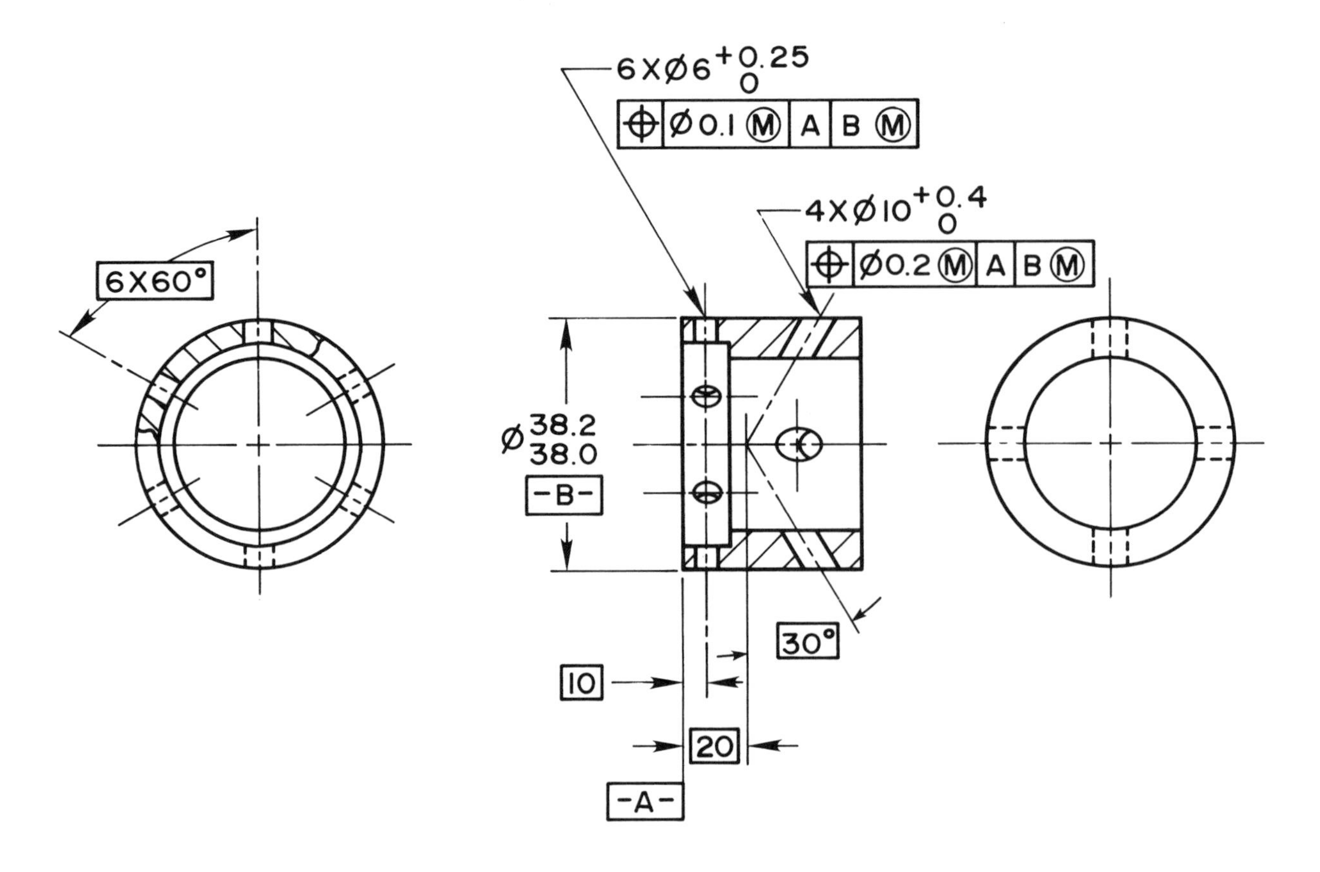

Example 7-23. Position tolerance of nonparallel holes.

POSITION TOLERANCE OF SPHERICAL FEATURES

A position tolerance may be used to control the location of a spherical feature relative to other features of a part. When dimensioning spherical features, the spherical diameter symbol preceeds the feature size dimension. The feature control frame is placed below the size dimension and the position tolerance zone is spherical in shape as shown in Example 7-24.

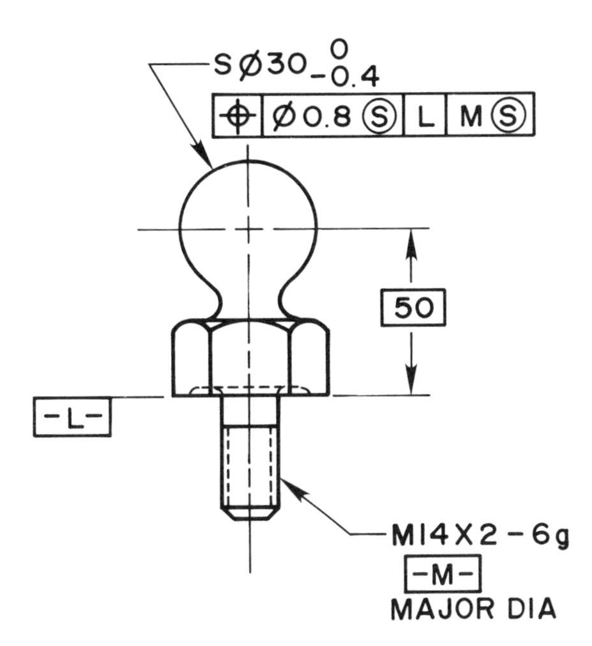

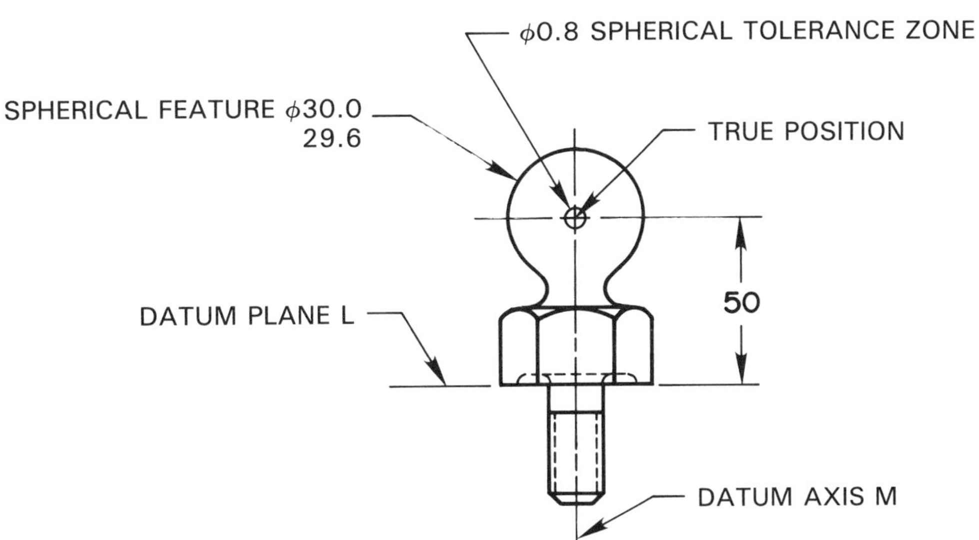

Example 7-24. Position tolerance of a spherical feature.

POSITION TOLERANCE LOCATING SYMMETRICAL FEATURES

SYMMETRY is a centerplane relationship of the features of an object. PERFECT SYMMETRY or true position occurs when the centerplanes of two or more related symmetrical features line up as shown in Example 7-25.

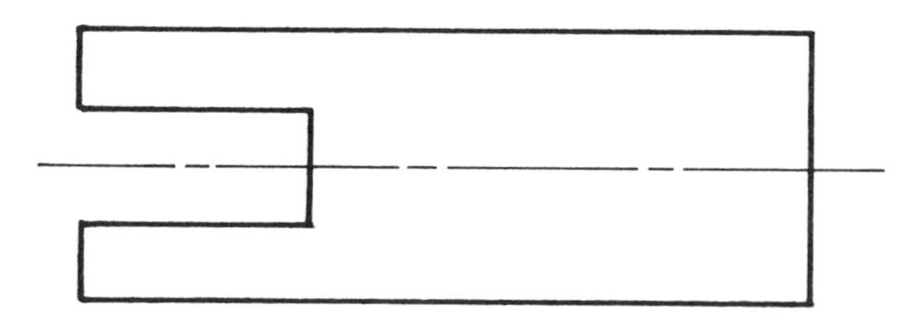

Example 7-25. Perfect symmetry is when the centerplanes of two features line up.

A position tolerance is used when it is required to locate one or more features symmetrically with respect to the centerplane of a datum feature. Refer to Example 7-26.

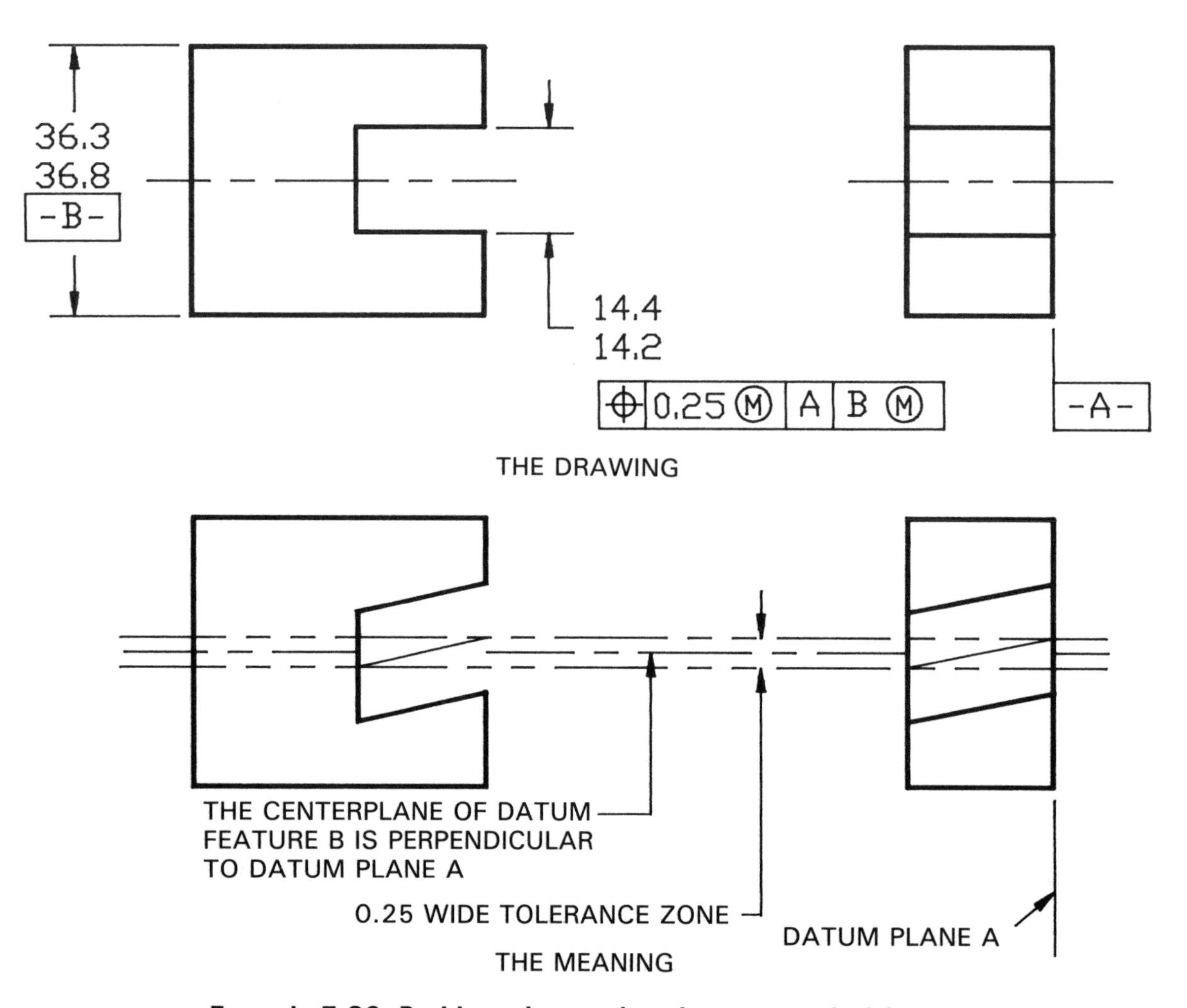

Example 7-26. Position tolerance locating symmetrical features.

The diameter symbol is omitted in front of the position tolerance in the feature control frame, because the given tolerance zone is the distance between two parallel planes equally divided on each side of true position rather than a cylindrical tolerance zone as described in previous applications. A material condition symbol MMC, RFS, or LMC must accompany the position tolerance as shown in Example 7-26.

Positional tolerancing of symmetrically shaped slots or tabs may be accomplished by identification of related datums, dimensioning the relationship between slots or tabs, providing the number of units followed by the size, and position tolerance feature control frame. The diameter symbol is omitted from the feature control frame and a material condition symbol is required, as shown in Example 7-27.

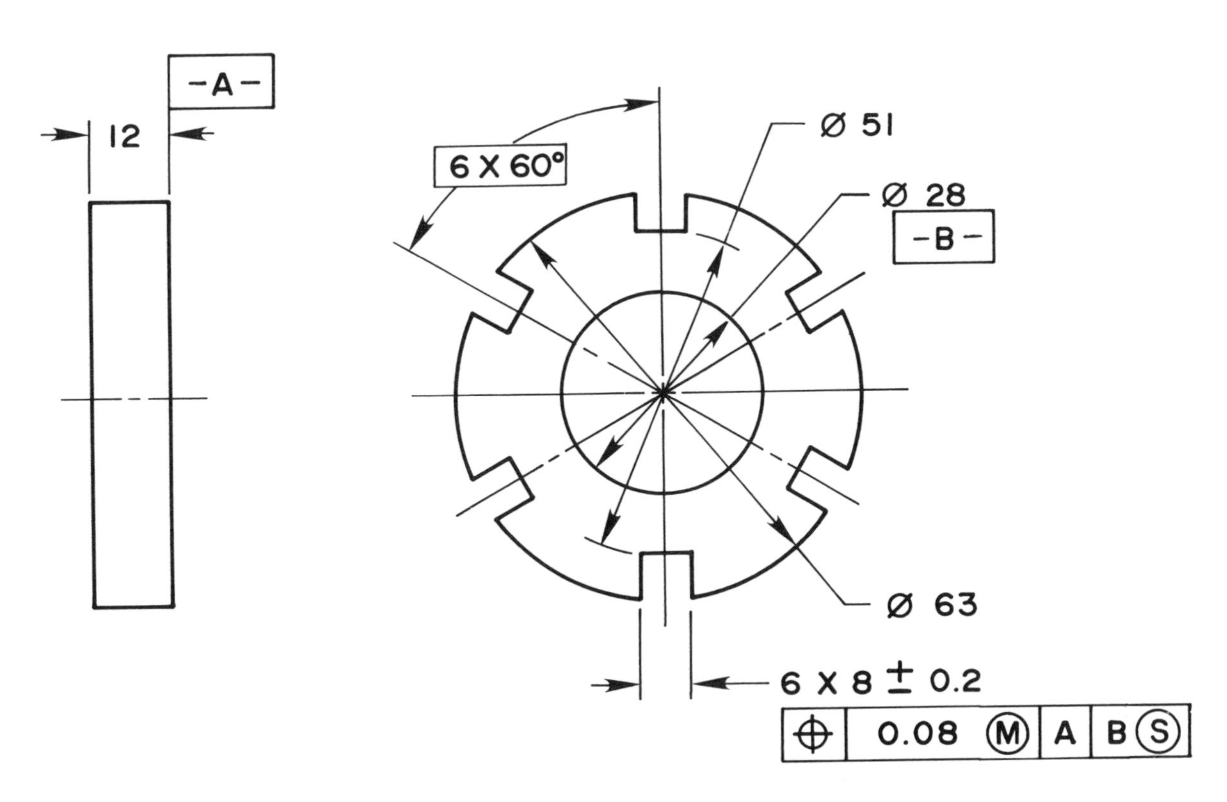

Example 7-27. Positional tolerance of slots or tabs.

FASTENERS

The application of geometric tolerancing such as orientation or location tolerances to threaded fasteners is applied to the axis of a cylinder established by the pitch diameter of the thread. Example 7-28a shows the elements of a screw thread.

If you want the geometric tolerance of the screw thread to be applied to the major diameter or the minor diameter, rather than the pitch diameter, then you need to place note MAJOR DIA or MINOR DIA, as appropriate, below the related feature control frame or datum feature symbol.

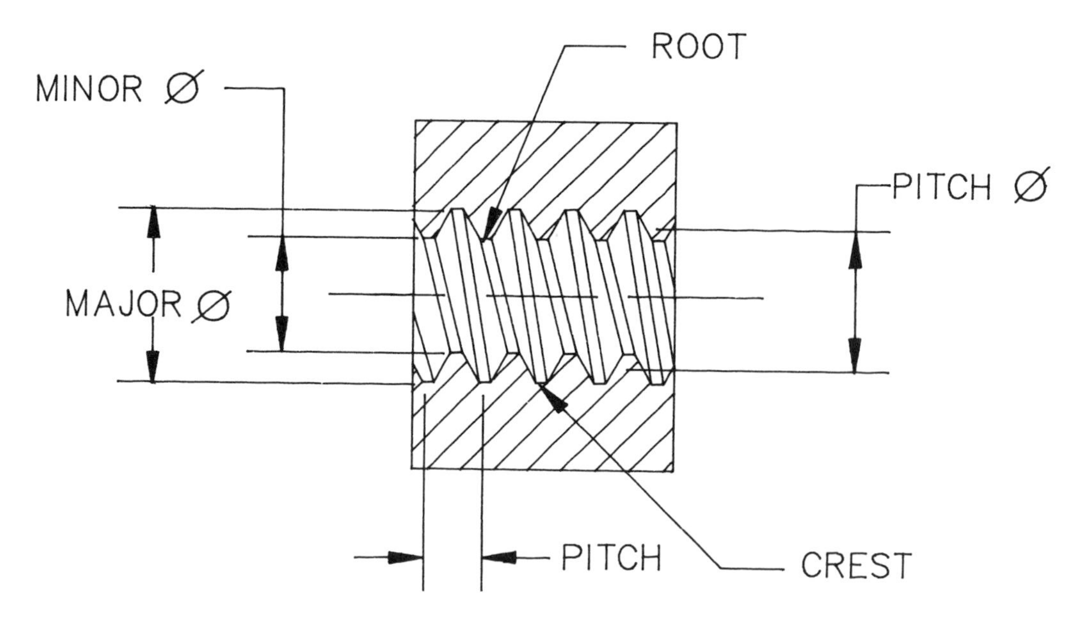

Example 7-28a. The elements of a screw thread.

FLOATING FASTENERS

The term FLOATING FASTENER applies to a situation where two or more parts are assembled with fasteners such as bolts and nuts, and all parts have clearance holes for the bolts. A floating fastener situation is shown in Example 7-28b. Notice Parts ''A'' and ''B'' are fastened together by a bolt, and a nut is required to hold the parts secure. When the holes in a pattern are the same diameters, the bolts used are the same diameters, and the same position tolerance for all holes is to be the same, then the position tolerance may be calculated using the formula:

MMC HOLE – MMC FASTENER (BOLT) = POSITION TOLERANCE FOR EACH PART

Note that each part is calculated separately.

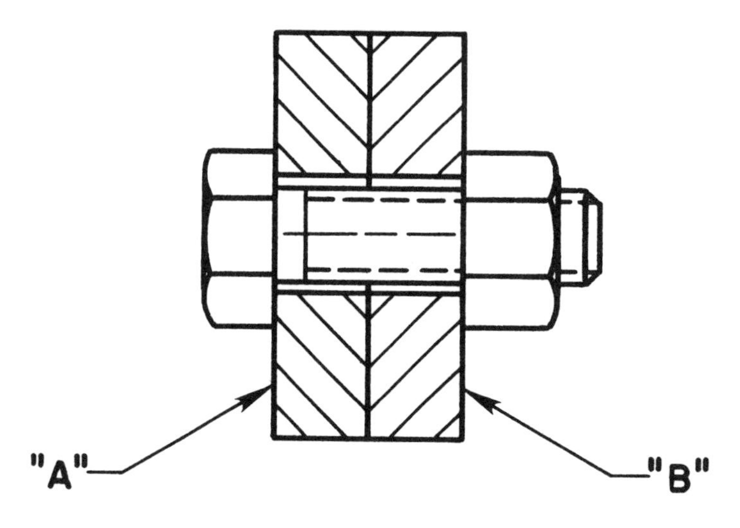

Example 7-28b. Floating fastener.

Given a situation where an M12 x 1,5 bolt is used to fasten together two identical parts with a hole diameter of 13.0/12.5 as shown in Example 7-29, the position tolerance required can be calculated as follows. The MMC of a bolt is considered as the nominal size, which is the same as the major diameter. The major diameter of the M 12 x 1.5 thread is 12 millimeters. Therefore:

MMC HOLE (12.5) — MMC BOLT (12) = POSITION TOLERANCE (0.5)

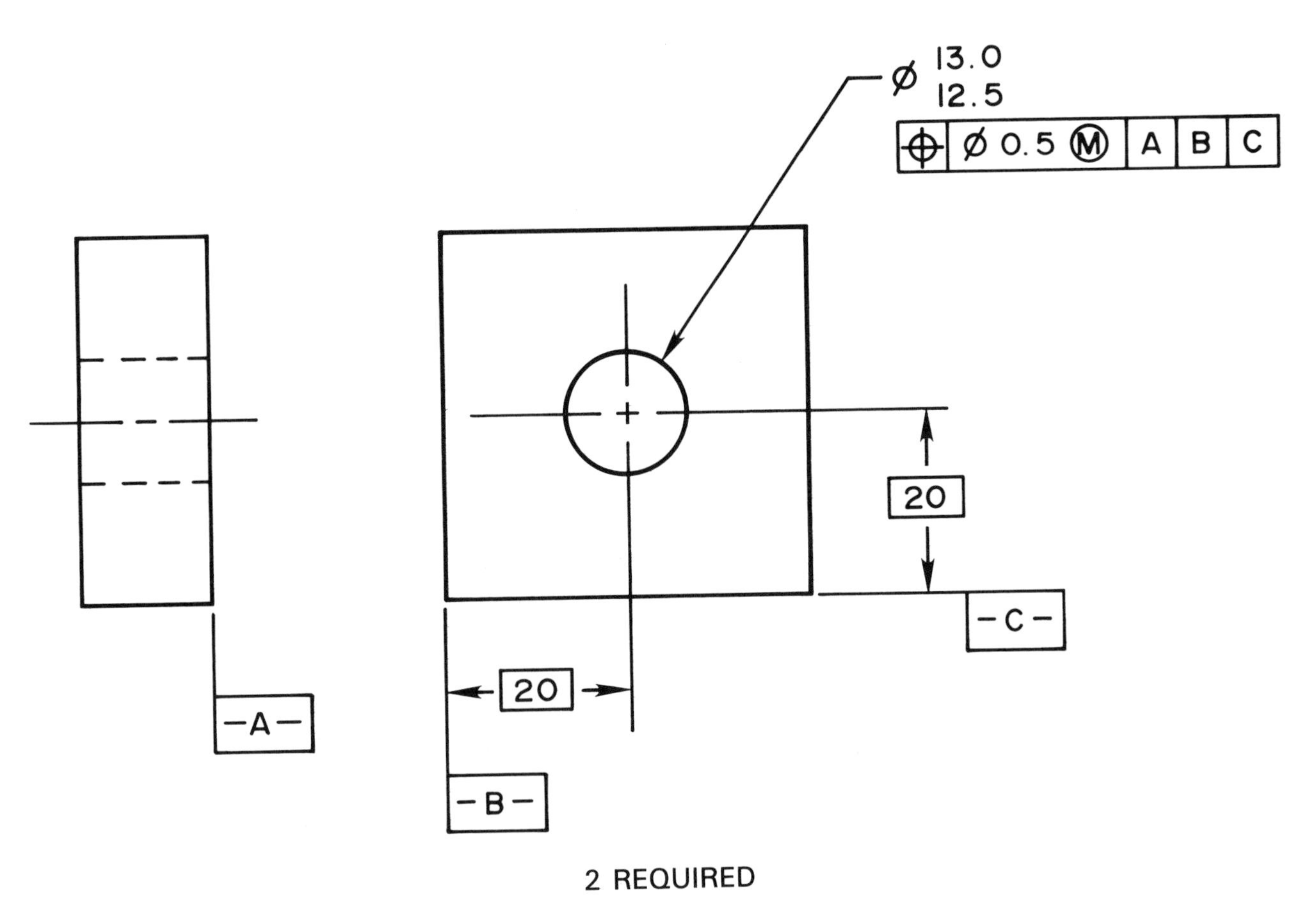

2 REQUIRED

Example 7-29. Calculating and showing the position tolerance for a floating fastener.

FIXED FASTENERS

The term FIXED FASTENER applies to a situation where one of the parts to be assembled has a held-in-place fastener such as a threaded hole for a bolt, screw, or stud. This applies to all holes of the same size in a pattern where the same position tolerance is specified. An example of a fixed fastener situation is shown in Example 7-30.

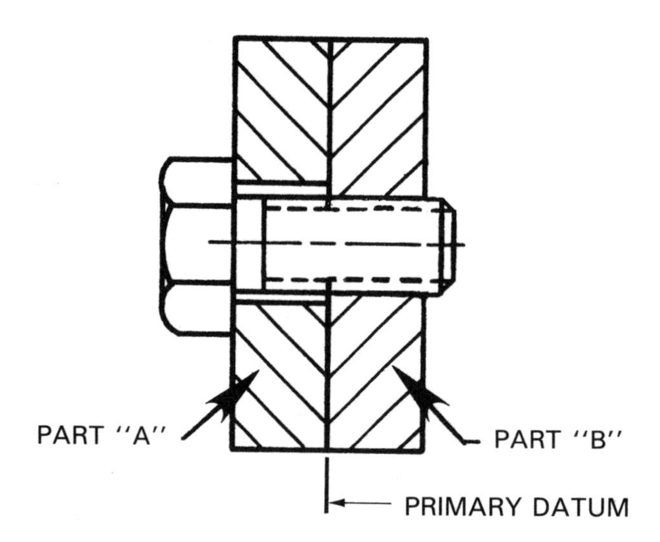

Example 7-30. Fixed fastener.

Notice in Example 7-30 that Part ''A'' has a through clearance hole and Part ''B'' is threaded. Part ''B'' acts as a nut. Therefore, a nut is not required as in the floating fastener situation. Notice that only Part ''A'' has clearance around the fastener. This means that half as much position tolerance is applied as compared to a floating fastener. The fixed fastener position tolerance may be calculated using the formula:

$$\frac{\text{MMC HOLE} - \text{MMC FASTENER (BOLT)}}{2} = \text{POSITION TOLERANCE FOR EACH PART}$$

Given a situation where an M14 x 2 bolt is used to fasten two parts together where Part ''A'' has a clearance hole diameter of 14.4/14.2, and Part ''B'' is threaded with M14 x 2 to accommodate the bolt, the position tolerance is calculated as follows:

$$\frac{\text{MMC HOLE (14.2)} - \text{MMC BOLT (14)}}{2} = \text{POSITION TOLERANCE (0.1)}$$

A drawing representing the position tolerance calculation for this fixed fastener is shown in Example 7-31.

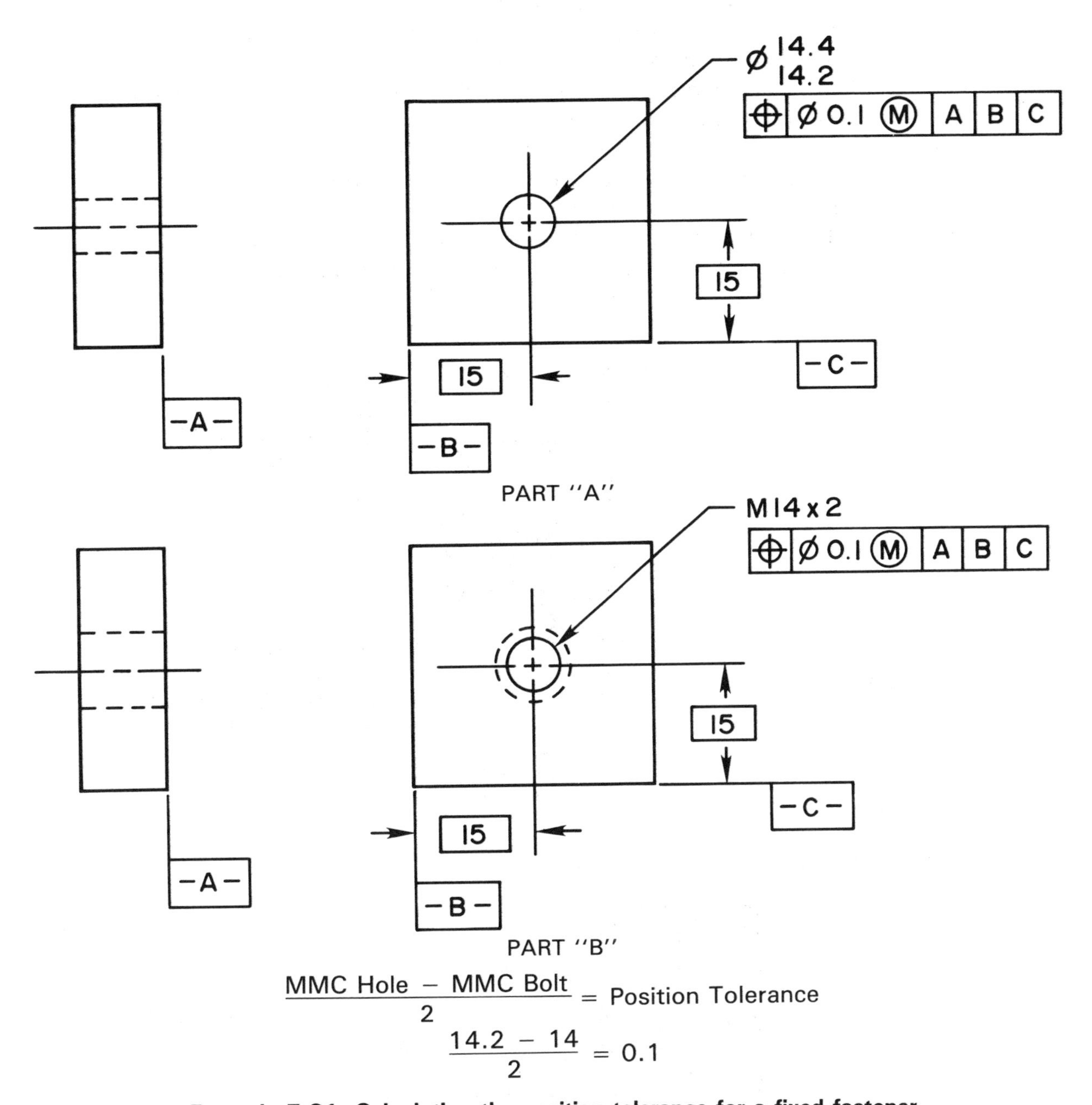

$$\frac{\text{MMC Hole} - \text{MMC Bolt}}{2} = \text{Position Tolerance}$$

$$\frac{14.2 - 14}{2} = 0.1$$

Example 7-31. Calculating the position tolerance for a fixed fastener.

Sometimes a designer may want to design the position tolerance between two or more parts in a fixed fastener situation with a greater amount of position tolerance applied to the threaded part than the unthreaded part. For example, 70% of the tolerance applied to the threaded part and 30% to the unthreaded part. We can calculate the revised position tolerance for Part "A" and Part "B" in Example 7-30 using the formula:

MMC HOLE (14.2) − MMC BOLT (14) = 0.2
0.2 x 30% (.30) = 0.06 POSITION TOLERANCE FOR PART A.
0.2 x 70% (.70) = 0.14 POSITION TOLERANCE FOR PART B.

PROJECTED TOLERANCE ZONE

In some situations where positional tolerance is used entirely in out-of-squareness it may be necessary to control perpendicularity and position above the part. The use of a PROJECTED TOLERANCE ZONE is recommended when variations in perpendicularity of threaded or press-fit holes could cause the fastener to interfere with the mating part. A projected tolerance zone is usually specified for fixed fastener situations such as the threaded hole for a bolt or screw, or the press fit hole of a pin application. The length of a projected tolerance zone may be specified as the distance the fastener extends into the mating part or the thickness of the part or the height of a press fit stud. A detailed example of the projected tolerance zone is shown in Example 7-32.

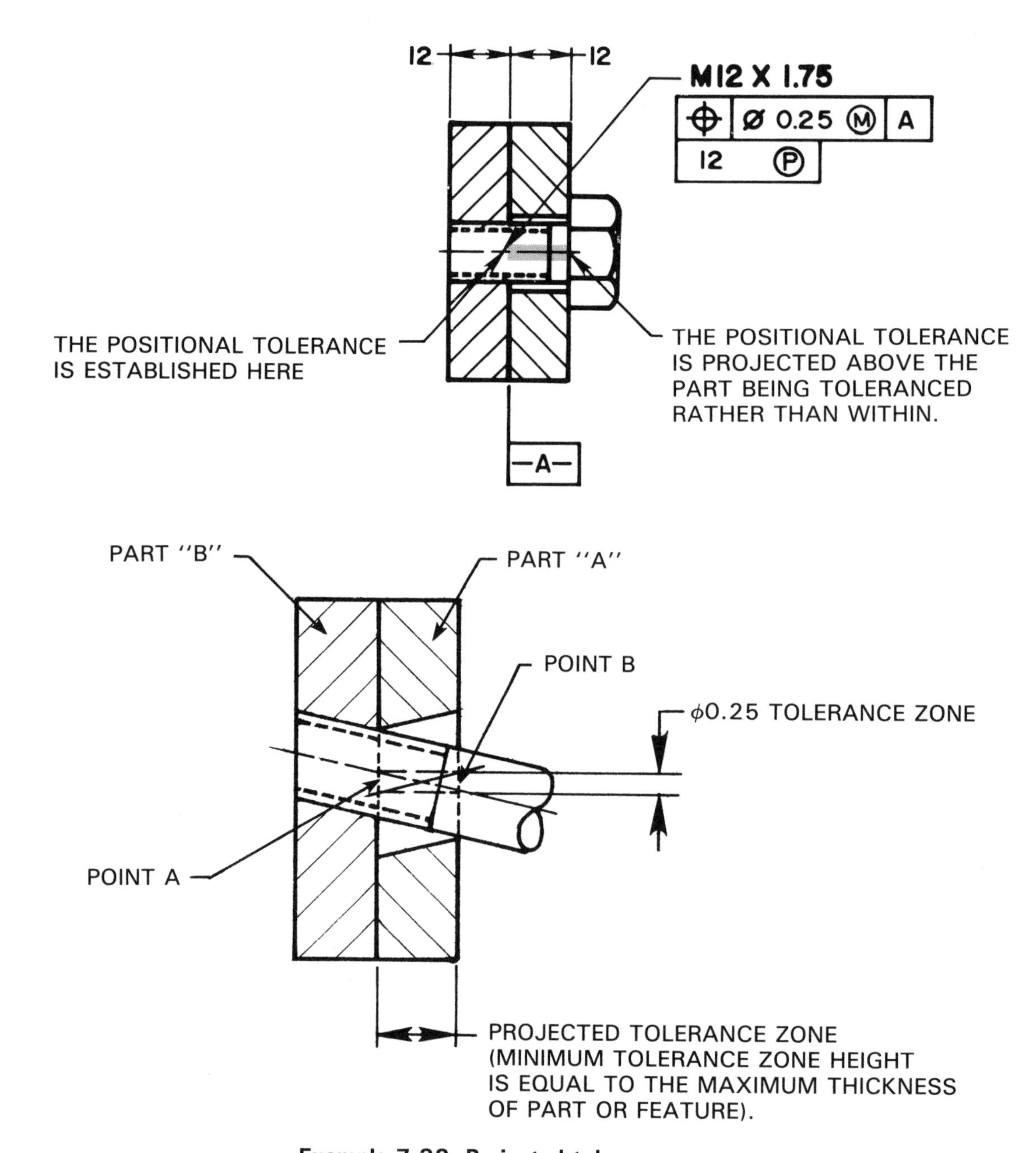

Example 7-32. Projected tolerance zone.

The projected tolerance zone representation on a drawing may be shown on a drawing a couple of different ways. The projected tolerance zone designation is displayed as a block below the feature control frame which is attached to the feature dimension in the view where the related datum controlling perpendicularity appears as a line. In this application, the projected tolerance zone extends away from the datum into the intended mating part. See Example 7-33.

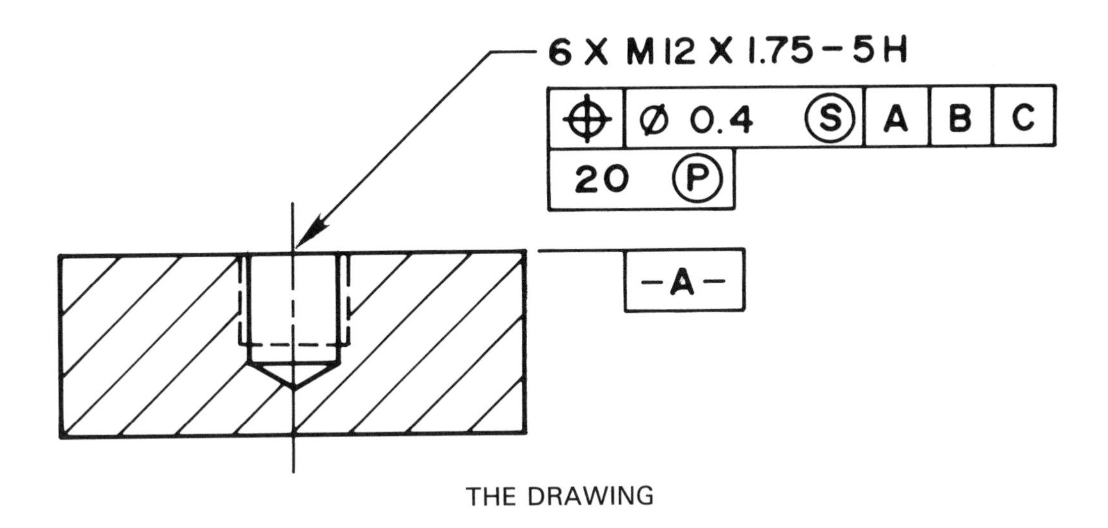

THE DRAWING

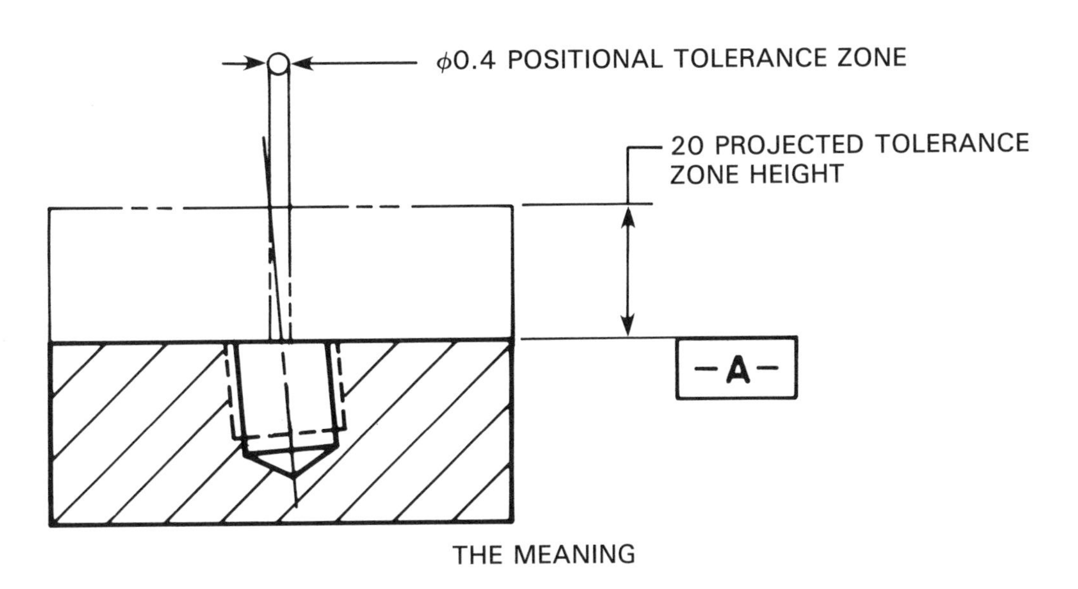

THE MEANING

Example 7-33. Projected tolerance zone representation.

To provide additional clarification the projected tolerance zone may be shown using a chain line in the view where the related datum appears as a line and the minimum height of the projection is dimensioned. Refer to Example 7-34. When this is done, the projected tolerance zone symbol is shown above in the blank below the feature control frame.

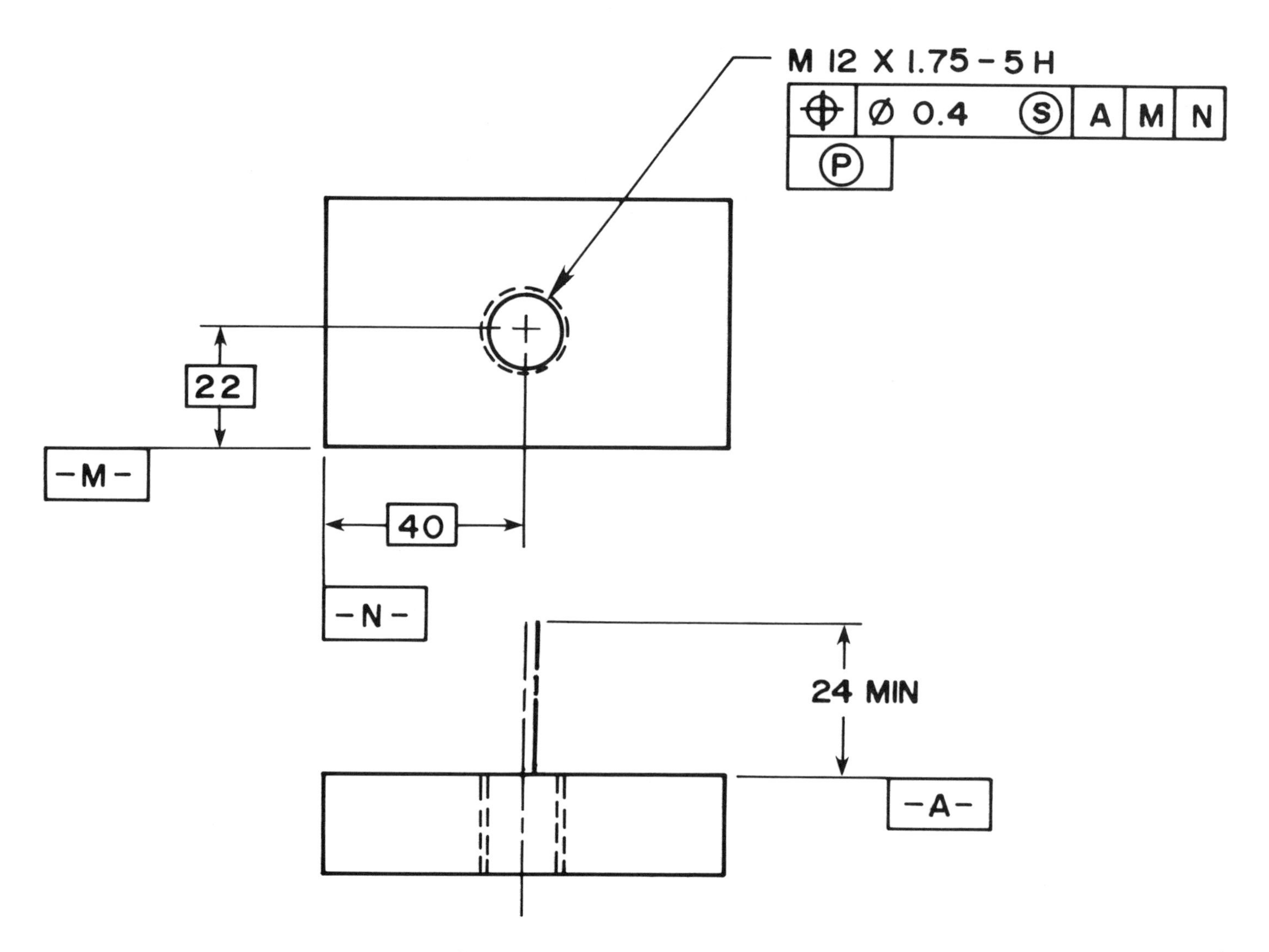

Example 7-34. Projected tolerance zone representation.

VIRTUAL CONDITION

The VIRTUAL CONDITION is a boundary that takes into consideration the combined effect of feature size at MMC and geometric tolerance. VIRTUAL CONDITION in essence establishes a working zone that is used to establish gage member sizes and the MMC size of mating parts or fasteners for mating parts. The virtual condition represents extreme conditions at MMC plus or minus the applicable geometric tolerance. This is used to determine clearance between mating parts. The virtual condition is calculated for situations involving internal or external features. When calculating the virtual condition of an internal feature, use the formula:

MMC SIZE OF THE FEATURE
− RELATED GEOMETRIC TOLERANCE
= VIRTUAL CONDITION

Given the part shown in Example 7-35, calculate the virtual condition.

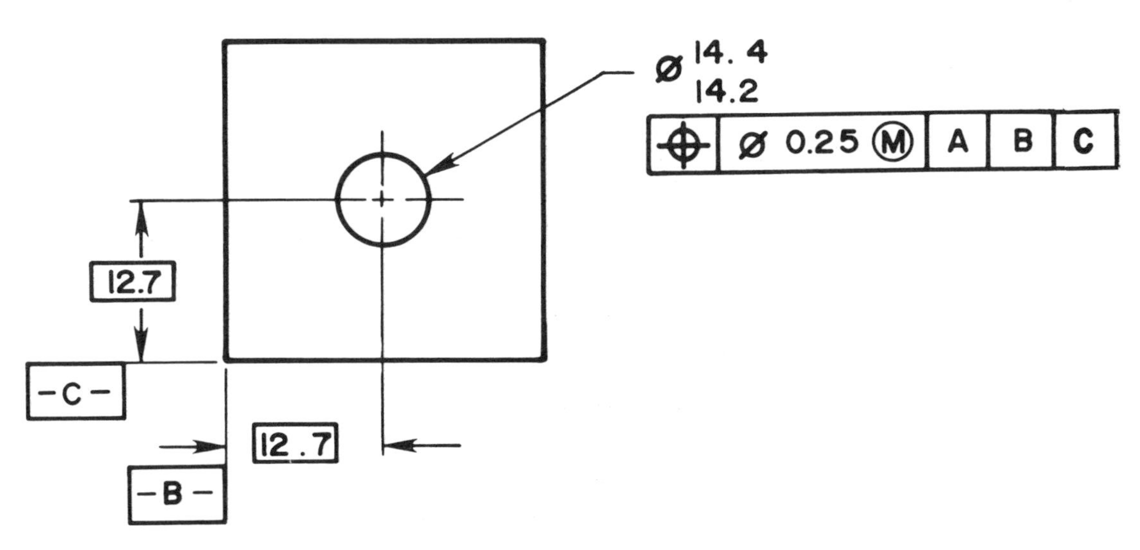

THE DRAWING

THE CALCULATION:

MMC HOLE = 14.2
− GEOMETRIC TOLERANCE = 0.25
VIRTUAL CONDITION = 13.95

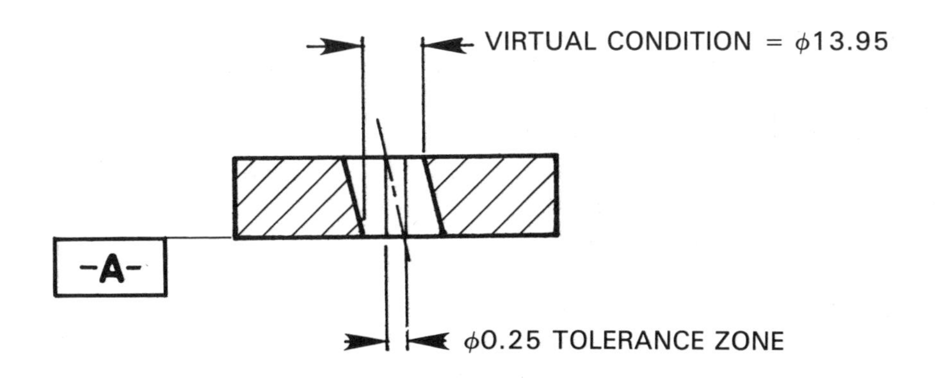

THE MEANING

Example 7-35. Virtual condition of an internal feature.

When calculating the virtual condition of an external feature, use the formula:

MMC SIZE OF THE FEATURE
+ RELATED GEOMETRIC TOLERANCE
= VIRTUAL CONDITION

Given the part shown in Example 7-36, calculate the virtual condition.

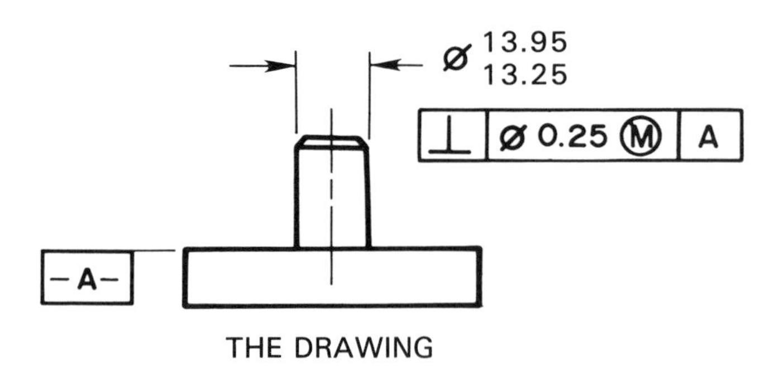

THE CALCULATION:

MMC PIN = 13.95
+ GEOMETRIC TOLERANCE = 0.25
VIRTUAL CONDITION = 14.20

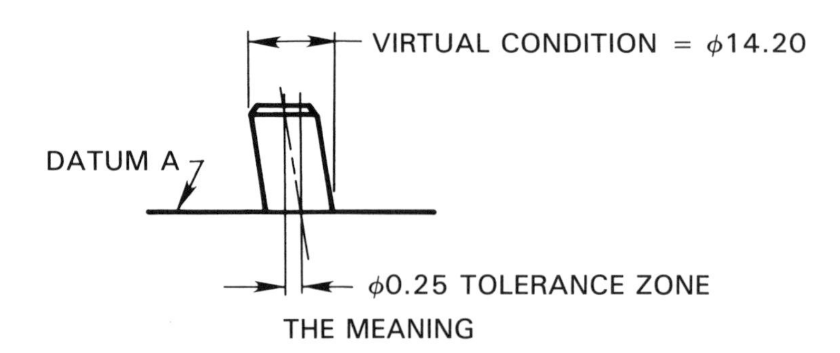

Example 7-36. Calculating the virtual condition of an external feature.

VIRTUAL CONDITION OF SIZE DATUMS

When the axis or center plane of a datum feature of size is controlled by a geometric tolerance the datum feature implies virtual condition even if a related datum reference is MMC. Refer to Example 7-36a and see the virtual condition of the feature at Datum D is ϕ24.70. The position tolerance of the 4 × ϕ8.0-8.5 holes in reference to Datum D at MMC implies a reference to Datum D at the virtual condition and not to Datum D at maximum material condition.

If you do not want the virtual condition to be applied to the MMC datum reference then you should consider controlling the datum feature with zero geometric tolerance at MMC. In the case of the drawing in Example 7-36a the position tolerance associated with Datum D could have been 0 at MMC in which case the maximum material condition (ϕ24.75) of the datum feature would have been equal to the virtual condition.

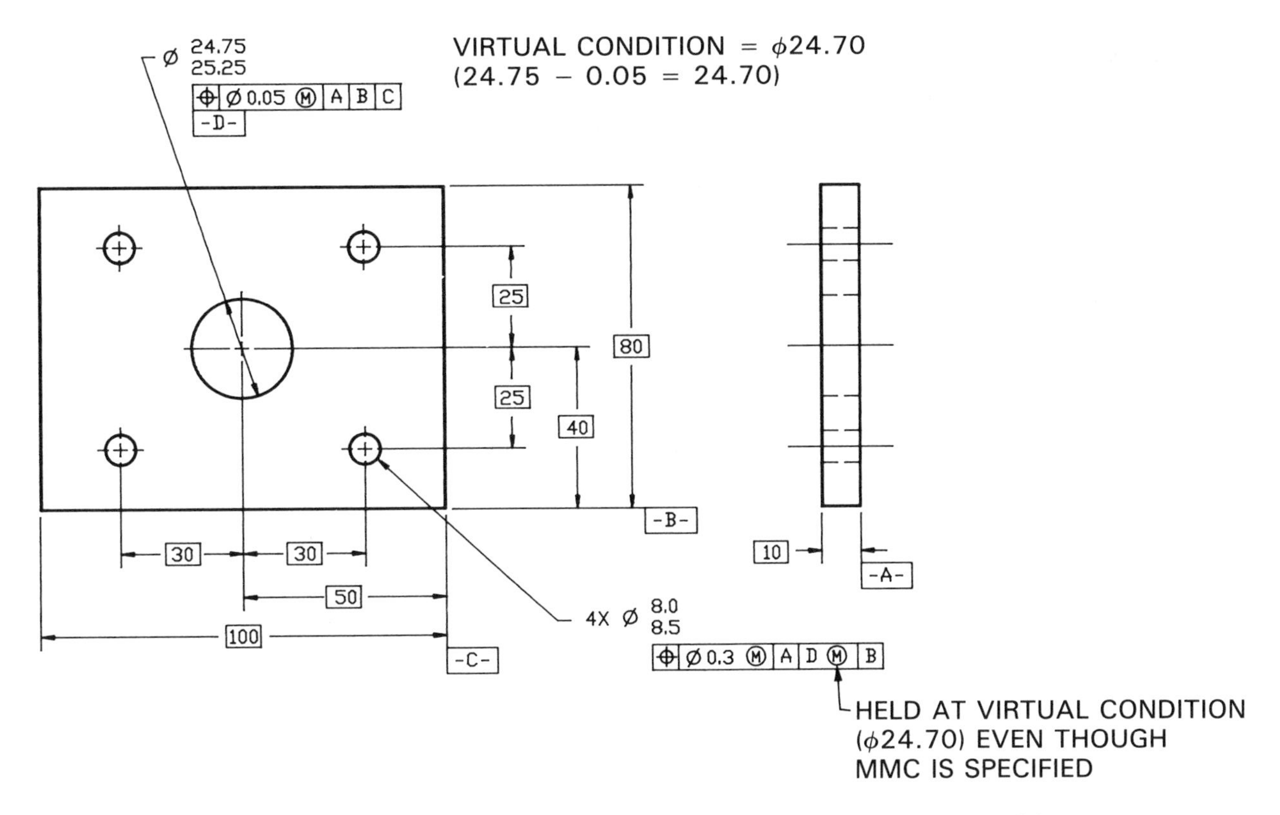

Example 7-36a. The application of datum features at virtual condition.

CONCENTRICITY TOLERANCE

CONCENTRICITY is used to establish a relationship between the axes of two or more cylindrical features of an object. The concentricity geometric characteristic symbol is shown detailed in a feature control frame in Example 7-37. Perfect concentricity exists when the axis of each cylindrical feature fall on the same line, as shown in Example 7-38.

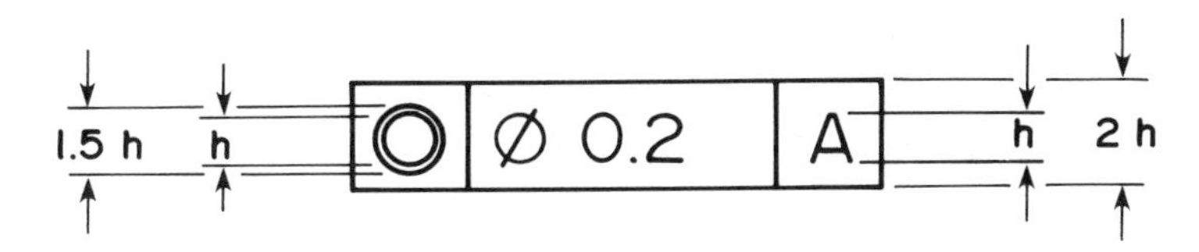

Example 7-37. Concentricity geometric characteristic symbol in a feature control frame.

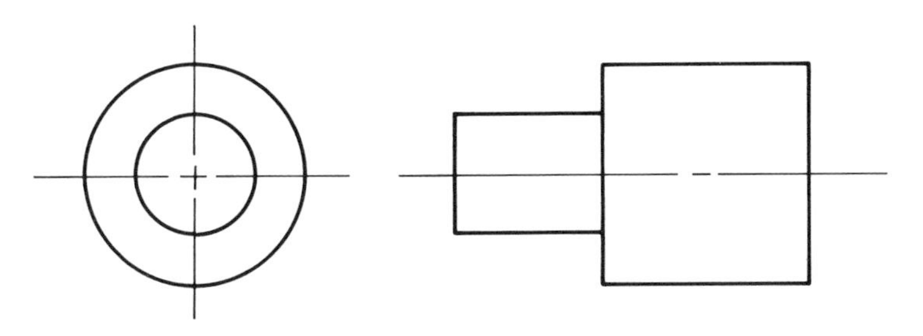

Example 7-38. Concentricity is when two or more cylindrical features have a common center axis.

CONCENTRICITY is the condition where the axis of all cross-sectional elements of a cylindrical surface are common the axis of a datum feature. The concentricity tolerance specifies a cylindrical (diameter) tolerance zone. The axis of this tolerance zone coincides with a datum axis and within which all cross-sectional axes of the control feature must lie, as shown in Example 7-39.

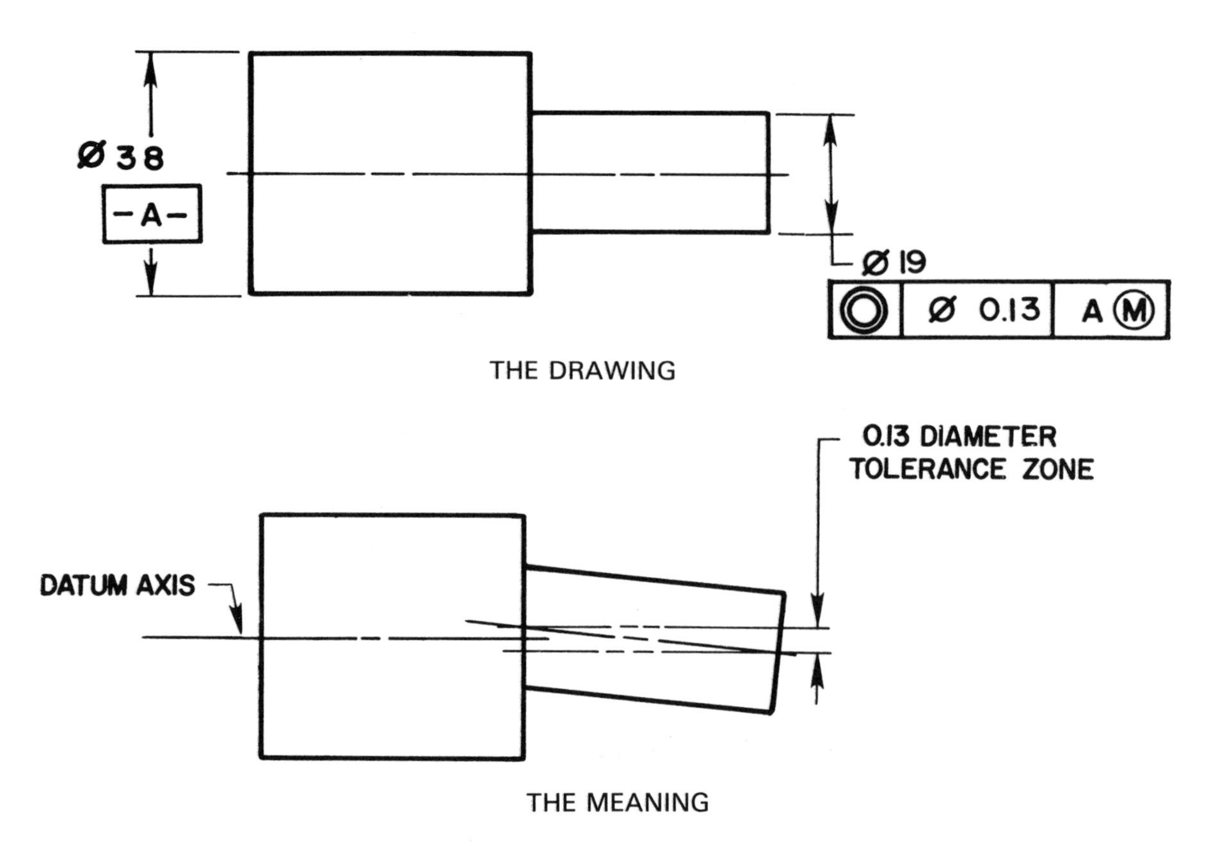

Example 7-39. Concentricity tolerance.

The specified concentricity tolerance and related datum reference apply only on an RFS basis. A concentricity tolerance requires the establishment and verification of axes unrelated to surface conditions. Irregularities in the form of the feature to be inspected may make it difficult to establish the axis of the concentric feature. Unless there is a need to control the axis, as in a dynamically balanced shaft, it is recommended that runout or positional tolerance be used.

POSITION TOLERANCING FOR COAXIALITY

Position tolerancing is recommended over concentricity tolerancing when the control of the axes of cylindrical features can be applied on a material condition basis. A coaxial relationship may be controlled by specifying a positional tolerance at MMC with the datum feature reference specified on an MMC or RFS basis depending on the design requirements. See Example 7-40.

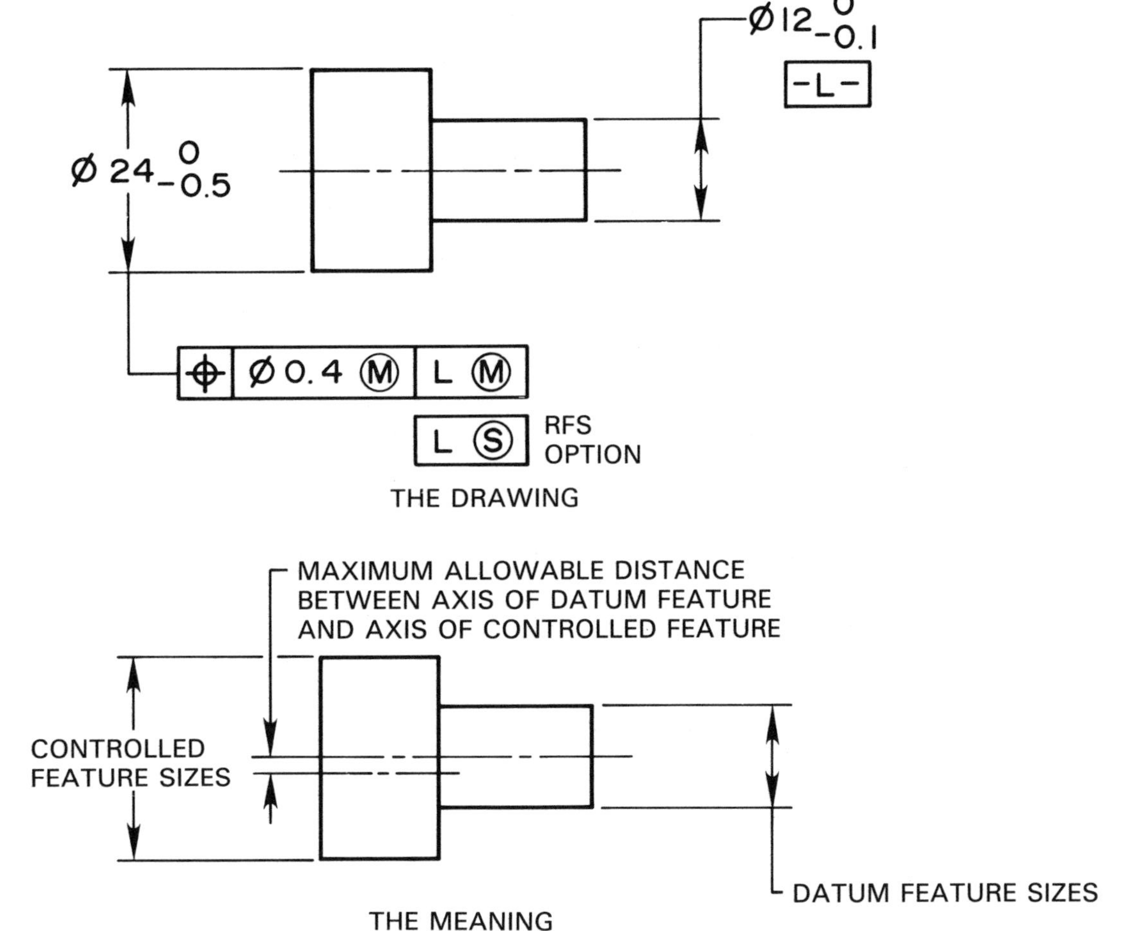

Example 7-40. Positional tolerancing for coaxiality.

When the datum feature is specified on an MMC basis, any departure of the datum feature from MMC may result in an additional displacement of the datum axis and the controlled feature axis. The maximum allowable distance between the axis of the datum feature and the axis of the controlled feature may be calculated at various produced sizes using the following formula:

$$a = \frac{\text{MMC controlled feature} - \text{Actual feature size} + \text{Geometric Tolerance.}}{2}$$

$$b = \frac{\text{MMC datum feature} - \text{Actual datum feature size.}}{2}$$

$a + b = c$ Where c = Maximum allowable distance between axis of datum feature and axis of controlled feature.

TEST 7, LOCATION TOLERANCES Name:____________________

1. Describe the purpose of location tolerances. ____________________
____
____
2. ____________________________ is used to define a zone within which the center, axis, or center plane of a feature of size is permitted to vary from true position.
3. Define true position. ____________________________
____
4. __________ dimensions are used to establish true position from specified datum features and between interrelated features.
5. Location tolerancing is specified by a ____________ or ____________ symbol, a ____________ and appropriate __________ references placed in a feature control frame.
6. When position tolerancing is used, the MMC, RFS, or LMC material condition symbols must be specified after the tolerance and applicable datum reference in the feature control frame. TRUE or FALSE?
7. Redraw the object below making a complete conversion from the conventional coordinate dimensioned drawing shown to a position toleranced drawing as follows:
   a. Calculate the position required to convert this drawing to a position tolerance drawing. (Show your calculations.)
   b. Use full scale.
   c. Use proper size symbols.

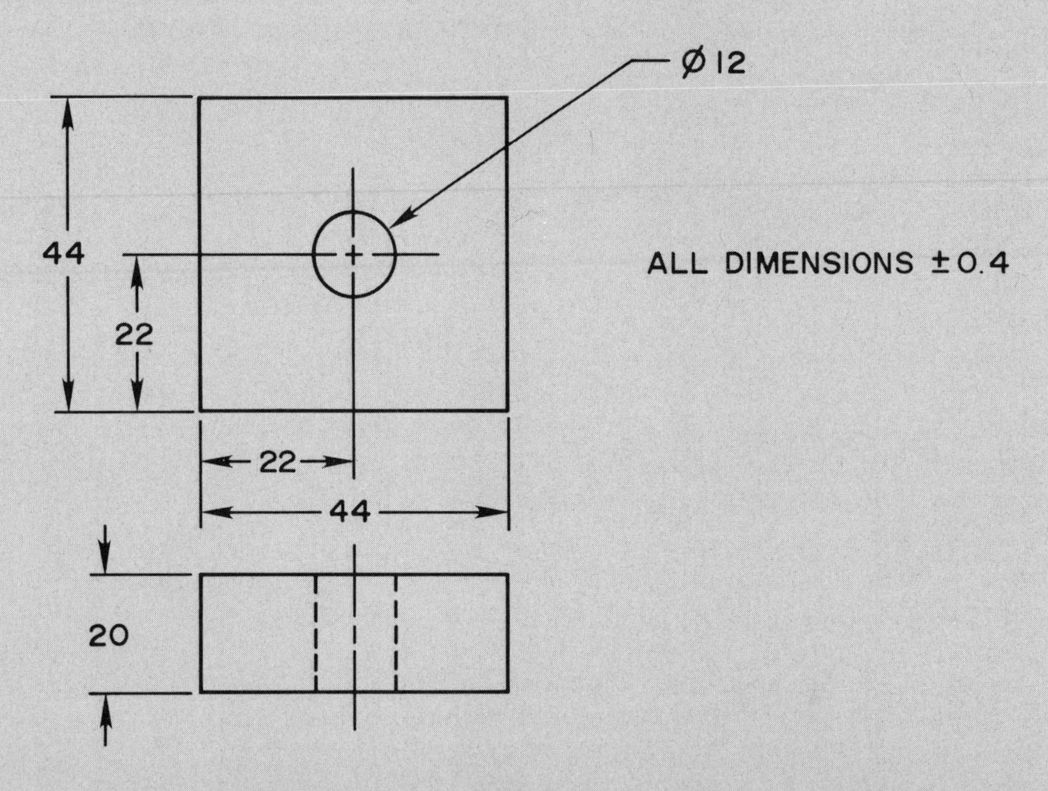

Space for drawing in problem 7.

Show your calculations here:

____

____

8. When locating features using position tolerancing, the basic dimensions may be drawn by placing the basic dimension symbol around each basic dimension, or specifying on the drawing or in a reference document the general note:

____

9. Given the following drawing, a reference chart showing a range of possible produced sizes, and four optional feature control frames that may be applied to the diameter dimension. Provide the position tolerance at each possible produced size for each feature control frame application.

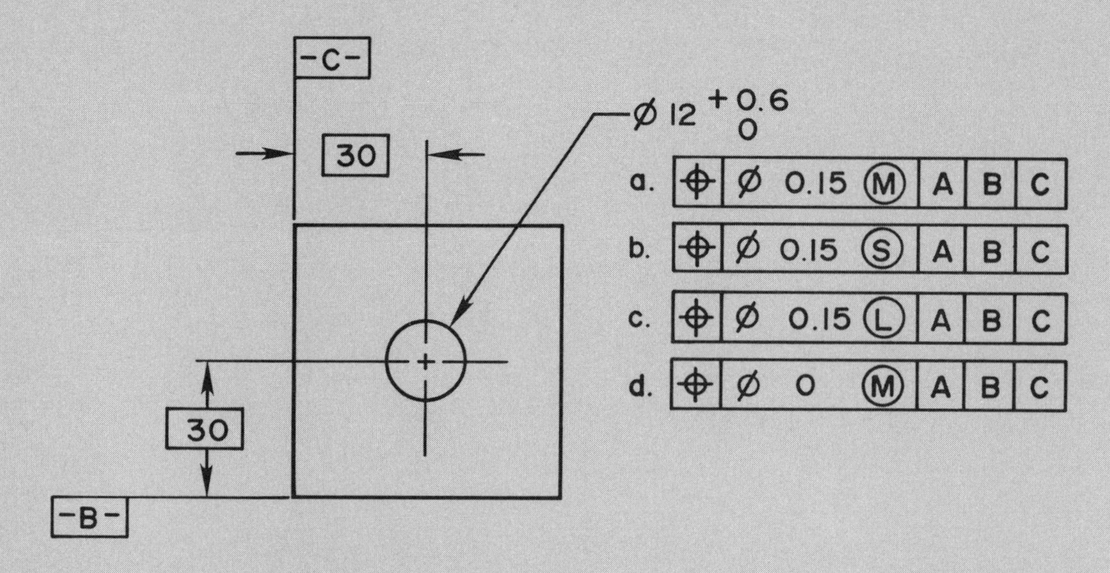

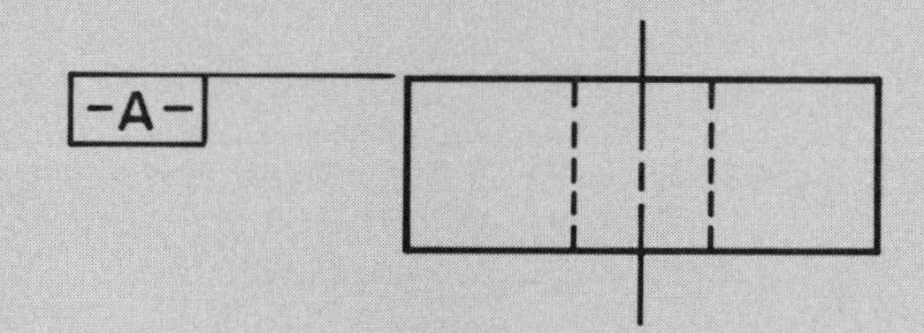

DIAMETER TOLERANCE ZONES ALLOWED

Possible Produced Sizes	a. MMC	b. RFS	c. LMC	d. 0 at MMC
12.0	______	______	______	______
12.2	______	______	______	______
12.4	______	______	______	______
12.6	______	______	______	______

10. Name the two types of dimensioning systems that are normally used when locating multiple features. ______________________________

____

11. The position tolerancing of slotted holes is accomplished by locating basic dimensions to the centers from datums and adding the feature control frame to the length and width dimensions. TRUE or FALSE?

12. Describe when composite positional tolerancing is used. ______________

____

____

____

13. When using composite positional tolerancing, the upper part of the feature control frame is referred to as the ______________________________ ______________ and specifies the larger tolerance for the pattern of features as a group, while the lower half of the frame is called the ______________________________ which specifies a smaller positional tolerance for individual features within the pattern.
14. Position tolerances may be used in situations where the axis of the holes are not parallel to each other and where they may be at an angle to the surface. TRUE or FALSE?
15. What is the shape of the positional tolerance zone when using position tolerancing to control the location of a spherical feature. ______________
16. Symmetry is a __________________ relationship of the features of an object.
17. Positional tolerancing of symmetrically shaped slots or tabs may be accomplished by:
    a. Identification of related datums.
    b. Dimensioning the relationship between features.
    c. Providing the number of units followed by the size.
    d. Placing the position tolerance feature control frame below the size dimension.
    e. All of the above.
18. Give the formula used to determine the position tolerance of a floating fastener. ____

    ____
19. Give the formula used to determine the position tolerance of a fixed fastener. ____

    ____

    ____
20. Under what conditions is a projected tolerance zone recommended?

    ____

    ____

    ____
21. Identify the two ways that a projected tolerance zone may be shown on a drawing. ____

    ____

    ____

    ____

    ____

    ____

    ____

22. Define virtual condition. ______________________________________

____

____

23. Give the formula used to calculate the virtual condition of an internal feature.____

____

24. Give the formula used to calculate the virtual condition of an external feature.____

____

25. ______________________ may be used to establish a relationship between the axes of two or more cylindrical features of an object.

26. A concentricity tolerance requires the establishment and varification of axes unrelated to surface conditions; therefore, unless there is a need to control the axis it is recommended that runout or position tolerance be used. TRUE or FALSE?

27. A coaxial relationship may be controlled by specifying a positional tolerance at MMC with the datum feature reference specified on an MMC or RFS basis depending on design requirements. TRUE or FALSE?

PRINT READING EXERCISES FOR CHAPTER 7

Name:________________________

The following print reading exercise is designed for use in programs for machining, welding, tool and die, dimensional inspection, and other manufacturing curriculums where the objective is the reading and interpretation of prints rather than the development of drafting skills. An actual industrial print is used with related questions that require you to read and interpret specific dimensioning and geometric tolerancing representations. The answers and interpretations should be based on the previously learned content of this book. The prints used are based on ANSI standards; however, company standards may differ slightly. When reading these prints or any other industrial prints, a degree of flexibility may be required to determine how individual applications correlate with the ANSI standard.

PRINT READING EXERCISE

Refer to the Hyster Company print of the FLYWHEEL-DSL found on page 218.

1. What type of coordinate dimensioning system is used to locate the M12 x 1.258 HOLES? ________________________

Refer to the Hyster Company print of the CASE-DIFF found on page 217.

2. Interpret the elements of the feature control frame associated with the ϕ157.2-156.7 dimension. ________________________

   ____

3. Interpret the elements of the feature control frame associated with the ϕ57.239-57.201 dimension. ________________________

   ____

   ____

   ____

4. Refer to the feature control frame associated with the ϕ57.239-57.201 dimension:

   a. Given the following list of possible produced sizes determine the position tolerance at each produced size:

Produced sizes	Position tolerance
57.239	________
57.231	________
57.223	________
57.215	________
57.207	________
57.201	________

5. Refer to the feature control frame associated with the ϕ11.90-11.40 dimension:

   a. Does the feature control frame also control the spotface? YES or NO? Explain: ______________________________

   ____

   ____

   b. Given the following list of possible produced sizes determine the position tolerance at each produced size:

Produced sizes	Position tolerance
11.90	________
11.80	________
11.70	________
11.60	________
11.50	________
11.40	________

Refer to the Curtis Associates print of the MOUNTING BRACKET (WORM GEAR DRIVE) found on page 220.

6. Interpret the feature control frame associated with the ϕ1.2593-1.2587 dimension. ______________________________

   ____

   ____

   ____

   ____

   ____

   ____

7. What is the primary purpose of using the method of position tolerancing associated with the ϕ1.2593-1.2587 dimension? ________

   ____

   ____

   ____

   ____

   ____

   ____

8. Interpret the feature control frame associated with the 4 x ϕ.334-.326 dimension. ______________________________

   ____

9. Given the following list of possible produced sizes associated with the ϕ.334-.326 dimension determine the position tolerance at each produced size:

Produced sizes	Position tolerance
.326	________
.328	________
.330	________
.332	________
.334	________

Refer to DIAL INDUSTRIES print of the BODY, CONNECTOR SAMPLE & HOLD FIXTURE found on page 221.

10. Interpret the feature control frame associated with the 1.000 ± .025 dimension. ____
____
____
____
____

11. Why is the diameter symbol omitted in front of the position tolerance in the feature control frame identified in question number 10?
____
____
____
____

12. Given the following list of possible produced sizes associated with the 1.750 ± .025 dimension determine the position tolerance at each produced size:

Produced sizes	Position tolerance
1.775	________
1.750	________
1.725	________

GEOMETRIC DIMENSIONING AND TOLERANCING FINAL EXAM

FINAL EXAM — Part I   Name: ____________________

Below left is a list of short descriptions with a list of words, or symbols at the right. Place the letter of the word or symbol that matches the description in the blank provided. Each letter may be used more than once. Some selections may not be used.

____ 1. The tolerance zone of a geometric characteristic that must be followed by a material condition symbol MMC, RFS, or LMC.

____ 2. The general term applied to a physical portion of a part.

____ 3. Condition in which a feature of size contains the maximum amount of material within the limits.

____ 4. Datum feature symbol.

____ 5. Considered a theoretically perfect dimension.

____ 6. The actual surface of an object that is used to establish a datum plane.

____ 7. All datum planes on a part intersecting at right angles are 90° basic by interpretation.

____ 8. A geometric tolerance or datum reference applies at any increment of size of the feature within its size tolerance.

____ 9. This geometric tolerance specifies a zone within which the required surface element or axis must lie.

____ 10. A geometric characteristic with a tolerance zone between two parallel planes and perpendicular to a datum.

____ 11. A geometric characteristic used to identify a location tolerance.

____ 12. A geometric characteristic with a tolerance zone between two parallel planes and parallel to a datum.

____ 13. A single element form control that establishes a tolerance zone between two concentric circles.

____ 14. A profile tolerance that is split equally on each side of the true profile.

____ 15. A geometric characteristic that establishes two perfectly concentric cylinders within which the actual surface must lie.

(A) ⊥

(B) Datum Target

(C) Ⓢ

(D) //

(E) ○

(F) Bilateral

(G) —

(H) LMC

(I) ⌖

(J) MMC

(K) ⌭

(L) basic

(M) ⌓

(N) Unilateral

(O) ↗

(P) Datum feature

(Q) [-A-]

(R) Three Plane Concept

(S) Feature

FINAL EXAM — Part II

1. The symbol below is called ______________________ and it would be drawn the size shown below: Fill in the dimensions as related to lettering height = h.

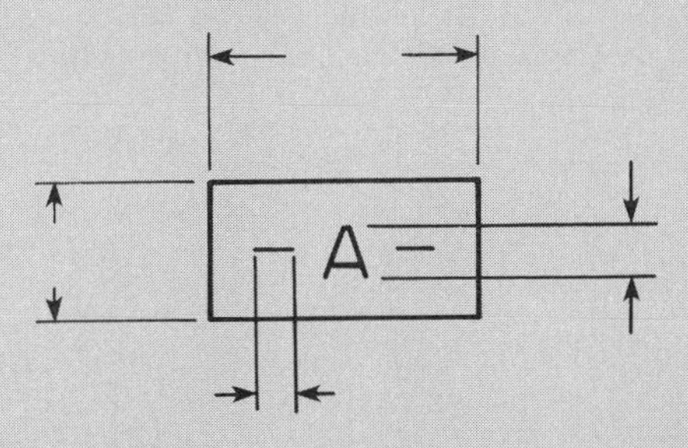

2. The symbol below is called ______________________ and it would be drawn the size below. Fill in the dimensions as related to lettering height = h.

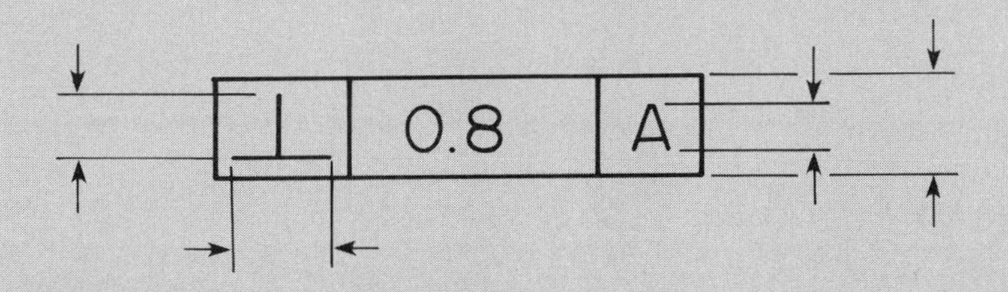

3. Identify the items in the symbol below and write your answers in the blanks on right.

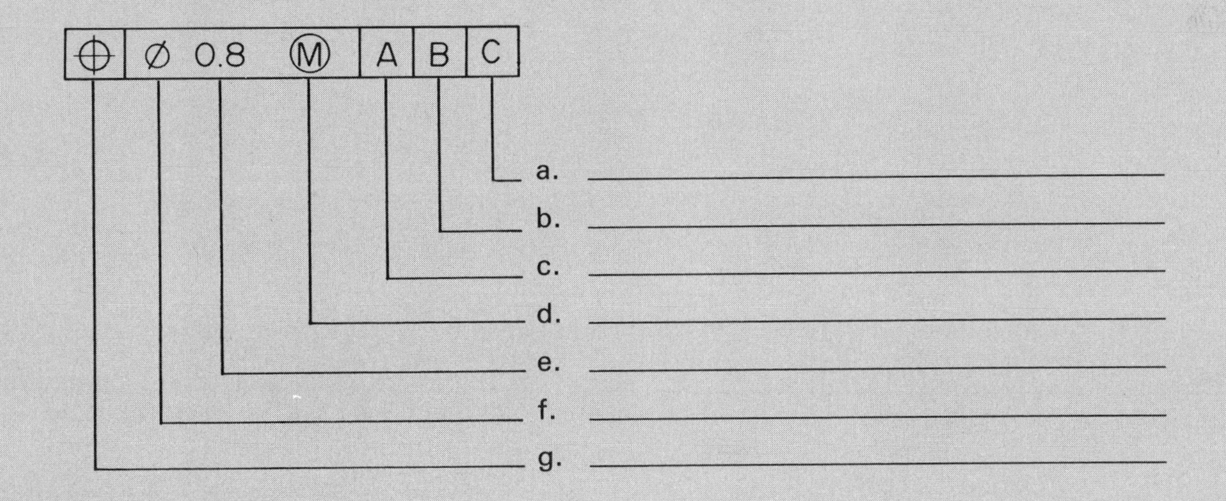

4. List the symbols for the geometric characteristics and indicate the symbol name.

5. What does this symbol represent?

A2 ______________________________

6. List three items that may be identified as datums.

____

____

____

7. What is the symbol for each of the items listed below?

Maximum Material Condition ____________________

Regardless of Feature Size ____________________

Projected Tolerance Zone ____________________

Least Material Condition ____________________

FINAL EXAM — Part III

1. Given the object shown below, answer the following questions:

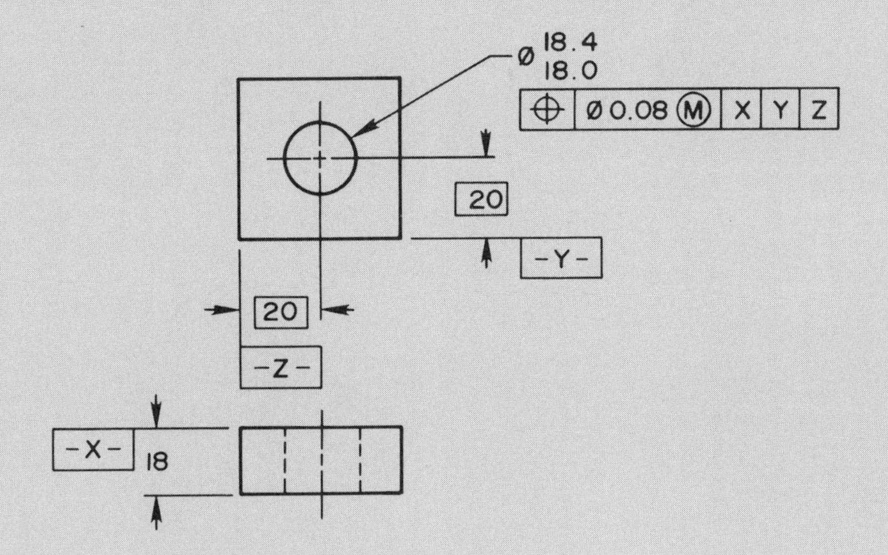

a. What is the MMC size of the hole? ____________

b. What is the LMC size of the hole? ____________

c. Provide the positional tolerance at the following produced sizes:

Produced sizes	Positional Tolerances
ϕ18	________
ϕ18.1	________
ϕ18.2	________
ϕ18.3	________
ϕ18.4	________

d. Show in the chart below how the positional tolerance would change at possible produced sizes if the following feature control frames were substituted:

a. ⌖ | Ø 0.08 Ⓢ |

b. ⌖ | Ø 0.08 Ⓛ |

c. ⌖ | Ø 0 Ⓜ |

Produced sizes	Positional Tolerances		
	a. RFS	b. LMC	c. O at MMC
ϕ18	________	________	________
ϕ18.1	________	________	________
ϕ18.2	________	________	________
ϕ18.3	________	________	________
ϕ18.4	________	________	________

2. Given the object shown below answer the following questions:

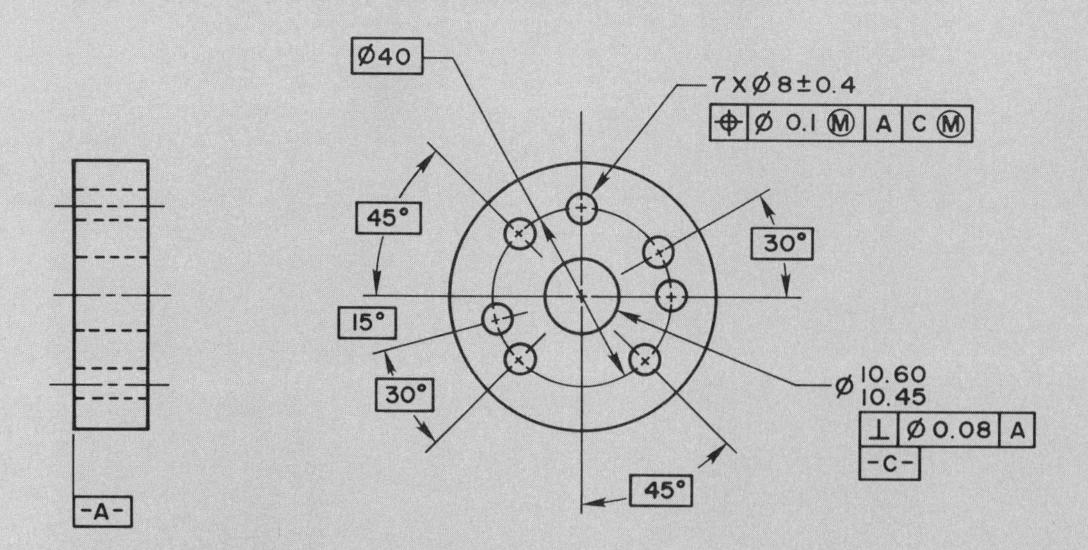

a. Are the holes located using rectangular or polar coordinate dimensioning?

____

b. What is the MMC of the small holes? __________

c. What is the virtual condition of the small holes? __________

d. Recalculate the positional tolerance for a floating fastener situation if the bolt used was specified as M7 × 1. __________

e. Recalculate the positional tolerance for a fixed fastener situation (equal distribution) if the bolt used was specified as M7 × 1.

____

f. What is the virtual condition of the ϕ10.60/10.45 hole? __________

g. Describe the datum -C-. ______________________

3. Provide a short, complete interpretation of the feature control frame associated with the following drawing.

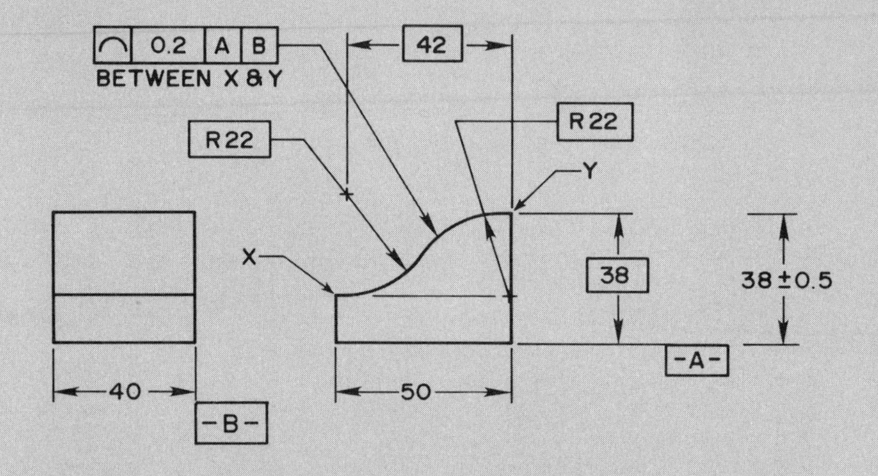

____

____

4. Provide a short, complete interpretation of the feature control frame associated with the following drawing.

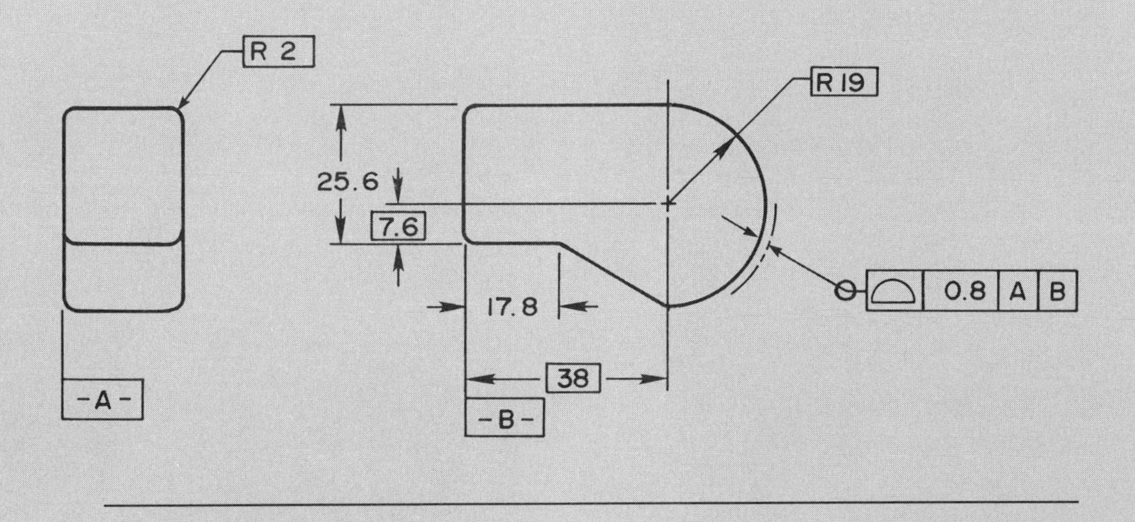

5. Provide a short, complete interpretation of feature control frame associated with the following drawing.

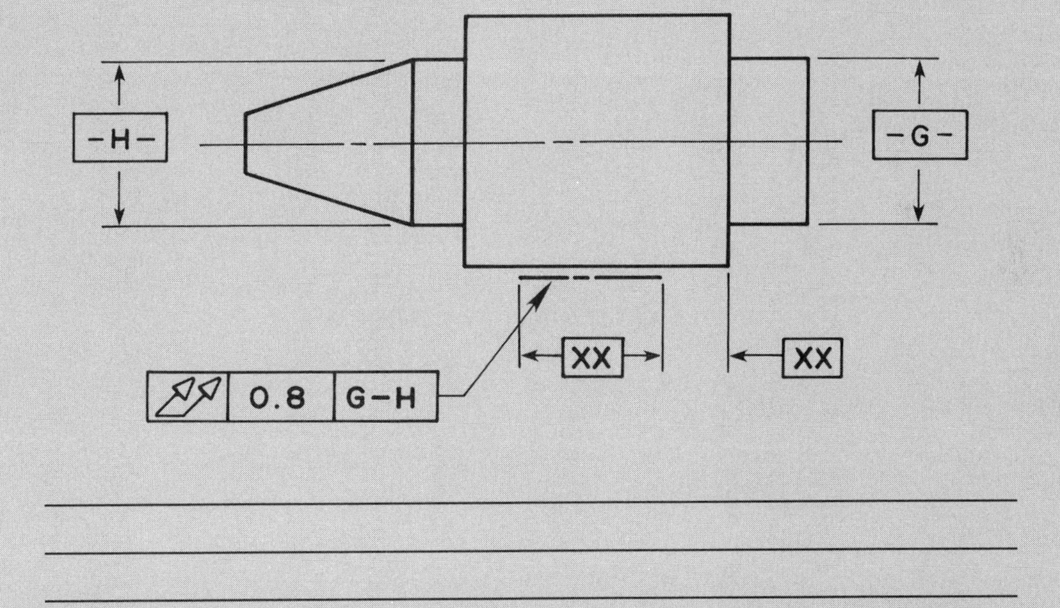

6. Provide a short, complete interpretation of the feature control frames associated with the following drawing.

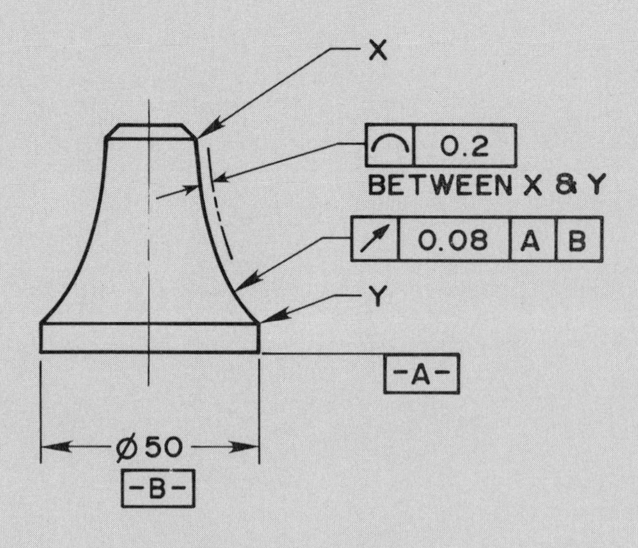

7. Provide a short complete interpretation of the feature control frame associated with the following drawing.

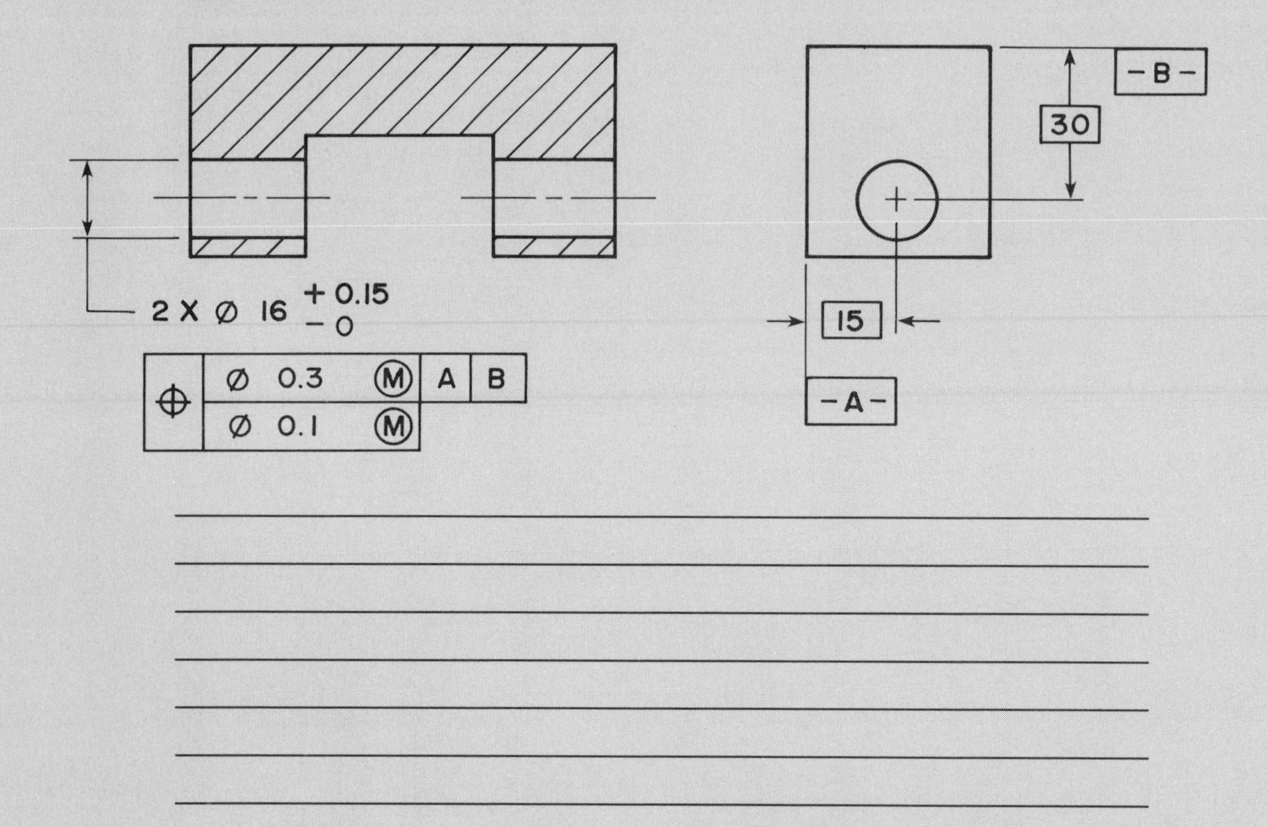

8. Provide a short, complete interpretation of the feature control frame associated with the following drawing.

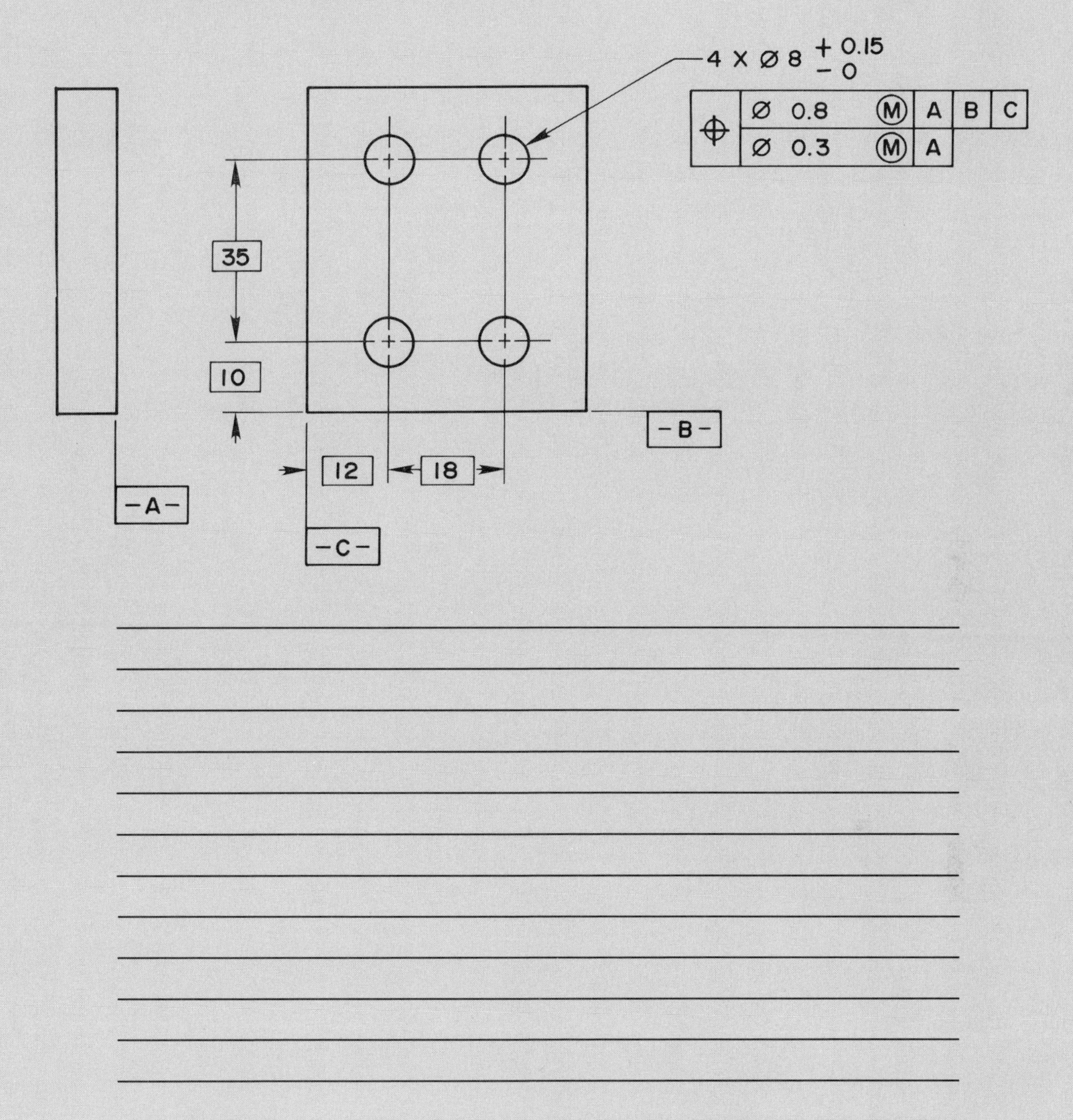

9. Provide a short, complete interpretation of feature control frame associated with the following drawing.

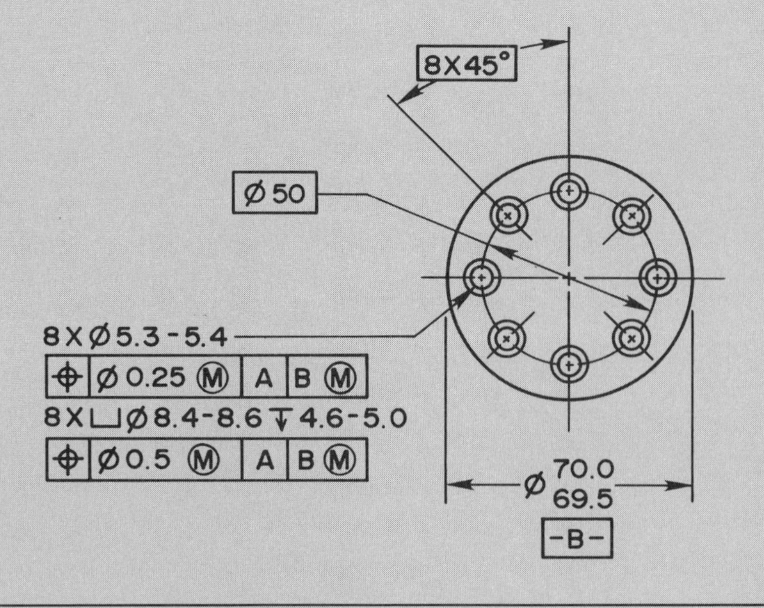

10. Provide a short, complete interpretation of feature control frame associated with the following drawing.

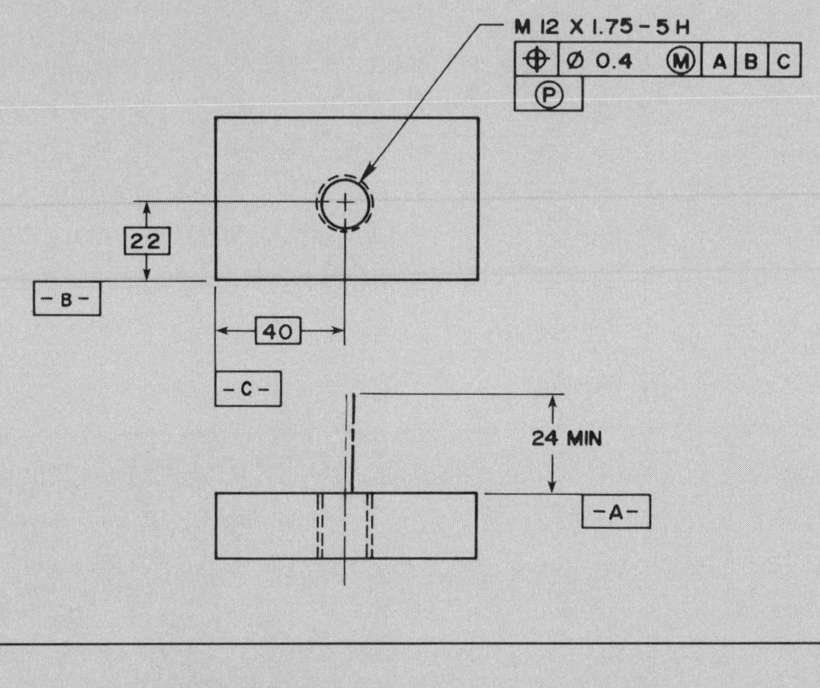

FINAL EXAM — Part IV

Circle T if the statement is true. Circle F if the statement is false.

1. T F Unit straightness may be used if the part must be controlled per unit of measure as well as over the total length.
2. T F Specific area flatness should be avoided on very large parts.
3. T F The zone descriptor of a circularity tolerance is diameter.
4. T F Cylindricity is identified by a radius tolerance zone that establishes two perfect concentric cylinders.
5. T F The profile of a line must be established between two given points on an object.
6. T F A parallelism tolerance zone must be between the size tolerance of the feature.
7. T F The perpendicularity of a shaft such as a stud or pin to a datum feature establishes a cylindrical diameter tolerance zone.
8. T F The terms EACH ELEMENT must be applied to a perpendicularity feature control frame.
9. T F An angularity tolerance must have a basic angular relationship to a datum.
10. T F The tolerance zone descriptor of a concentricity tolerance is R.
11. T F Concentricity is used to establish the relationship between the axis of two or more cylindrical features of an object.
12. T F The profile tolerance zone may be bilateral or unilateral.
13. T F Surface straightness may violate perfect form at MMC.
14. T F Geometric tolerances and related references imply RFS unless otherwise specified (except position).
15. T F A concentricity tolerance should be used if there is a need to control the axis as in a dynamically balanced shaft, otherwise it is recommended that a runout or positional tolerance be used.
16. T F A coaxial relationship may be controlled by a positional tolerance at MMC with the datum reference at MMC or RFS.
17. T F LMC is often used to control minimum edge distance.
18. T F True position is the theoretically exact location of a feature.
19. T F The datum reference frame exists in theory; the theoretical reference frame is simulated by positioning the part on datum features to adequately relate the part to the datum reference frame and to restrict motion of the part relative to the reference frame.
20. T F When reference is made to the datum reference frame the primary datum should be given first followed by the secondary and tertiary, this is referred to as datum precedence.

FINAL EXAM — Part V

Given the following specifications, calculate the required unknown values.

1. Given: A shaft with a diameter of 36 ± 0.2

   Calculate:

   a. Tolerance ________

   b. MMC ________

   c. LMC ________

2. Given: A hole with a diameter of $36.5 \begin{smallmatrix} +0.3 \\ -0.1 \end{smallmatrix}$

   Calculate:

   a. Tolerance ________

   b. MMC ________

   c. LMC ________

   d. Allowance with shaft in Question 1. Show formula and calculations

   ____

   ____

   ____

   e. Clearance with shaft in Question 1. Show formula and calculations.

   ____

   ____

   ____

3. Given: Two parts to be bolted together with a 5 ± 0.4 hole through each part and an M4.5 x 0.75 Hex socket head cap screw and hex nut for fastening.

   Calculate: Show formula and calculations.

   a. Positional tolerance. ____________________

   ____

   ____

   b. Virtual condition. ____________________

   ____

   ____

4. Given: Two parts to be bolted together. One part has a hole with a diameter of 16.5-16.1. The other part has a threaded hole, M16 x 2, located to align with the hole through the first part. An M16 x 2 Standard Metric Heavy Hex Screw is used to fasten the two parts together. Calculate the positional tolerance for each part based on the following:
   a. Equally distributed positional tolerance.
      Show formula and calculations.

      ____

      ____

      ____

   b. Provide 60% of the positional tolerance to the threaded part. Determine position tolerance for each part. Show formulas and calculations.

      ____

      ____

      ____

5. Given: A pin with a diameter of $8 \, {}^{\;0}_{-0.5}$ is held perpendicular to datum surface -A- by ϕ0.2.

   Calculate:

   a. MMC Pin. ____________

   b. LMC Pin. ____________

   c. Virtual Condition. Show formula and calculations.

      ____

      ____

      ____

APPENDICES

SPECIAL SYMBOLS

All symbols are drawn recommended size based on a .125 inch lettering height.

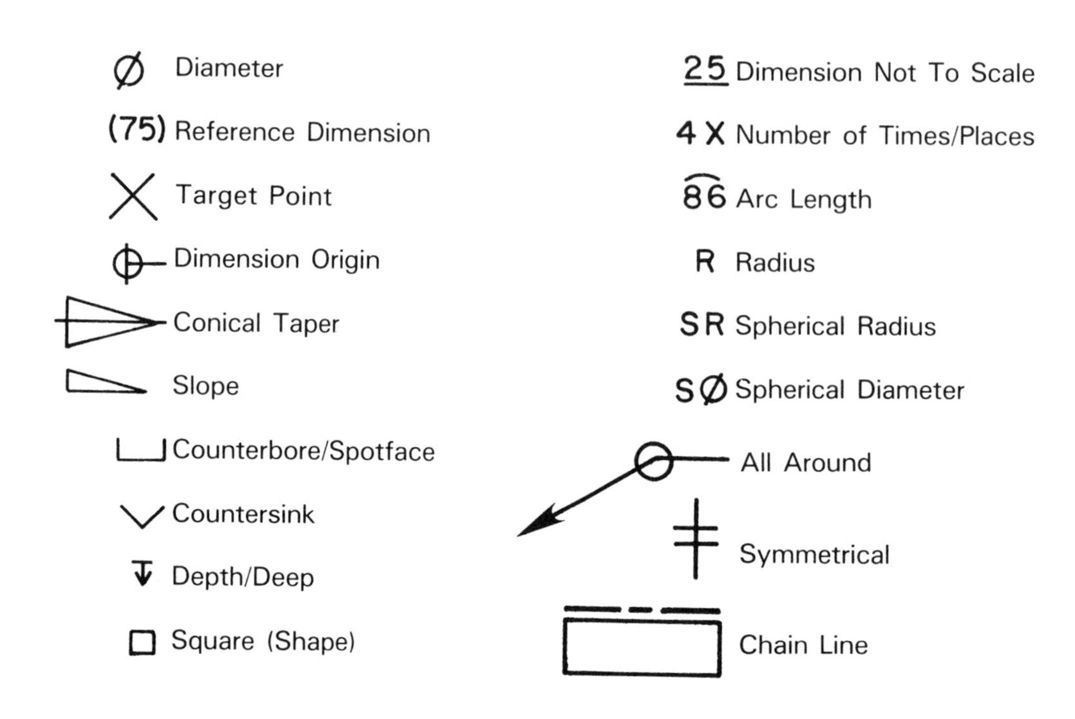

SOME COMMON SYMBOL USES

1. 8X Ø 8.4
   ⌴ Ø 12.7
   ↧ 5.2

2. 6 X Ø $12^{+0.27}_{0}$

3. Ø 6.2 THRU
   ⌵ Ø 12 X 82°

4. ⌖ 45

CONVERSION TABLE
METRIC TO ENGLISH

WHEN YOU KNOW	MULTIPLY BY: * = Exact		TO FIND
	VERY ACCURATE	APPROXIMATE	
LENGTH			
millimeters	0.0393701	0.04	inches
centimeters	0.3937008	0.4	inches
meters	3.280840	3.3	feet
meters	1.093613	1.1	yards
kilometers	0.621371	0.6	miles
WEIGHT			
grains	0.00228571	0.0023	ounces
grams	0.03527396	0.035	ounces
kilograms	2.204623	2.2	pounds
tonnes	1.1023113	1.1	short tons
VOLUME			
milliliters		0.2	teaspoons
milliliters	0.06667	0.067	tablespoons
milliliters	0.03381402	0.03	fluid ounces
liters	61.02374	61.024	cubic inches
liters	2.113376	2.1	pints
liters	1.056688	1.06	quarts
liters	0.26417205	0.26	gallons
liters	0.03531467	0.035	cubic feet
cubic meters	61023.74	61023.7	cubic inches
cubic meters	35.31467	35.0	cubic feet
cubic meters	1.3079506	1.3	cubic yards
cubic meters	264.17205	264.0	gallons
AREA			
square centimeters	0.1550003	0.16	square inches
square centimeters	0.00107639	0.001	square feet
square meters	10.76391	10.8	square feet
square meters	1.195990	1.2	square yards
square kilometers		0.4	square miles
hectares	2.471054	2.5	acres
TEMPERATURE			
Celsius	*9/5 (then add 32)		Fahrenheit

CONVERSION TABLE
ENGLISH TO METRIC

WHEN YOU KNOW	MULTIPLY BY: * = Exact		TO FIND
	VERY ACCURATE	APPROXIMATE	
LENGTH			
inches	* 25.4		millimeters
inches	* 2.54		centimeters
feet	* 0.3048		meters
feet	* 30.48		centimeters
yards	* 0.9144	0.9	meters
miles	* 1.609344	1.6	kilometers
WEIGHT			
grains	15.43236	15.4	grams
ounces	* 28.349523125	28.0	grams
ounces	* 0.028349523125	.028	kilograms
pounds	* 0.45359237	0.45	kilograms
short ton	* 0.90718474	0.9	tonnes
VOLUME			
teaspoons		5.0	milliliters
tablespoons		15.0	milliliters
fluid ounces	29.57353	30.0	milliliters
cups		0.24	liters
pints	* 0.473176473	0.47	liters
quarts	* 0.946352946	0.95	liters
gallons	* 3.785411784	3.8	liters
cubic inches	* 0.016387064	0.02	liters
cubic feet	* 0.028316846592	0.03	cubic meters
cubic yards	* 0.764554857984	0.76	cubic meters
AREA			
square inches	* 6.4516	6.5	square centimeters
square feet	* 0.09290304	0.09	square meters
square yards	* 0.83612736	0.8	square meters
square miles		2.6	square kilometers
acres	* 0.40468564224	0.4	hectares
TEMPERATURE			
Fahrenheit	* 5/9 (after subtracting 32)		Celsius

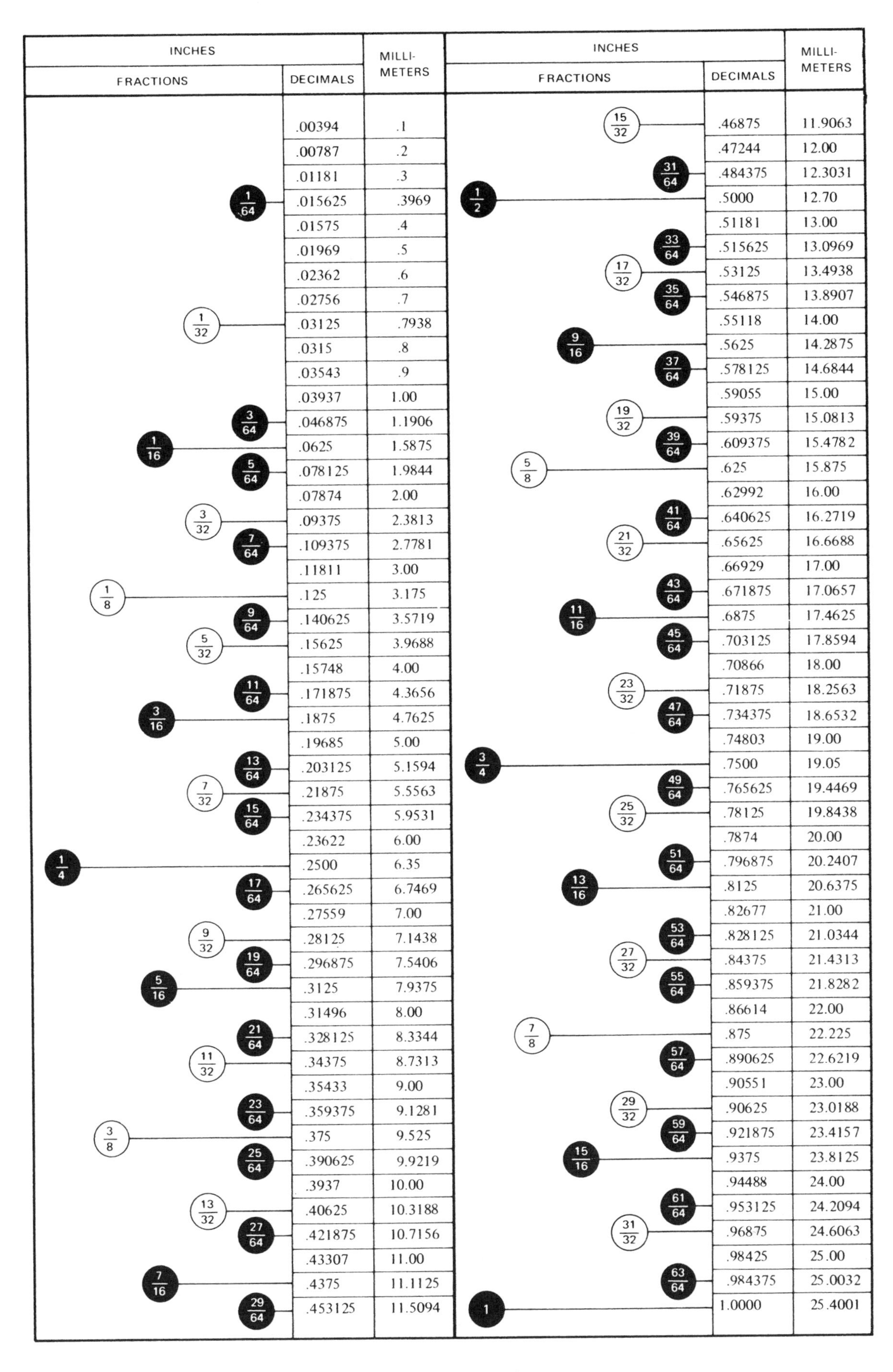

INCHES		MILLIMETERS
FRACTIONS	DECIMALS	
	.00394	.1
	.00787	.2
	.01181	.3
1/64	.015625	.3969
	.01575	.4
	.01969	.5
	.02362	.6
	.02756	.7
1/32	.03125	.7938
	.0315	.8
	.03543	.9
	.03937	1.00
3/64	.046875	1.1906
1/16	.0625	1.5875
5/64	.078125	1.9844
	.07874	2.00
3/32	.09375	2.3813
7/64	.109375	2.7781
	.11811	3.00
1/8	.125	3.175
9/64	.140625	3.5719
5/32	.15625	3.9688
	.15748	4.00
11/64	.171875	4.3656
3/16	.1875	4.7625
	.19685	5.00
13/64	.203125	5.1594
7/32	.21875	5.5563
15/64	.234375	5.9531
	.23622	6.00
1/4	.2500	6.35
17/64	.265625	6.7469
	.27559	7.00
9/32	.28125	7.1438
19/64	.296875	7.5406
5/16	.3125	7.9375
	.31496	8.00
21/64	.328125	8.3344
11/32	.34375	8.7313
	.35433	9.00
23/64	.359375	9.1281
3/8	.375	9.525
25/64	.390625	9.9219
	.3937	10.00
13/32	.40625	10.3188
27/64	.421875	10.7156
	.43307	11.00
7/16	.4375	11.1125
29/64	.453125	11.5094
15/32	.46875	11.9063
	.47244	12.00
31/64	.484375	12.3031
1/2	.5000	12.70
	.51181	13.00
33/64	.515625	13.0969
17/32	.53125	13.4938
35/64	.546875	13.8907
	.55118	14.00
9/16	.5625	14.2875
37/64	.578125	14.6844
	.59055	15.00
19/32	.59375	15.0813
39/64	.609375	15.4782
5/8	.625	15.875
	.62992	16.00
41/64	.640625	16.2719
21/32	.65625	16.6688
	.66929	17.00
43/64	.671875	17.0657
11/16	.6875	17.4625
45/64	.703125	17.8594
	.70866	18.00
23/32	.71875	18.2563
47/64	.734375	18.6532
	.74803	19.00
3/4	.7500	19.05
49/64	.765625	19.4469
25/32	.78125	19.8438
	.7874	20.00
51/64	.796875	20.2407
13/16	.8125	20.6375
	.82677	21.00
53/64	.828125	21.0344
27/32	.84375	21.4313
55/64	.859375	21.8282
	.86614	22.00
7/8	.875	22.225
57/64	.890625	22.6219
	.90551	23.00
29/32	.90625	23.0188
59/64	.921875	23.4157
15/16	.9375	23.8125
	.94488	24.00
61/64	.953125	24.2094
31/32	.96875	24.6063
	.98425	25.00
63/64	.984375	25.0032
1	1.0000	25.4001

Metric – inch equivalents.

ENGLISH	METRIC
LENGTH	
12 inches = 1 foot	1 kilometer = 1000 meters
36 inches = 1 yard	1 hectometer = 100 meters
3 feet = 1 yard	1 decameter = 10 meters
5,280 feet = 1 mile	1 meter = 1 meter
16.5 feet = 1 rod	1 decimeter = 0.1 meter
320 rods = 1 mile	1 centimeter = 0.01 meter
6 feet = 1 fathom	1 millimeter = 0.001 meter
WEIGHT	
27.34 grains = 1 dram	1 tonne = 1,000,000 grams
438 grains = 1 ounce	1 kilogram = 1000 grams
16 drams = 1 ounce	1 hectogram = 100 grams
16 ounces = 1 pound	1 dekagram = 10 grams
2000 pounds = 1 short ton	1 gram = 1 gram
2240 pounds = 1 long ton	1 decigram = 0.1 gram
25 pounds = 1 quarter	1 centigram = 0.01 gram
4 quarters = 1 cwt	1 milligram = 0.001 gram
VOLUME	
8 ounces = 1 cup	1 hectoliter = 100 liters
16 ounces = 1 pint	1 decaliter = 10 liters
32 ounces = 1 quart	1 liter = 1 liter
2 cups = 1 pint	1 deciliter = 0.1 liter
2 pints = 1 quart	1 centiliter = 0.01 liter
4 quarts = 1 gallon	1 milliliter = 0.001 liter
8 pints = 1 gallon	1000 milliliter = 1 liter
AREA	
144 sq. inches = 1 sq. foot	100 sq. millimeters = 1 sq. centimeter
9 sq. feet = 1 sq. yard	100 sq. centimeters = 1 sq. decimeter
43,560 sq. ft. = 160 sq. rods	100 sq. decimeters = 1 sq. meter
160 sq. rods = 1 acre	10,000 sq. meters = 1 hectare
640 acres = 1 sq. mile	

TEMPERATURE		
FAHRENHEIT		CELSIUS (centigrade)
32 degrees F	Water freezes	0 degrees C
68 degrees F	Reasonable room temperature	20 degrees C
98.6 degrees F	Normal body temperature	37 degrees C
173 degrees F	Alcohol boils	78.34 degrees C
212 degrees F	Water boils	100 degrees C

SYMBOL FOR:	ANSI Y14.5	ISO
DIMENSION ORIGIN	⌖→	NONE
FEATURE CONTROL FRAME	⌖ | Ø0.5Ⓜ | A | B | C	⌖ | Ø0.5Ⓜ | A | B | C
CONICAL TAPER	⊳	⊳
SLOPE	◺	◺
COUNTERBORE/SPOTFACE	⌴	NONE
COUNTERSINK	⌵	NONE
DEPTH/DEEP	↧	NONE
SQUARE (SHAPE)	□	□
DIMENSION NOT TO SCALE	<u>15</u>	<u>15</u>
NUMBER OF TIMES/PLACES	8X	8X
ARC LENGTH	⌒105	NONE
RADIUS	R	R
SPHERICAL RADIUS	SR	NONE
SPHERICAL DIAMETER	SØ	NONE

(ANSI-Y14.5M-1982)

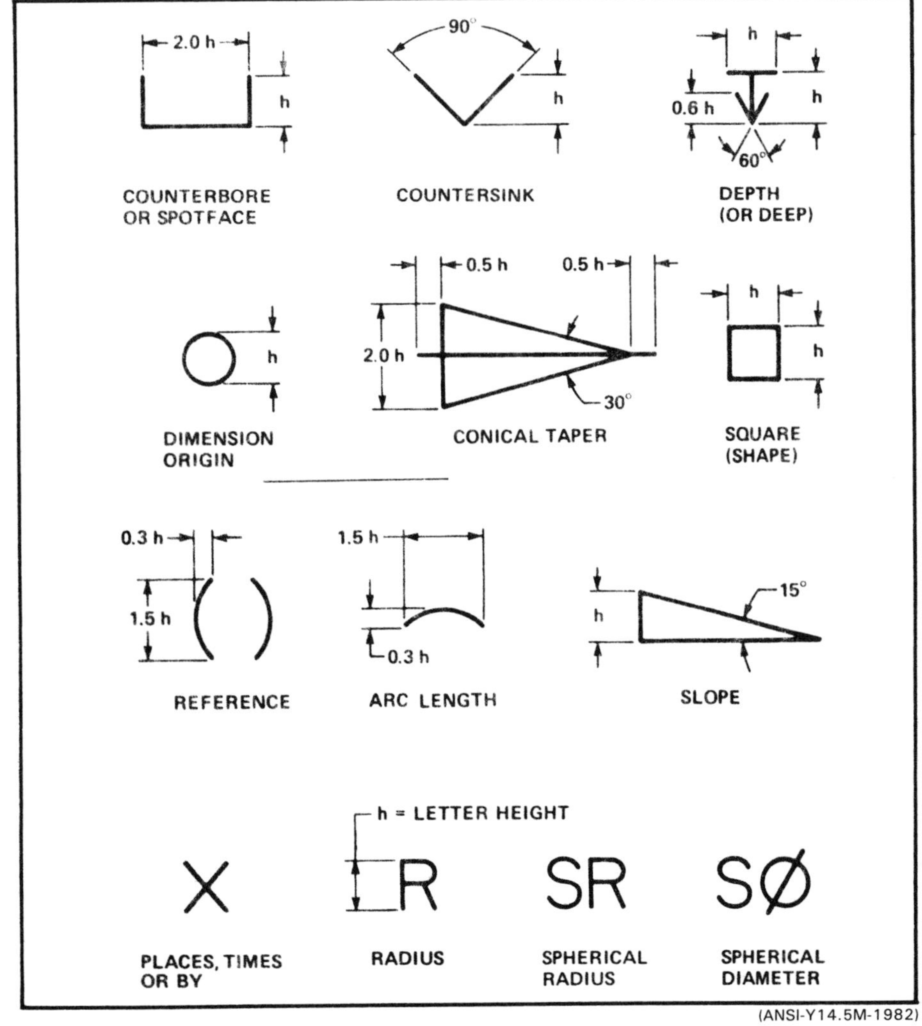

(ANSI-Y14.5M-1982)

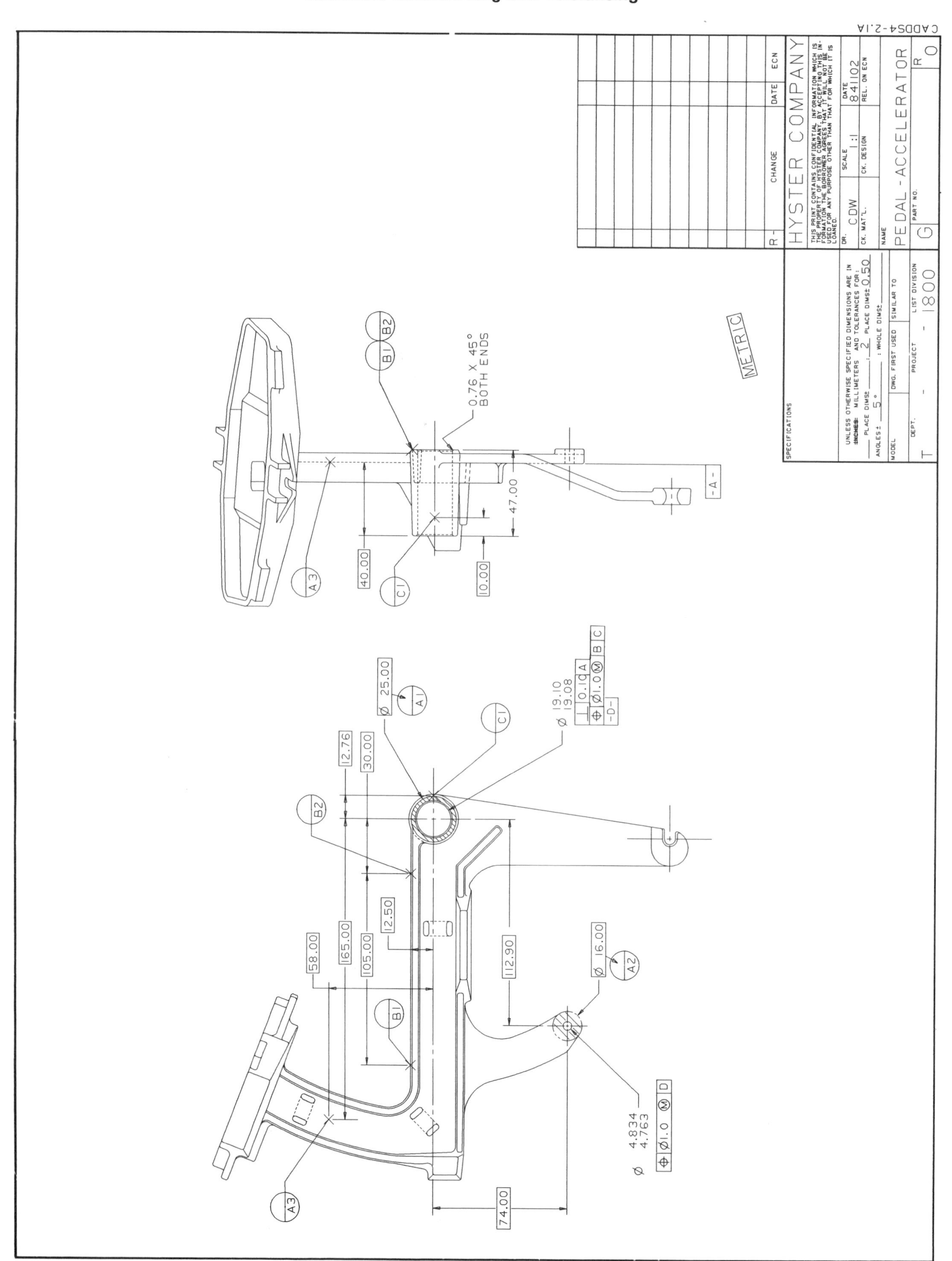
CADDS4-2.1A
R-
CHANGE
DATE
ECN
HYSTER COMPANY
THIS PRINT CONTAINS CONFIDENTIAL INFORMATION WHICH IS THE PROPERTY OF HYSTER COMPANY. BY ACCEPTING THIS INFORMATION THE BORROWER AGREES THAT IT WILL NOT BE USED FOR ANY PURPOSE OTHER THAN THAT FOR WHICH IT IS LOANED.
DR.
CDW
SCALE
1:1
DATE
841102
CK. MAT'L.
CK. DESIGN
REL. ON ECN
NAME
PEDAL-ACCELERATOR
PART NO.
G
R
0
SPECIFICATIONS
UNLESS OTHERWISE SPECIFIED DIMENSIONS ARE IN MILLIMETERS AND TOLERANCES FOR:
PLACE DIMS±
2 PLACE DIMS± 0.50
ANGLES± 5°
WHOLE DIMS±
MODEL
DWG. FIRST USED
SIMILAR TO
DEPT.
T
PROJECT
LIST DIVISION
1800
METRIC
0.76 X 45°
BOTH ENDS
-A-
47.00
40.00
10.00
A3
C1
B1
B2
Ø 25.00
A1
12.76
30.00
Ø 19.10
19.08
0.1 A
Ø1.0 Ⓜ B C
-D-
165.00
105.00
58.00
12.50
112.90
Ø 16.00
A2
Ø 4.834
4.763
Ø1.0 Ⓜ D
74.00

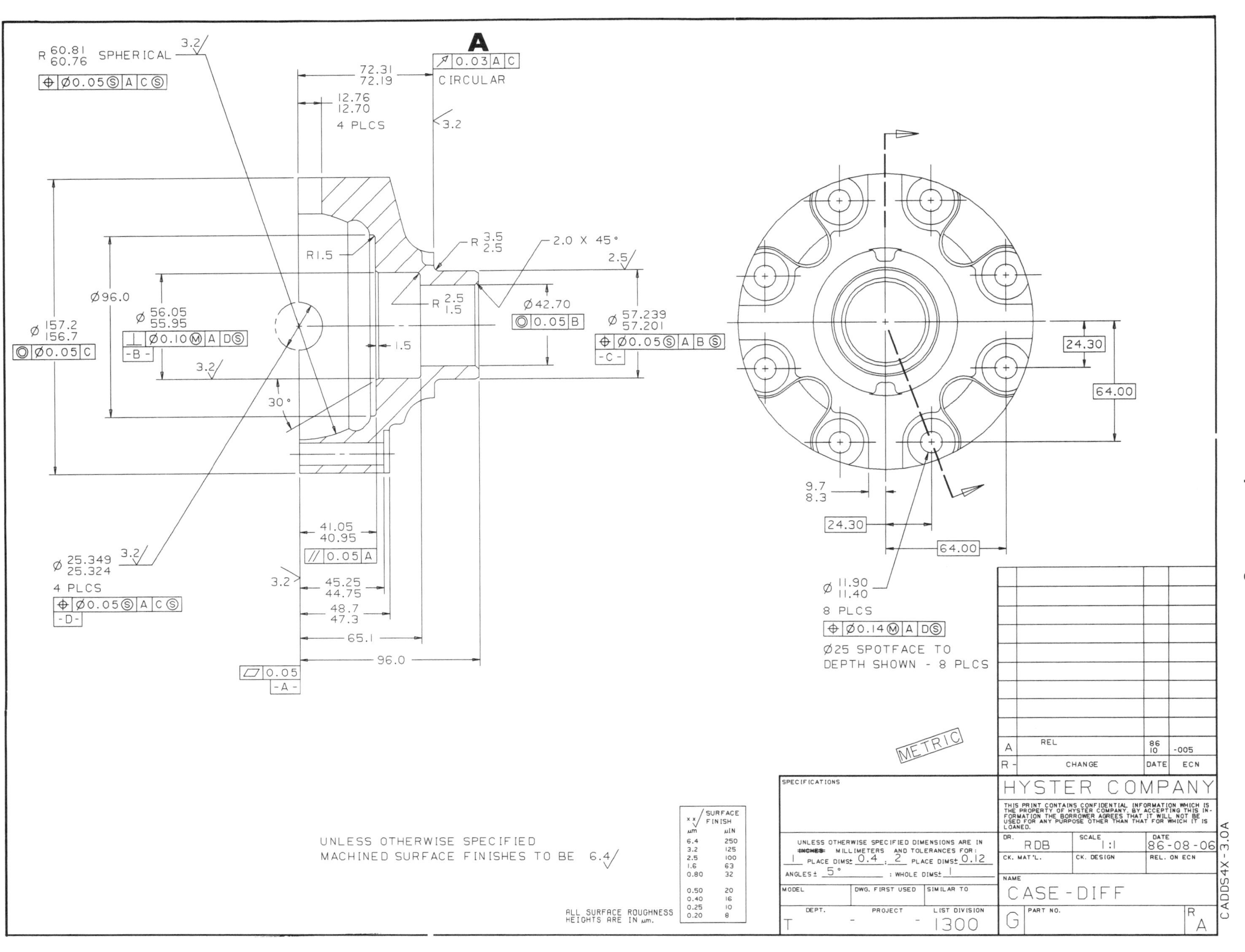

µm	µIN
6.4	250
3.2	125
2.5	100
1.6	63
0.80	32
0.50	20
0.40	16
0.25	10
0.20	8

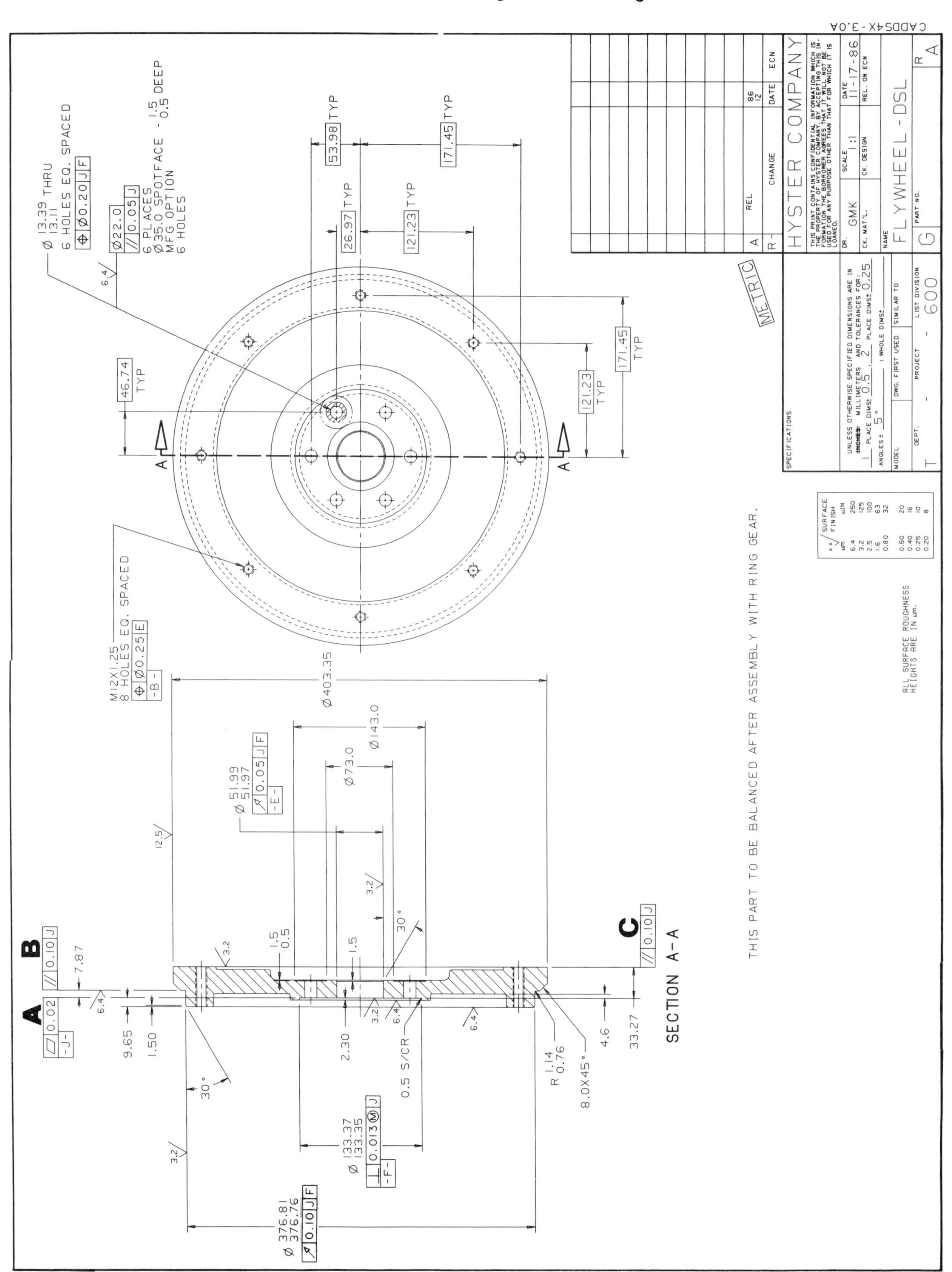

HYSTER COMPANY
FLYWHEEL - DSL
METRIC
SECTION A-A
THIS PART TO BE BALANCED AFTER ASSEMBLY WITH RING GEAR.
ALL SURFACE ROUGHNESS HEIGHTS ARE IN µm.
6 HOLES EQ. SPACED
8 HOLES EQ. SPACED
M12X1.25
MFG OPTION
6 PLACES
6 HOLES
Ø35.0 SPOTFACE
Ø403.35
Ø143.0
Ø73.0
0.5 S/CR
8.0X45°
CADDS4X-3.0A

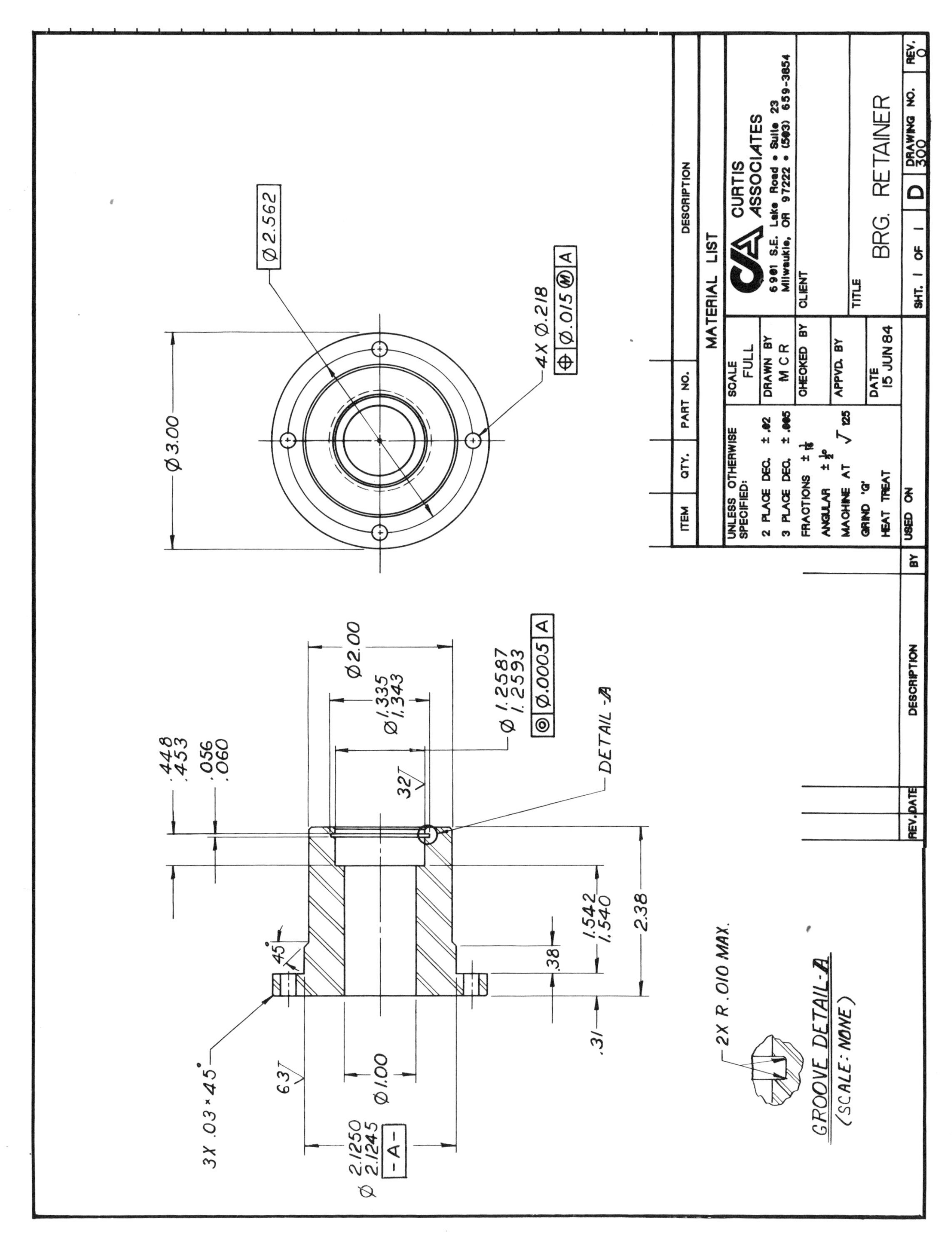

Ø 2.562
Ø 3.00
4X Ø.218
Ø.015 M A
Ø 2.00
Ø 1.335
1.343
Ø 1.2587
1.2593
Ø.0005 A
DETAIL-A
.448
.453
.056
.060
32
45°
.38
1.542
1.540
2.38
.31
3X .03 × 45°
63
Ø1.00
Ø 2.1250
2.1245
-A-
2X R .010 MAX.
GROOVE DETAIL-A
(SCALE: NONE)
ITEM
QTY.
PART NO.
DESCRIPTION
MATERIAL LIST
UNLESS OTHERWISE SPECIFIED:
2 PLACE DEC. ± .02
3 PLACE DEC. ± .005
FRACTIONS ± 1/16
ANGULAR ± 1/2°
MACHINE AT 125
GRIND 'G'
HEAT TREAT
SCALE
FULL
DRAWN BY
M C R
CHECKED BY
APPVD. BY
DATE
15 JUN 84
CURTIS
ASSOCIATES
6901 S.E. Lake Road • Suite 23
Milwaukie, OR 97222 • (503) 659-3854
CLIENT
TITLE
BRG. RETAINER
USED ON
SHT. 1 OF 1
D
DRAWING NO.
300
REV.
0
REV.
DATE
DESCRIPTION
BY

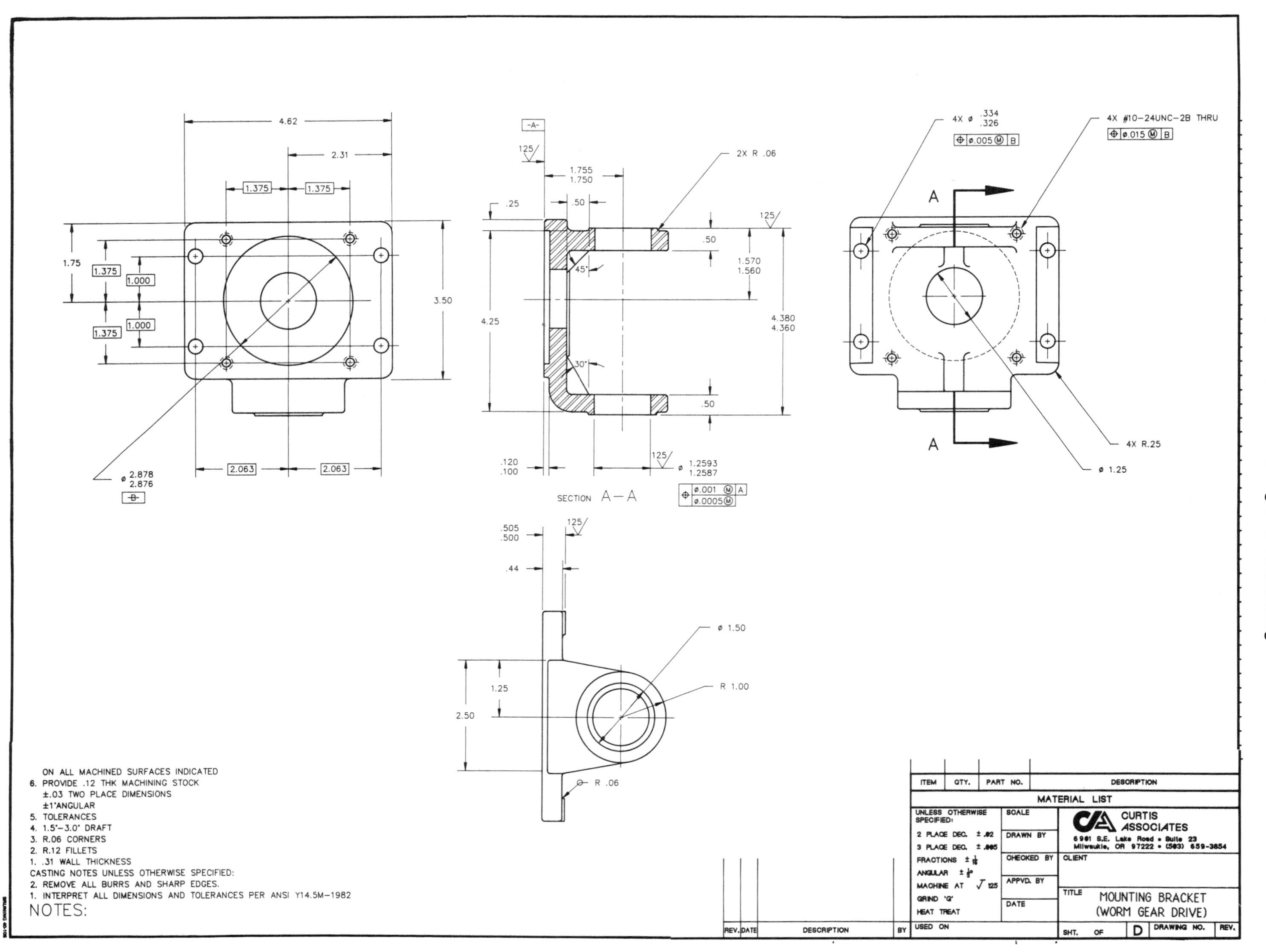
4X #10-24UNC-2B THRU
4X ø .334 .326
2X R .06
4X R.25
ø 1.25
ø 2.878 2.876
-B-
-A-
SECTION A—A
ø 1.50
R 1.00
R .06
ON ALL MACHINED SURFACES INDICATED
6. PROVIDE .12 THK MACHINING STOCK
±.03 TWO PLACE DIMENSIONS
±1°ANGULAR
5. TOLERANCES
4. 1.5°-3.0° DRAFT
3. R.06 CORNERS
2. R.12 FILLETS
1. .31 WALL THICKNESS
CASTING NOTES UNLESS OTHERWISE SPECIFIED:
2. REMOVE ALL BURRS AND SHARP EDGES.
1. INTERPRET ALL DIMENSIONS AND TOLERANCES PER ANSI Y14.5M-1982
NOTES:
MATERIAL LIST
CURTIS ASSOCIATES
6901 S.E. Lake Road • Suite 23
Milwaukie, OR 97222 • (503) 659-3854
TITLE MOUNTING BRACKET
(WORM GEAR DRIVE)
D

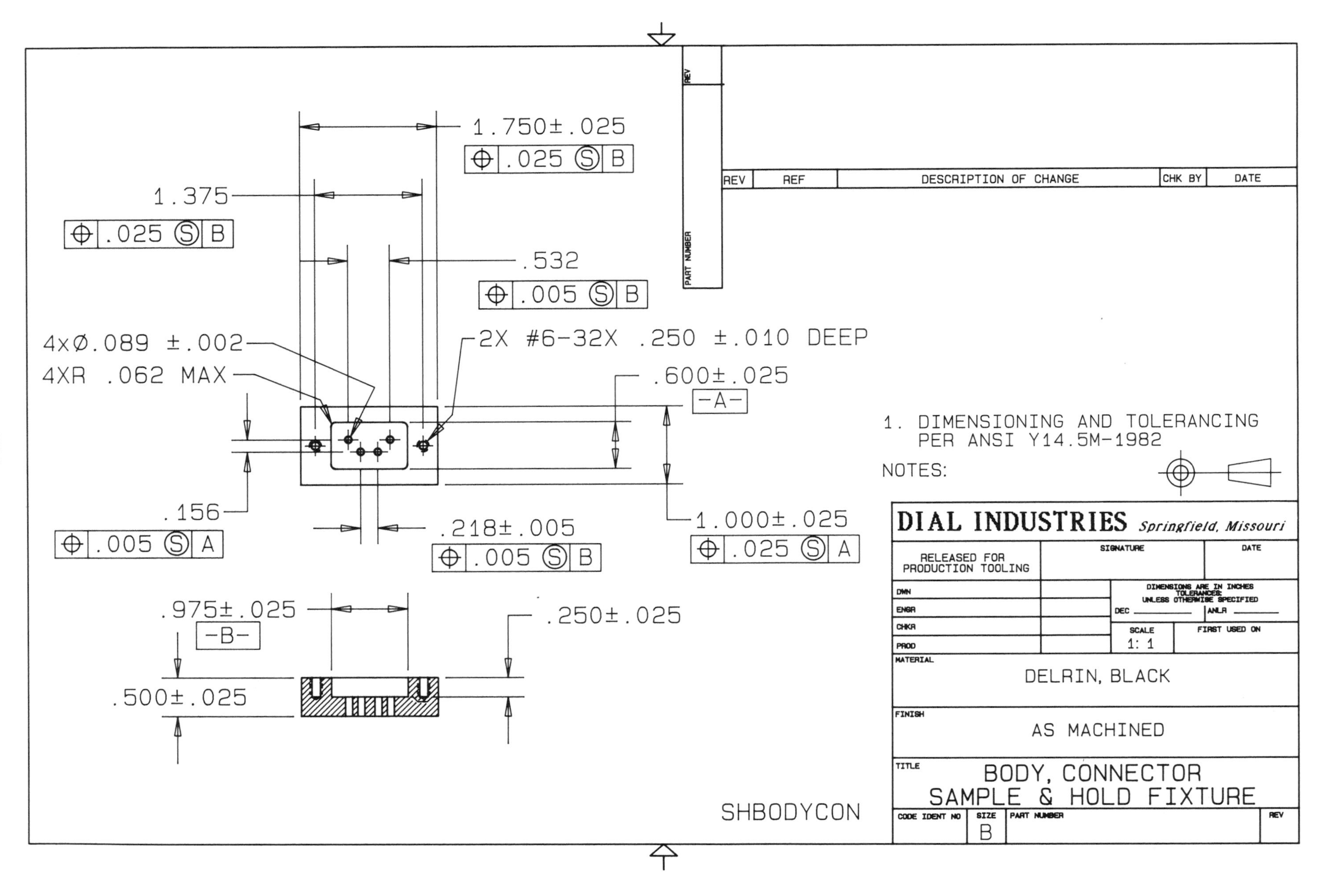

REV
PART NUMBER
REV
REF
DESCRIPTION OF CHANGE
CHK BY
DATE
1.750±.025
.025 S B
1.375
.025 S B
.532
.005 S B
4XØ.089 ±.002
4XR .062 MAX
2X #6-32X .250 ±.010 DEEP
.600±.025
-A-
1. DIMENSIONING AND TOLERANCING PER ANSI Y14.5M-1982
NOTES:
.156
.005 S A
.218±.005
.005 S B
1.000±.025
.025 S A
.975±.025
-B-
.250±.025
.500±.025
SHBODYCON
DIAL INDUSTRIES Springfield, Missouri
RELEASED FOR PRODUCTION TOOLING
SIGNATURE
DATE
DWN
ENGR
CHKR
PROD
DIMENSIONS ARE IN INCHES TOLERANCES: UNLESS OTHERWISE SPECIFIED
DEC
ANLR
SCALE 1: 1
FIRST USED ON
MATERIAL
DELRIN, BLACK
FINISH
AS MACHINED
TITLE
BODY, CONNECTOR
SAMPLE & HOLD FIXTURE
CODE IDENT NO
SIZE B
PART NUMBER
REV

GEOMETRIC DIMENSIONING AND TOLERANCING

DRAFTING PROBLEMS

Geometric dimensioning and tolerancing drafting problems may be assigned as needed to meet individual course length and requirements. A variety of problems are presented as engineering sketches. This is similar to the way drafters are assigned drawings in many real-life industrial situations. Keep in mind that engineering sketches are often rough. For example, a hole is called out as a ϕ, but the sketch may not look like a circle. Engineering sketches do not always place dimensions in proper locations. Follow the dimensions and instructions carefully. Do not assume that the final drawing will look exactly like the sketch. Given the following engineering sketches, you should:

1. Use manual or Computer Aided Drafting as specified by course requirements.
2. Select a drawing scale that clearly displays the features and dimensions.
3. Drafting vellum or polyester film should be sized in accordance with recommended practices such as: scale, number of views, amount of dimensions and notes, and necessary clear space. Do not crowd the drawing.
3. Prepare formal drawings using properly selected multiviews (orthographic projection). The number of views needed depends on the requirements of each drafting problem.
4. Use proper sectioning techniques as necessary.
5. Place dimensions and geometric tolerancing as specified by ANSI Y14.5M-1982. Use unidirectional dimensioning unless otherwise specified. Dimensions are presented on the engineering sketches in METRIC or INCHES. Give the appropriate inch or metric general note. Provide the general note: INTERPRET DIMENSIONS AND TOLERANCES PER ANSI Y14.5M-1982. Additional general notes are given with each problem.

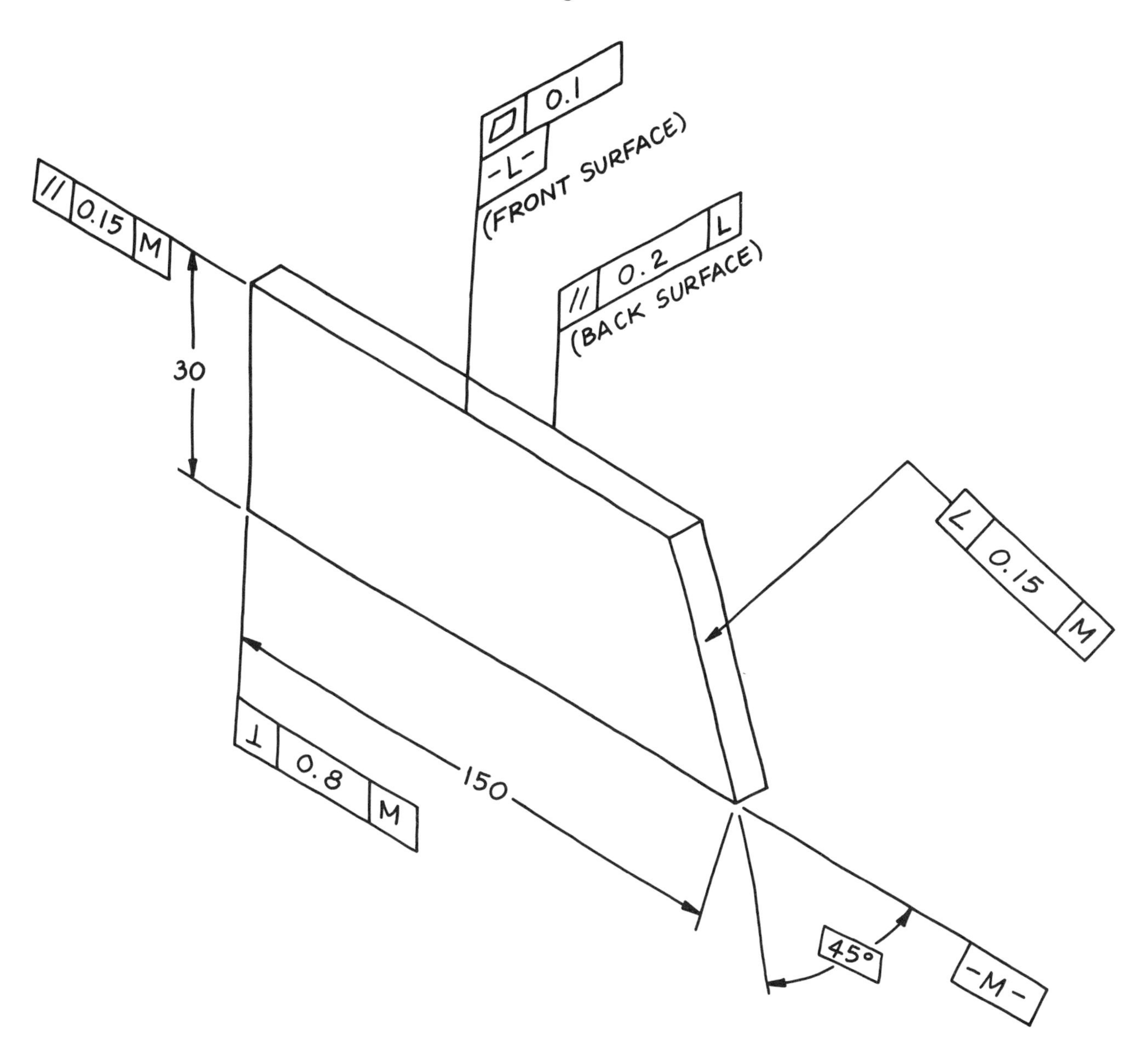

DRAFTING PROBLEM 1

METRIC

GT100

NAME: GAUGE BLOCK.

MATERIAL: SAE 4320. 8 mm THICK.

FINISH ALL OVER 0.20 μM.

REMOVE ALL BURRS AND SHARP EDGES.

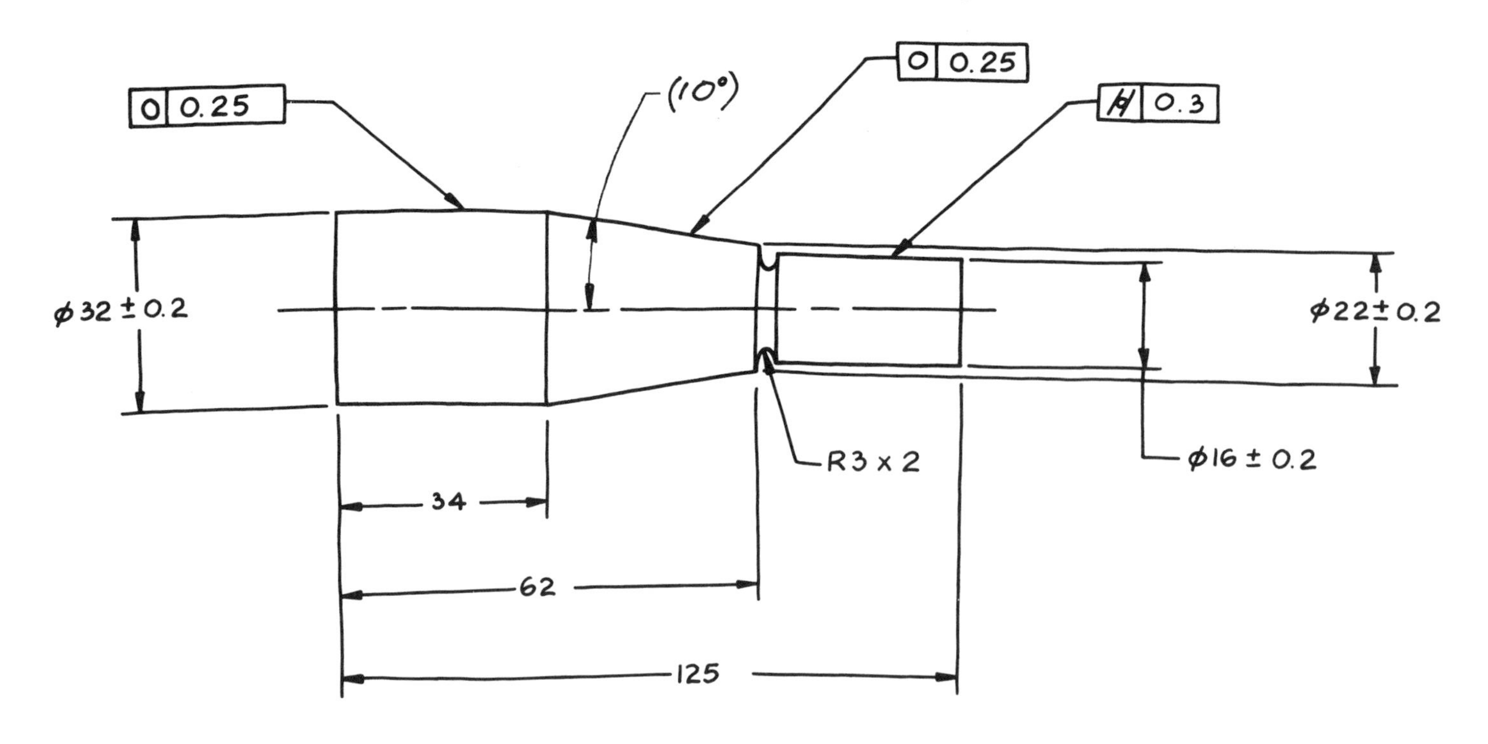

DRAFTING PROBLEM 2

METRIC

GT 200

NAME: VALVE PIN.

MATERIAL: PHOSPHOR BRONZE.

FINISH ALL OVER 0.2 μM.

REMOVE ALL BURRS AND SHARP EDGES.

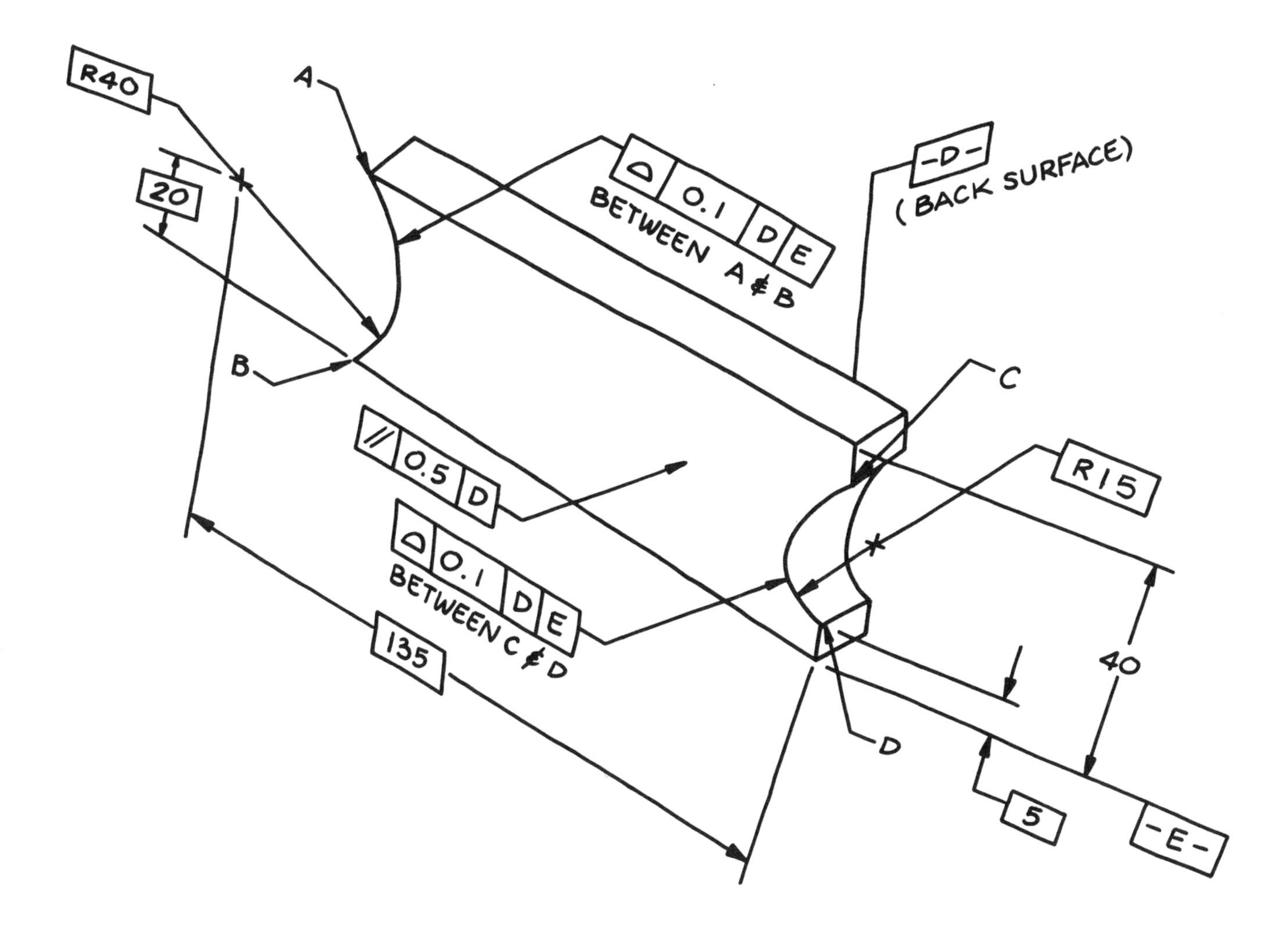

DRAFTING PROBLEM 3

METRIC

GT 300

NAME: SPRING CLIP.

MATERIAL: SAE 1060. 10mm THICK.

FINISH ALL OVER 0.50 μM.

REMOVE ALL BURRS AND SHARP EDGES.

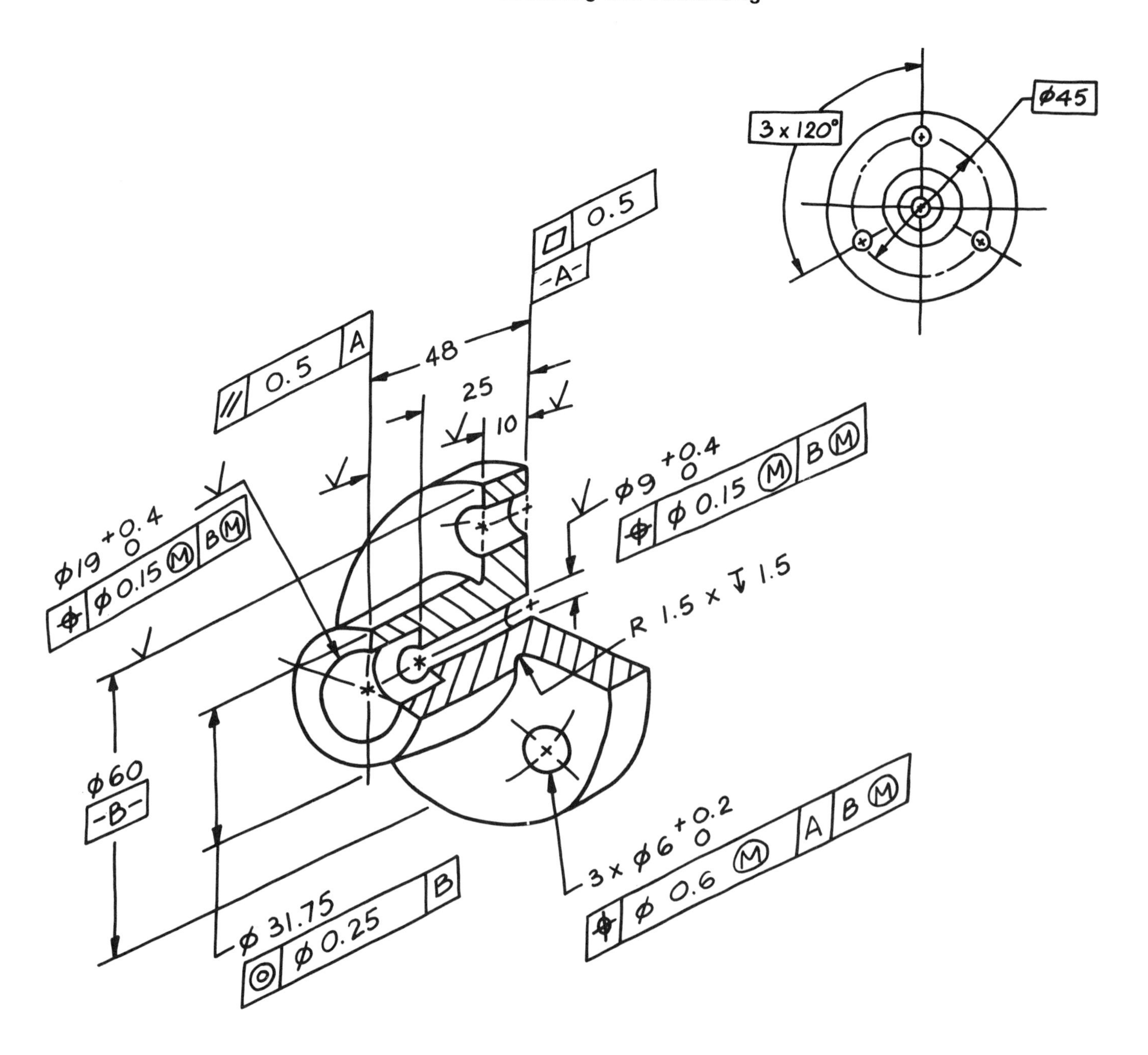

DRAFTING PROBLEM 4

METRIC

GT 400

NAME: HUB.

MATERIAL: SAE 1045.

FINISH ALL OVER 1.6 μM.

REMOVE ALL BURRS AND SHARP EDGES.

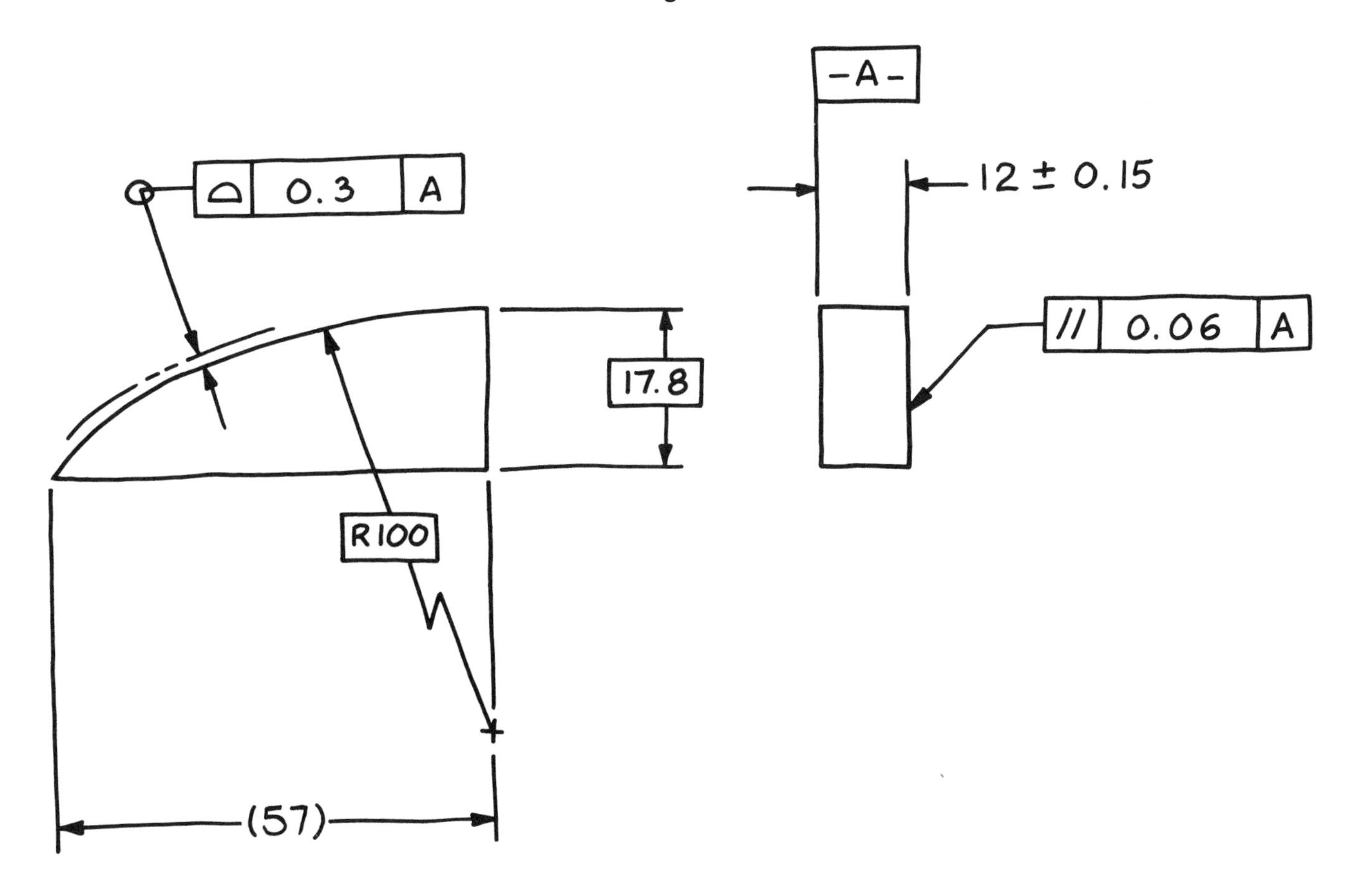

DRAFTING PROBLEM 5

METRIC

GT 500

NAME: INSERT.

MATERIAL: SAE 4640.

FINISH ALL OVER 0.2 μM.

REMOVE ALL BURRS AND SHARP EDGES.

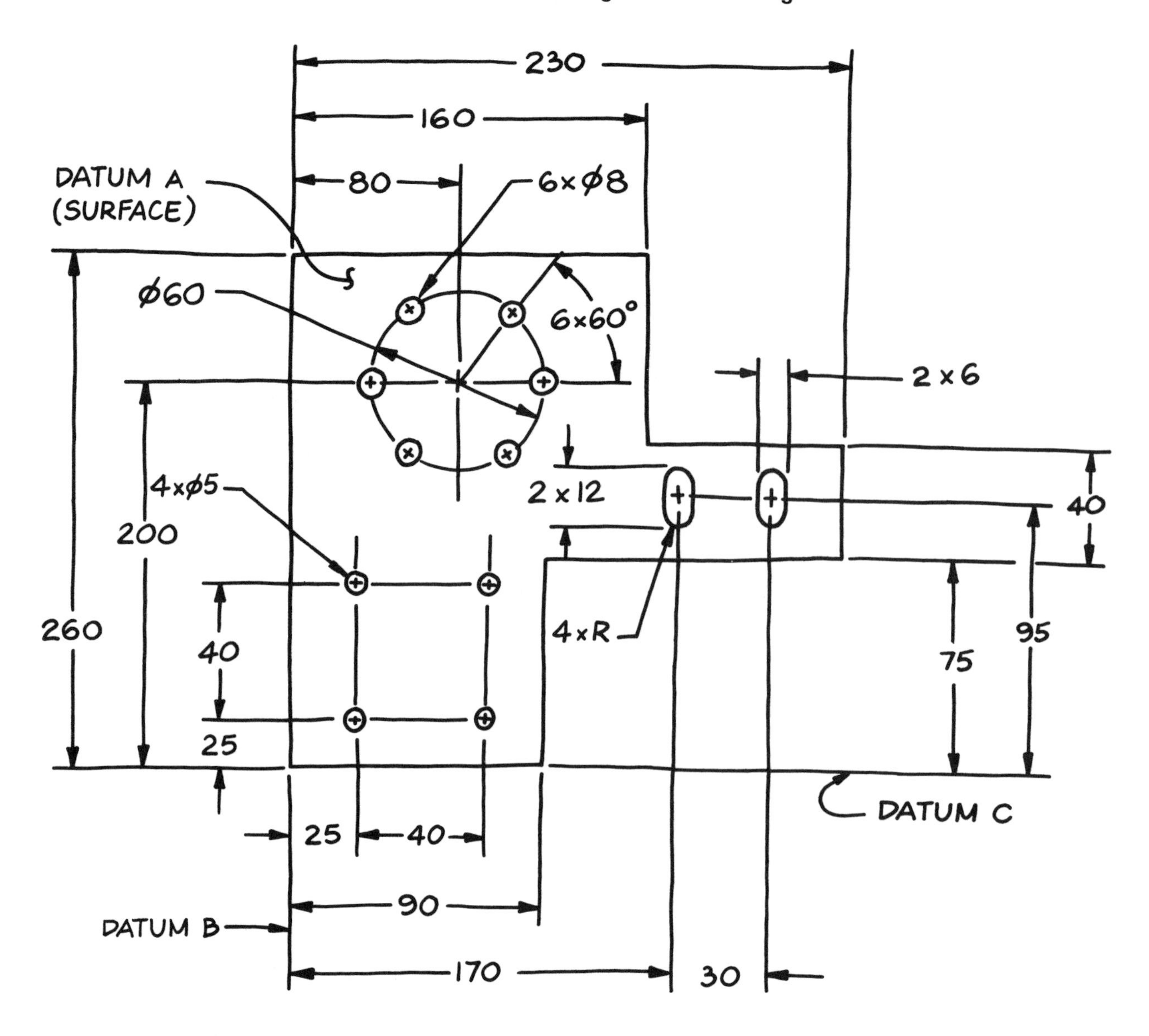

DRAFTING PROBLEM 6

METRIC

GT 600

NAME: MODULAR CHASSIS PLATE.

MATERIAL: SAE 30308. 1.5 THICK.

FINISH ALL OVER 0.2 μM.

REMOVE ALL BURRS AND SHARP EDGES.

ADDITIONAL INSTRUCTIONS:

1. Two slots (12 x 6 with full radius each end) shall have a position tolerance applied to both the length (12) and the width (6) of 0.5 MMC to datums A, B, and C.
2. The 6 x ⌀8 and the 4 x ⌀5 holes shall be position toleranced to datums A, B, and C by 0.25 at MMC.

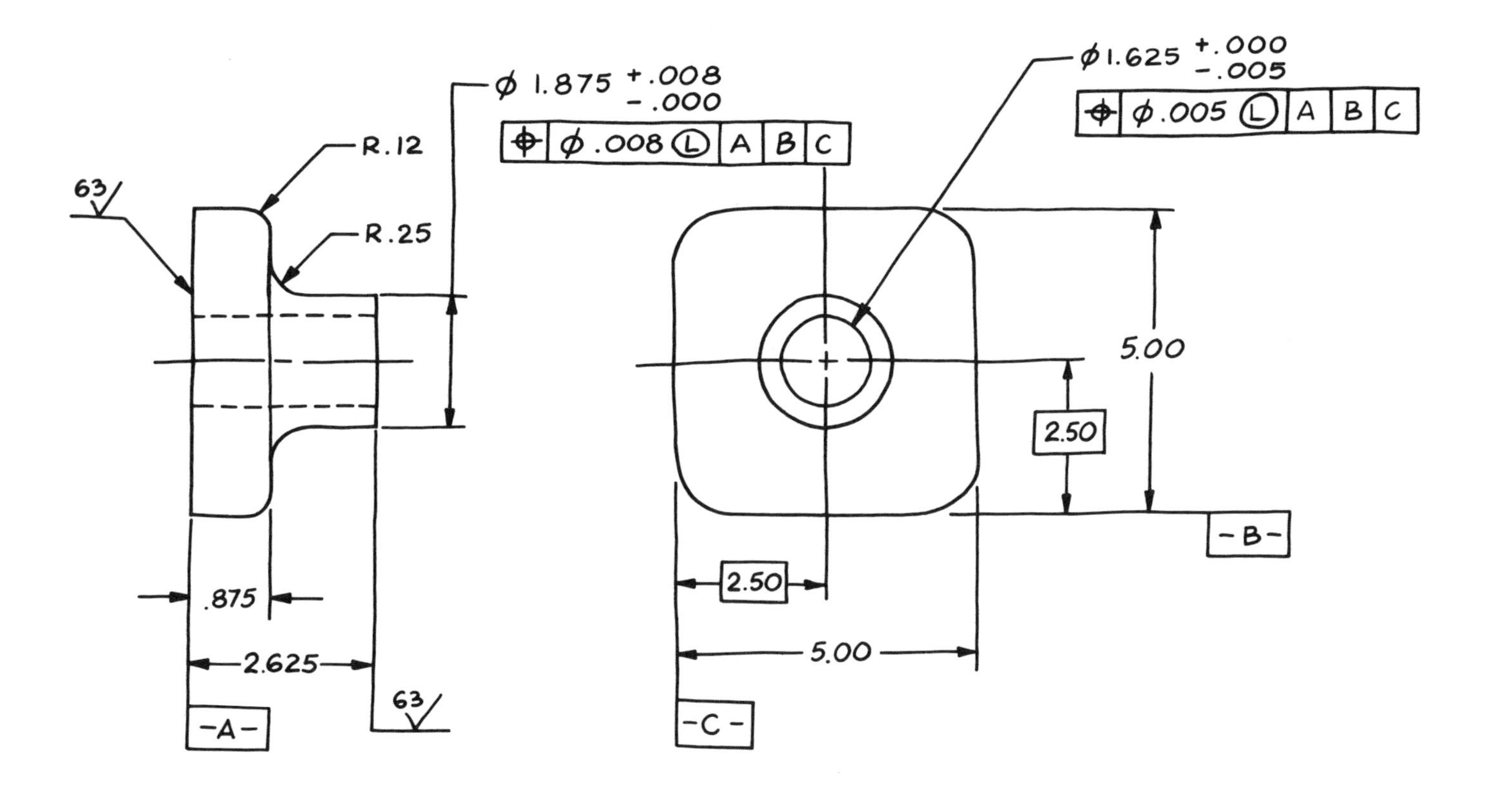

DRAFTING PROBLEM 7

INCHES

GT 700

NAME: THRUST WASHER.

MATERIAL: SAE 5150.

REMOVE ALL BURRS AND SHARP EDGES.

CORNERS ARE R .25 UNLESS OTHERWISE SPECIFIED.

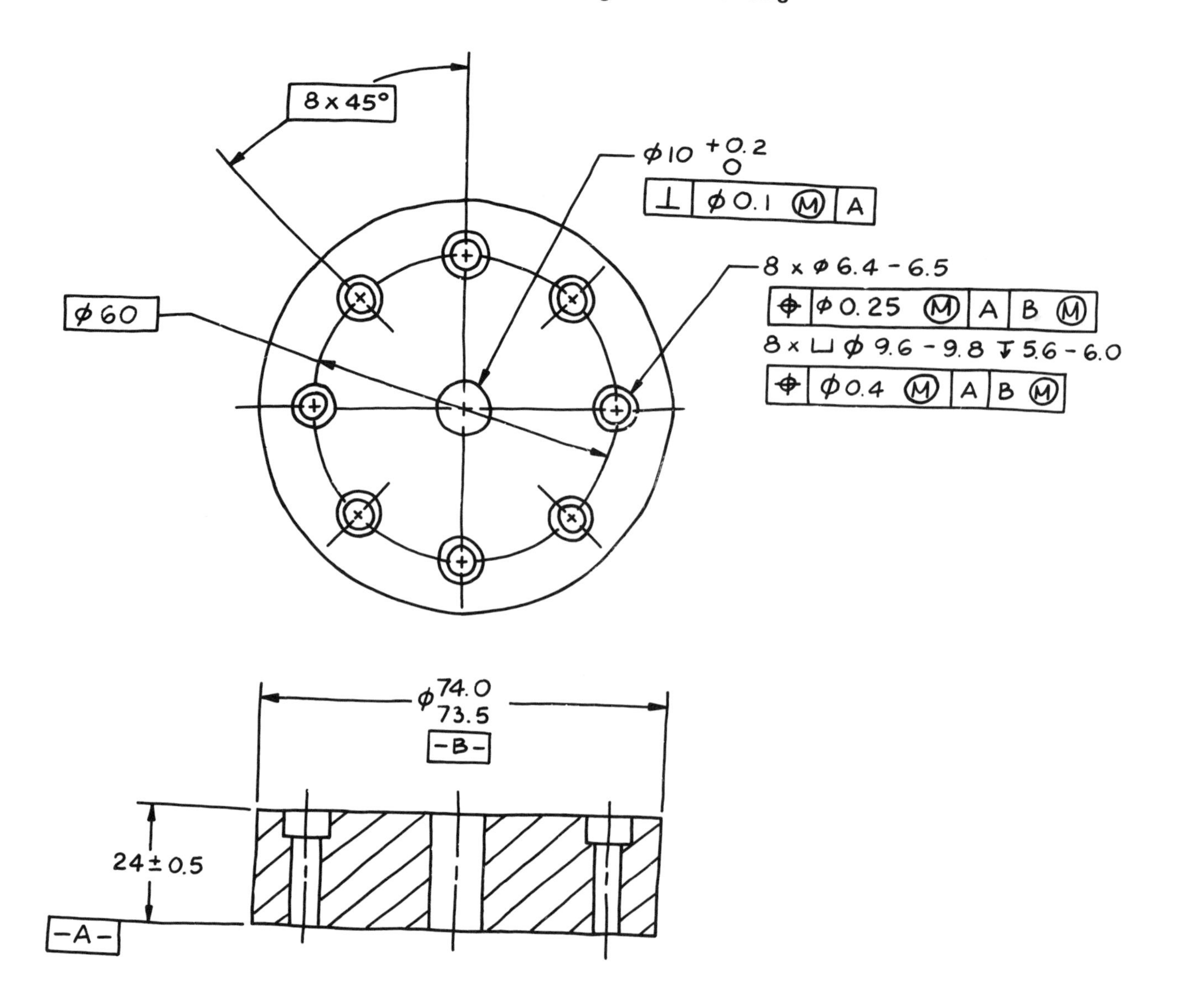

DRAFTING PROBLEM 8

METRIC

GT 800

NAME: MOUNTING PLATE.

MATERIAL: SAE 4140.

FINISH ALL OVER 0.80 μM.

REMOVE ALL BURRS AND SHARP EDGES.

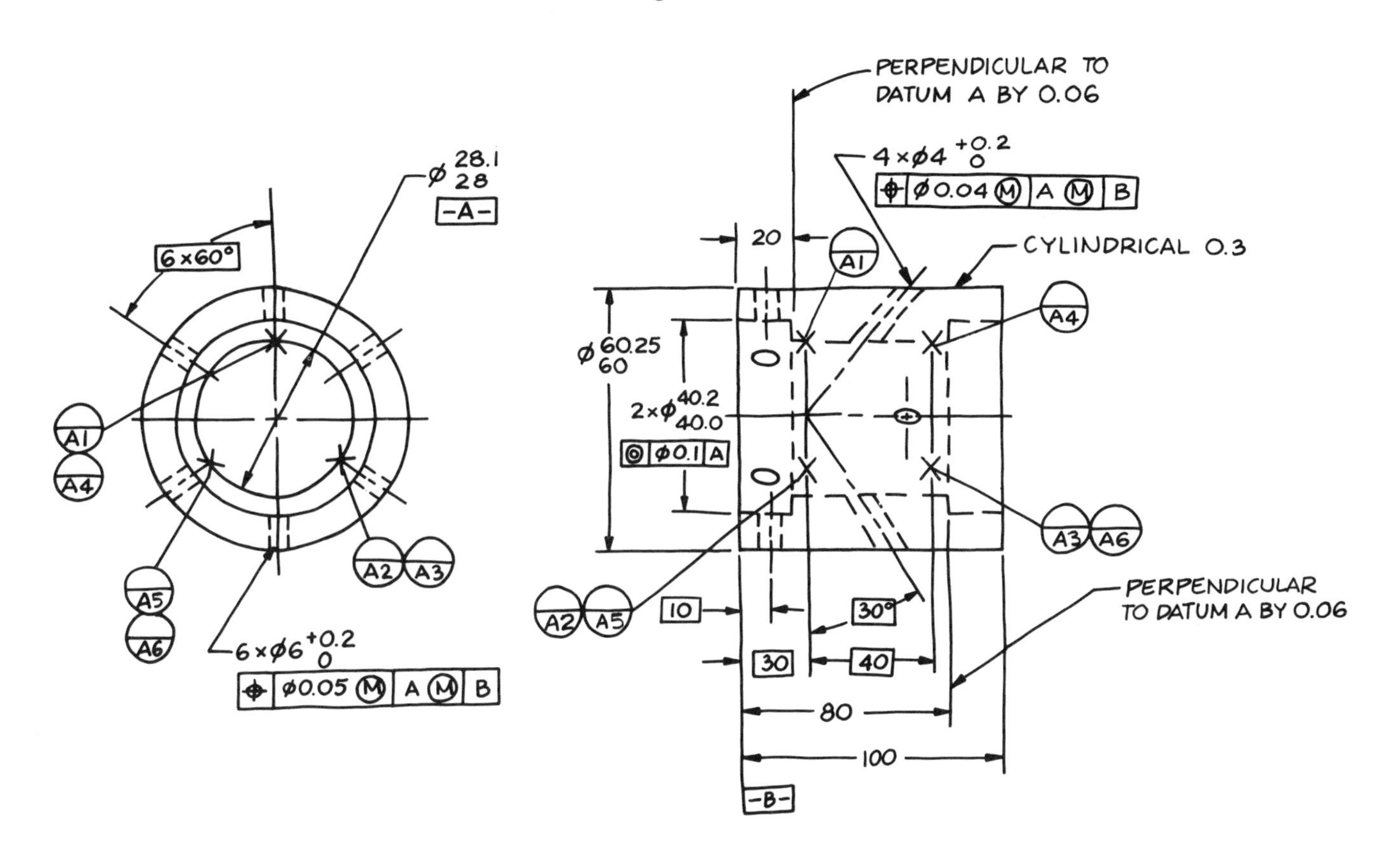

DRAFTING PROBLEM 9

METRIC

GT 900

NAME: OSCILLATOR.

MATERIAL: PHOSPHOR BRONZE.

FINISH ALL OVER 0.25 μM.

REMOVE ALL BURRS AND SHARP EDGES.

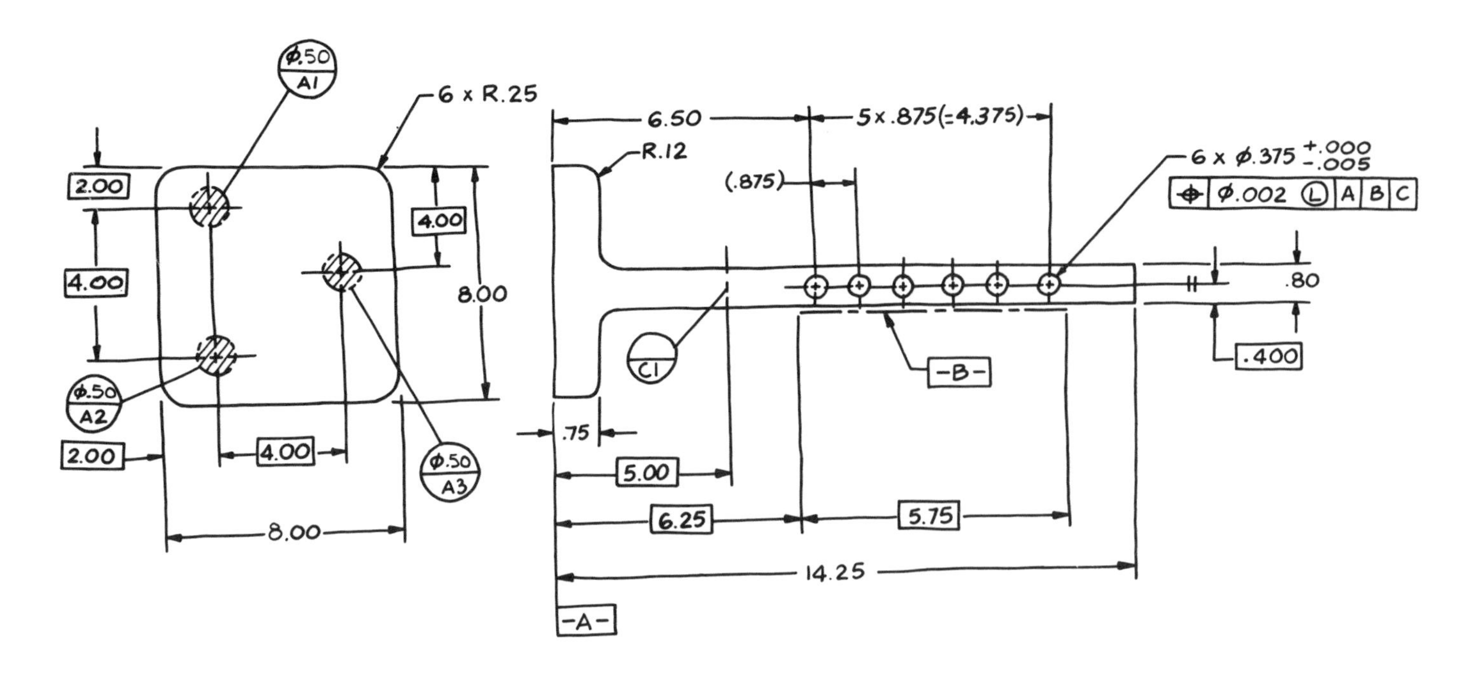

DRAFTING PROBLEM 10

INCHES

GT 1000

NAME: EXTENSION SUPPORT.

MATERIAL: ALUMINUM.

FINISH ALL OVER 125 μIN.

REMOVE ALL BURRS AND SHARP EDGES.

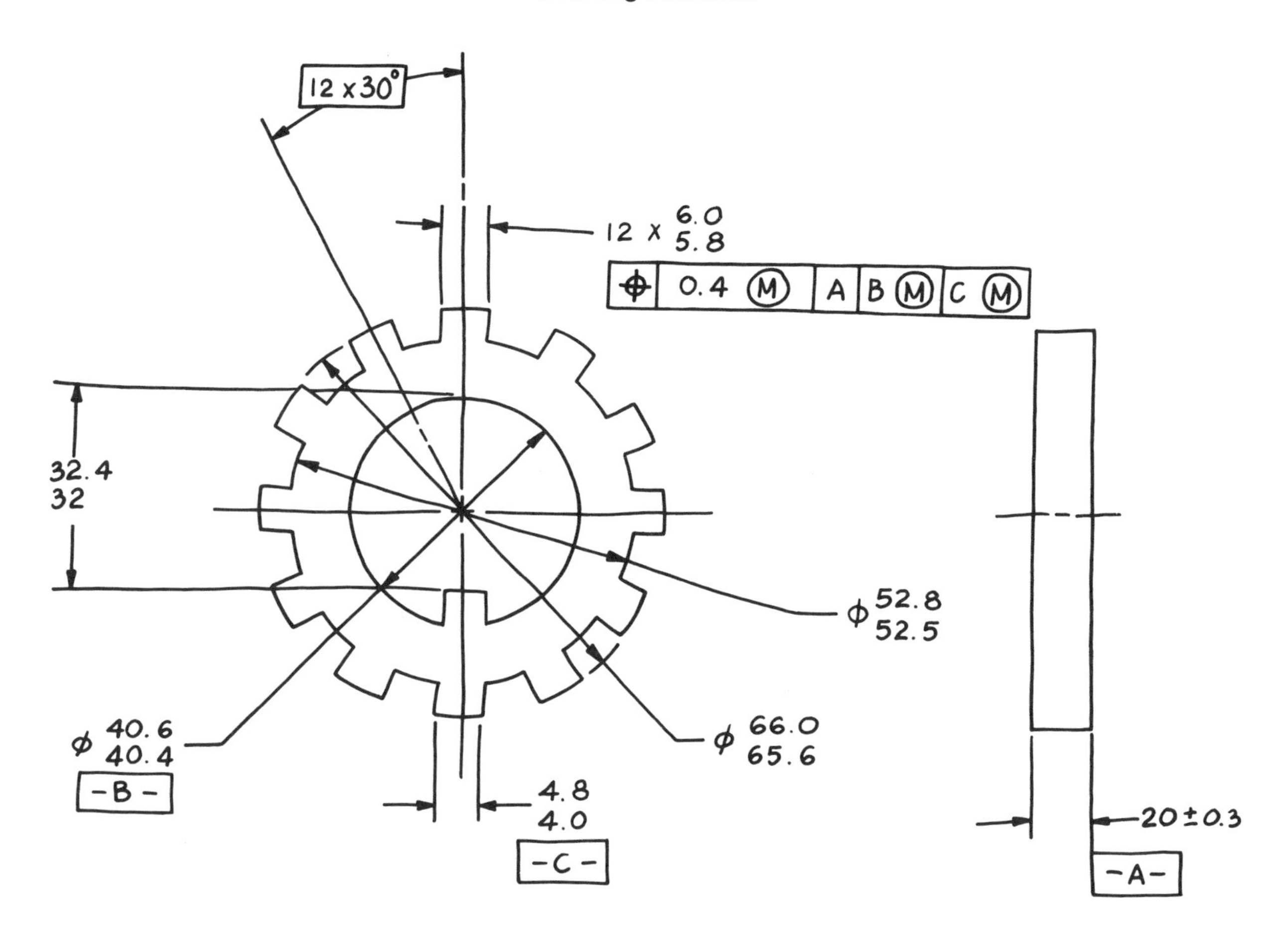

DRAFTING PROBLEM 11

METRIC

GT 1100

NAME: SPLINE PLATE.

MATERIAL: SAE 3135.

FINISH ALL OVER 1.6 μM.

REMOVE ALL BURRS AND SHARP EDGES.

DRAFTING PROBLEM 12

INCHES

GT 1200

NAME: LATCH BRACKET.

MATERIAL: TITANIUM.

FINISH ALL OVER 32 μIN.

REMOVE ALL BURRS AND SHARP EDGES.

ADDITIONAL INSTRUCTIONS:

1. Label all datums.
2. Use dimensions shown on sketch, but not necessarily the placement shown.
3. Provide angularity tolerance of .005 to datum C for 70° angle.
4. Hold surface profile of .005 between Points X and Y at both surfaces controlled by the R 1.25 dimension.
5. Position the ϕ.750 + .005/−.002 and the counterbore ϕ1.380 to datums A, B, and C by .003 MMC.
6. The surfaces labeled 1, 2, 3, and 4 shall be held symmetrical with datum centerplane D by .004. (Note: the numbers 1, 2, 3, and 4 will not be shown on your drawing).
7. Provide a coaxial position tolerance to locate the ϕ.187 feature with a coaxial ϕ tolerance of .002 at MMC relative to datums A, B, and C within which the holes, together, must lie. Also control the coaxial ϕ.001 tolerance at MMC within which the axes of the holes must lie relative to each other.
8. Provide a coaxial position tolerance to locate the ϕ.86-.89 feature with a coaxial ϕ tolerance of .005 at MMC relative to datums E, A, and C within which the holes, together, must lie. Also control the coaxial ϕ.002 tolerance at MMC within which the axes of the holes must lie relative to each other.
9. Datum B perpendicular to Datum A by .002.
10. Datum C perpendicular to Datums A and B by .002.

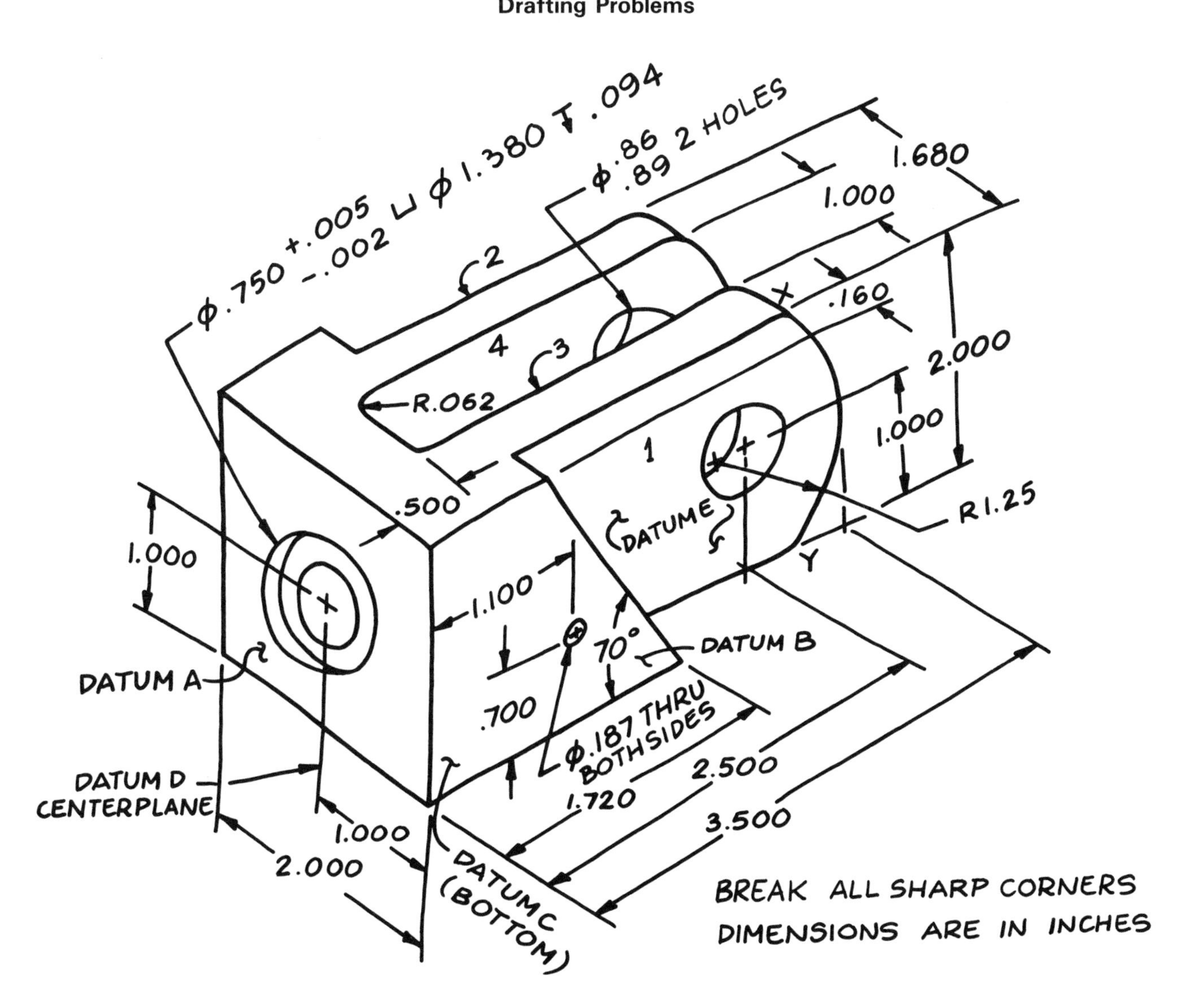

DRAFTING PROBLEM 12

INDEX

F

G

H

I

L

M

N

O

P

R

S